Proceedings of the

Fourth Meeting on CPT and Lorentz Symmetry

Proceedings of the

Fourth Meeting on CPT and Lorentz Symmetry

Bloomington, USA 8 – 11 August 2007

Editor

V. Alan Kostelecký

Indiana University, USA

World Scientific

NEW JERSEY • LONDON • SINGAPORE • BEIJING • SHANGHAI • HONG KONG • TAIPEI • CHENNAI

Published by

World Scientific Publishing Co. Pte. Ltd.
5 Toh Tuck Link, Singapore 596224
USA office: 27 Warren Street, Suite 401-402, Hackensack, NJ 07601
UK office: 57 Shelton Street, Covent Garden, London WC2H 9HE

British Library Cataloguing-in-Publication Data
A catalogue record for this book is available from the British Library.

CPT AND LORENTZ SYMMETRY
Proceedings of the Fourth Meeting

ISBN-13 978-981-277-950-2
ISBN-10 981-277-950-7

Printed in Singapore by B & JO Enterprise

PREFACE

This proceedings volume contains papers presented during the Fourth Meeting on CPT and Lorentz Symmetry, which took place between Wednesday August 8 and Saturday August 11, 2007 in the Physics Department at Indiana University, Bloomington. The primary focus of the event was experimental and theoretical investigations of these spacetime symmetries.

The papers in this volume were solicited from all the individuals who delivered presentations at the meeting. The schedule included invited talks, contributed talks, and posters. The papers are categorized accordingly and are ordered within each category following their timing during the event. My appreciation goes to all contributors for their efforts in preparing timely manuscripts.

The meeting featured descriptions of a wide variety of experimental studies of CPT and Lorentz symmetry, including ones involving astrophysical observations, clock-comparison measurements, cosmological birefringence, electromagnetic resonant cavities, gravitational tests, matter interferometry, muon behavior, neutrino oscillations, oscillations and decays of neutral mesons, particle-antiparticle comparisons, post-newtonian gravity, space-based missions, spectroscopy of hydrogen and antihydrogen, and spin-polarized matter. Theoretical treatments of CPT and Lorentz symmetry included physical effects at the level of the Standard Model, General Relativity, and beyond, possible origins and mechanisms, and associated classical and quantum issues in field theory, particle physics, gravity, and string theory.

I would like to thank the many people who helped to organize and run the meeting. Advice and assistance with various day-to-day matters were provided by Brett Altschul, Quentin Bailey, Robert Bluhm, Jorge Díaz, Jason Henline, Matt Mewes, and Jay Tasson, among numerous others. Several crucial organizational challenges were skillfully met by Jordan Tillett and Neil Russell, whose efforts were key to the smooth operation and success of the event.

Alan Kostelecký
October 2007

CONTENTS

IMPROVED TESTS OF LORENTZ AND CPT SYMMETRY USING NOBLE-GAS MASERS

ALEX GLENDAY, DAVID F. PHILLIPS AND RONALD L. WALSWORTH
Harvard-Smithsonian Center for Astrophysics
Cambridge, MA 02138, U.S.A.

We discuss recent improvements of the $^{129}Xe/^{3}He$ Zeeman maser used to make measurements constraining Lorentz and CPT violation. Experimental investigations of Lorentz and CPT symmetry provide important tests of the framework of the Standard Model of particle physics and theories of gravity. Previous measurements using the $^{129}Xe/^{3}He$ Zeeman maser set stringent limits on rotation- and boost-dependent Lorentz and CPT violation involving the neutron, consistent respectively with no effect at the level of 10^{-31} GeV and 10^{-27} GeV. Recent upgrades to the system should lead to an order-of-magnitude improvement in sensitivity to Lorentz and CPT violation.

1. Introduction

Lorentz symmetry is a fundamental feature of modern descriptions of nature. Lorentz transformations include both spatial rotations and boosts. Therefore, experimental investigations of rotation and boost symmetry provide important tests of the framework of the Standard Model of particle physics and single metric theories of gravity.

Clock comparisons[1] provide sensitive tests of these symmetries by bounding the frequency variation of a given clock as its orientation changes, e.g., with respect to the fixed stars. In practice, the most precise limits are obtained by comparing the frequencies of two co-located clocks as they rotate with the Earth and as they revolve with the Earth around the Sun. Atomic clocks are typically used, involving the electromagnetic signals emitted or absorbed on hyperfine or Zeeman transitions.

Here we discuss recent upgrades to our two-species $^{129}Xe/^{3}He$ Zeeman maser to enable an improved test of Lorentz and CPT symmetry. Using a previous version of this device we placed a limit on rotation-dependent Lorentz and CPT violation involving the neutron of 10^{-31} GeV,[2] improving by more than an order of magnitude on the best previous measurement.[3,4]

With the same device we performed the first clean test for the fermion sector of the symmetry of spacetime under boost transformations, placing a limit on boost-dependent Lorentz and CPT violation involving the neutron of 10^{-27} GeV.[5]

2. Motivation

Our atomic clock tests of Lorentz and CPT symmetry are interpreted in terms of the Standard-Model Extension (SME) developed by Kostelecký and others.[7] The SME parameterizes arbitrary coordinate-independent Lorentz violation. Since violation of CPT symmetry (the product of Charge conjugation, Parity inversion, and Time reversal) must come with Lorentz violation,[8] the SME also parameterizes general CPT violation. Observable Lorentz and CPT violation could be a remnant of Planck-scale physics. One attractive origin is spontaneous Lorentz breaking in a fundamental theory,[9] but other sources are possible.[10]

The SME provides a widely-accepted formalism for the interpretation and comparison of experimental searches for Lorentz and CPT violation, and has been applied to many systems, including mesons, photons, and leptons. For example, the SME admits Lorentz-violating couplings of noble-gas nuclei to expectation values of tensor fields. (Some of these couplings also violate CPT.) Each of the tensor fields may have an unknown magnitude and orientation in space, to be limited or detected by experiment. Our atomic-clock comparisons using the ^{129}Xe/^{3}He Zeeman maser have provided the most stringent laboratory tests of Lorentz and CPT symmetry (and hence rotation and boost invariance) in any system.[2,5]

3. ^{129}Xe/^{3}He maser upgrades for Lorentz/CPT tests

We provide here a brief review of the design and operation of the two-species ^{129}Xe/^{3}He maser, and an overview of recent upgrades that should enable an order-of-magnitude improvement in sensitivity to violations of Lorentz and CPT symmetry of the neutron. A schematic of the ^{129}Xe/^{3}He maser is shown in Fig. 1. Co-located ensembles of ^{129}Xe and ^{3}He atoms at pressures of hundreds of mbar are held in a double-chamber glass cell placed in a homogeneous magnetic field of ~ 6 G. Both species have spin-1/2 nuclei and the same sign nuclear magnetic dipole moment, but no higher-order electric or magnetic nuclear multipole moments. In one chamber of the glass cell, the pump bulb, the noble gas atoms are nuclear-spin-polarized into a population inversion by spin-exchange collisions with optically-pumped

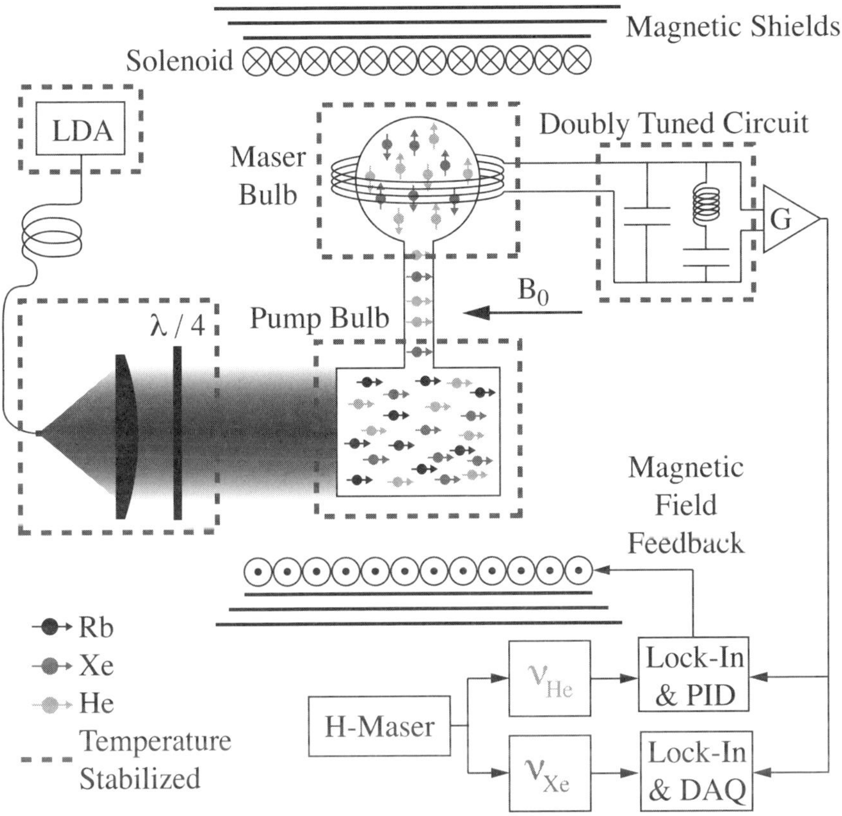

Figure 1. Schematic of the ^{129}Xe/^{3}He Zeeman maser (not to scale).

Rb vapor.[11] The noble gas atoms diffuse down a transfer tube into the second chamber, the maser bulb, which is surrounded by an inductive coil connected to a circuit resonant both at the ^{3}He and ^{129}Xe Zeeman frequencies (19.6 kHz and 7.1 kHz, respectively). For a sufficiently high flux of population-inverted nuclear magnetization, active maser oscillation of both species can be maintained indefinitely. By comparing one of the noble-gas masers (usually ^{129}Xe in practice) to a stable frequency reference (a hydrogen maser) which does not couple to SME tensor fields to leading order, we can stabilize the magnetic field and then use the other noble gas maser (usually ^{3}He) as a sensor for new physics coupling to spin, such as SME fields. The detection of SME fields relies on the fact that the gyromagnetic ratios of ^{3}He and ^{129}Xe differ by a factor of ~ 2.75, but the coupling to

SME fields should be the same for both noble gas species to leading order. We also assume that the SME fields are static and homogenous on scales of at least as long as a year and large as the solar system so that we can detect modulations in the maser frequencies as the Earth rotates and orbits the Sun.

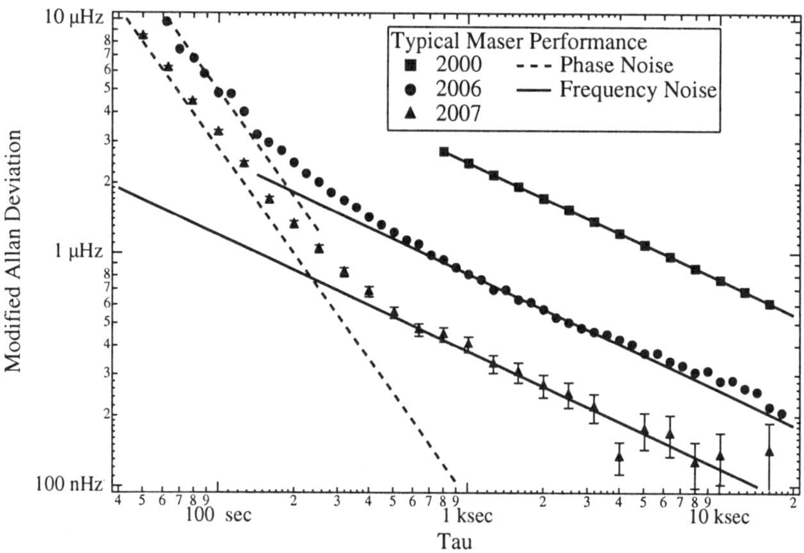

Figure 2. Examples of Allan deviation plots from the last several years, showing improvements in frequency stability of the $^{129}Xe/^{3}He$ Zeeman maser due to increased maser amplitudes (hence lower phase noise), longer coherence times (extends phase-to-frequency-noise crossover), and reduced long-term drift in system parameters.

To make a sensitive measurement of a Lorentz violating background field, the noble-gas masers must have excellent absolute frequency sensitivity, which requires narrow absolute linewidths (i.e., long atomic coherence times) and high signal-to-noise. They also need long term frequency stability to realize that sensitivity on the time scale of the rotation of the apparatus through the Lorentz violating background field. (In our experiments this has been performed using the Earth's rotation.) Very narrow absolute linewidths are possible due to the generally weak interactions of noble gas atoms with the walls and during atomic collisions. The ^{3}He and ^{129}Xe ensembles can have long Zeeman coherence (T_2) times of hundreds of seconds in magnetic fields of a few gauss with high absolute homogeneity.

Recent optimizations of the noble gas pressures and double bulb cell cell geometry to maximize maser amplitudes and coherence times as well as an increase in Zeeman frequency (i.e., applied magnetic field) to increase maser power led to significant improvements in frequency stability, as shown in Fig. 2. Importantly, the upgraded $^{129}Xe/^{3}He$ maser now regularly provides frequency stability on timescales of a day that is free of 1/f drift, with a modified Allan deviation that continues to decrease as $\tau^{-0.5}$ for consecutive measurement intervals τ. Thus no advantage would be gained by rotating the apparatus for tests of Lorentz symmetry (and problems associated with rotation can be avoided).

The frequency stability of the $^{129}Xe/^{3}He$ maser can be degraded by frequency shifts not corrected for by comagnetometery: including intra- and inter-noble gas species magnetization, noble gas collisions with polarized Rb, collisions with the cell wall, cavity pulling, imperfect spatial overlap of the masing ensembles, and the extended nature of the masing ensembles. All of these frequency shifts are temperature sensitive and hence good control of local environmental temperatures as well as heating power delivered by the optical pumping laser are critical for long term frequency stability. To improve the temperature stability of the double bulb cell we redesigned the oven in which the cell is contained (i) to use separate boron nitride (high thermal conductivity and mass) ceramic blocks around each bulb of the cell; and (ii) to heat and cool each block with blown air not directly in contact with the cell. With this system we achieved thermal stability of 1 m^{o}C per day for each temperature-controlled region. In addition, we added thermal stabilization to both the laser (an array of free running diodes) and the optics to stabilize the heat load from the absorbed optical pumping light.

Several systematic frequency shifts have been greatly reduced in the upgraded maser system. For example, the pump bulb (see Fig. 1) must be at a high temperature ($\sim 130\ ^{o}$C) with high Rb density to enable rapid noble gas population inversion; however, the upgraded temperature control system allows the maser bulb and transfer tube between the two bulbs to be up to 80 ^{o}C cooler with low Rb density and no laser light and hence negligible loss of polarization and Zeeman frequency shifts due to Rb-noble gas collisions. Also, the use of a spherical geometry maser bulb and optimization of noble gas pressures through systematic variation has greatly reduced frequency shifts due to intra-noble gas species magnetization (which is exactly eliminated for a sphere with uniform magnetization).

Although the double bulb geometry separates the pumping and masing actions, atoms in the pump bulb can still weakly couple to the inductive coil

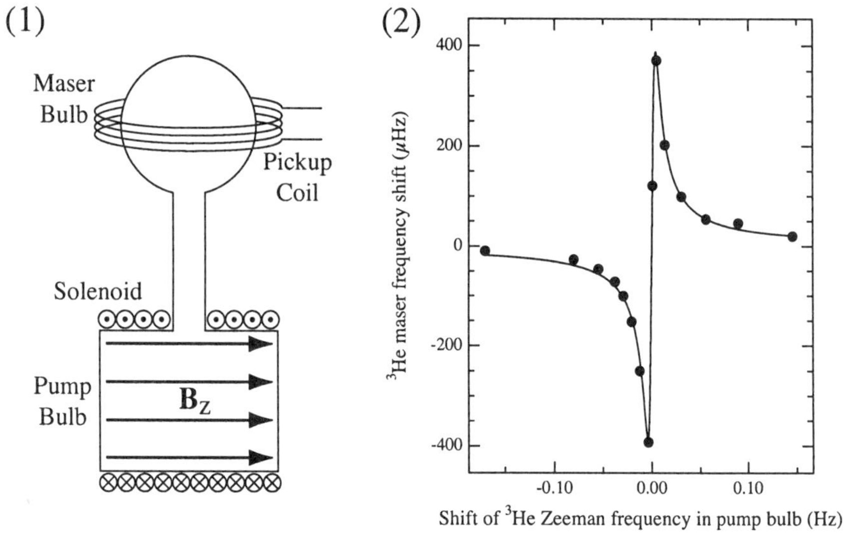

Figure 3. (1) A solenoid wrapped around the pump bulb is used to locally shift the noble gas Zeeman frequency away from that in the maser bulb. (2) Measurements of ^{3}He maser frequency as a function of ^{3}He Zeeman frequency in the pump bulb (shifted by the solenoid magnetic field). Detuning the Zeeman frequency in the pump bulb from that in the maser bulb by several linewidths prevents atoms in the pump bulb from participating in the maser ensemble.

surrounding the maser bulb, and thereby contribute to the maser ensemble and overall maser frequency (see Fig. 3). This effect is unwanted for several reasons: collisional frequency shifts due to polarized Rb in the pump bulb can affect the maser frequency; also small magnetic field gradients would lead to large frequency shifts given the distance between bulbs. To avoid these systematic effects we added a local magnetic field to the pump bulb (as shown in Fig. 3) to detune the noble gas Zeeman frequencies in the pump bulb many linewidths away from the maser bulb Zeeman frequencies so that the pump bulb atoms are not part of the masing ensembles.

Improved frequency stability of the ^{129}Xe/^{3}He maser requires stringent stability of the tuned circuit resonant frequencies to avoid cavity pulling of the maser frequencies. Improvements in coil design and construction greatly reduced the parasitic capacitance of the coil and the stray capacitive couplings of the coil to other parts of the circuit and environment; temperature stabilization of the double-tuned resonator was also significantly improved. As shown in Fig. 4, cavity-pulling shifts from the double-tuned resonator should not limit maser frequency stability on timescales of hours to a day.

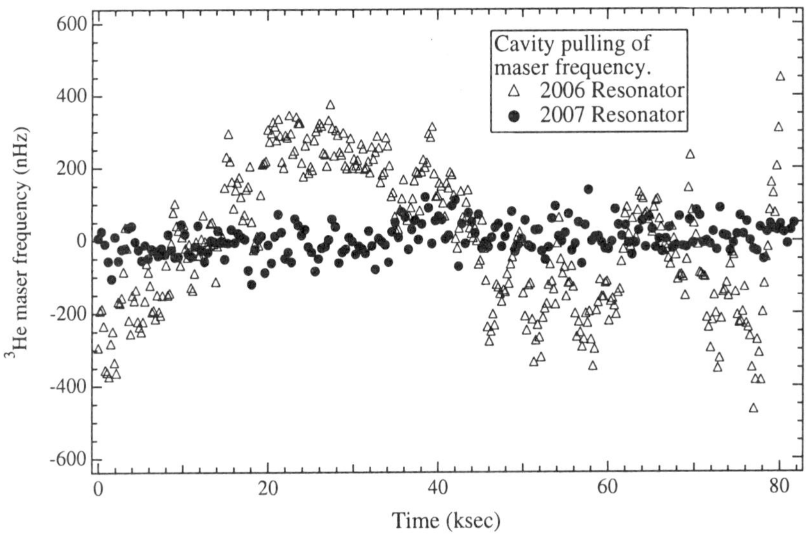

Figure 4. Upgraded stability of the double-tuned resonator has enabled improved long-term maser frequency stability, as illustrated in this comparison of one day's worth of data for the 2006 and 2007 resonators..

4. Conclusions

Precision comparisons of atomic clocks provide sensitive tests of Lorentz and CPT symmetry, thereby probing extensions to the Standard Model in which these symmetries can be spontaneously broken. Previous measurements using the two-species ^{129}Xe/^{3}He Zeeman maser constrain rotation-violation of the neutron at the 10^{-31} GeV level and boost-dependent violations at the 10^{-27} GeV level. We expect about an order of magnitude improvement in sensitivity to Lorentz/CPT violation of the neutron using an upgraded ^{129}Xe/^{3}He maser. Improved temperature control of the pump and maser regions, optimized noble gas pressures and cell geometry, increased Zeeman frequency, proper spatial definition of masing ensembles, and improved stability of the double-tuned resonator should enable this goal to be achieved over the next couple of years.

Acknowledgments

We gratefully acknowledge our collaborators on the maser upgrade work described above: Federico Canè, Matthew Rosen and Marc Browning. Support for the ^{129}Xe/^{3}He maser upgrade was provided by NASA and NSF.

References

1. V.A. Kostelecký and C.D. Lane, Phys. Rev. D **60**, 116010 (1999).
2. D. Bear, R.E. Stoner, R.L. Walsworth, V.A. Kostelecký, and C.D. Lane, Phys. Rev. Lett. **85**, 5038 (2000); *ibid.*, **89**, 209902 (2002).
3. C.J. Berglund, L.R. Hunter, D. Krause, Jr., E.O. Prigge, M.S. Ronfeldt, and S.K. Lamoreaux, Phys. Rev. Lett. **75**, 1879 (1995).
4. L. R. Hunter, C.J. Berglund, M.S. Ronfeldt, E.O. Prigge, D. Krause, Jr., and S.K. Lamoreaux, "A Test of Local Lorentz Invariance Using Hg and Cs Magnetometers," in V.A. Kostelecký, ed., *CPT and Lorentz Symmetry*, World Scientific, Singapore, 1999, p. 180.
5. F. Canè, D. Bear, D.F. Phillips, M.S. Rosen, C.L. Smallwood, R.E. Stoner and R.L. Walsworth, Phys. Rev. Lett. **93**, 230801 (2004).
6. D.F. Phillips, M.A. Humphrey, E.M. Mattison, R.E. Stoner, R.F. C. Vessot, and R.L. Walsworth, Phys. Rev. D **63**, 111101 (2001).
7. D. Colladay and V.A. Kostelecký, Phys. Rev. D **55**, 6760 (1997); **58**, 116002 (1998); V.A. Kostelecký and R. Lehnert, Phys. Rev. D **63**, 065008 (2001); V.A. Kostelecký, Phys. Rev. D **69**, 105009 (2004).
8. O.W. Greenberg, Phys. Rev. Lett. **89**, 231602 (2002); Phys. Lett. B **567**, 179 (2003).
9. V.A. Kostelecký and S. Samuel, Phys. Rev. D **39**, 683 (1989); Phys. Rev. Lett. **63**, 224 (1989); Phys. Rev. D **40**, 1886 (1989); V.A. Kostelecký and R. Potting, Nucl. Phys. B **359**, 545 (1991); Phys. Rev. D **51**, 3923 (1995).
10. For reviews of approaches to Lorentz and CPT violation, see, for example, V.A. Kostelecký, ed., *CPT and Lorentz Symmetry I, II, III*, World Scientific, Singapore, 1999, 2002, 2005.
11. T.E. Chupp *et al.*, Phys. Rev. A **38**, 3998 (1988); G.D. Cates *et al.*, Phys. Rev. A **45**, 4631 (1992).
12. K.F. Woodman *et al.*, J. Navig. **40**, 366 (1987).
13. R.E. Stoner and R.L. Walsworth, Phys. Rev. A **66**, 032704 (2002).
14. R. Bluhm *et al.*, Phys. Rev. Lett. **88**, 090801 (2002).
15. D. Kleppner, H.M. Goldenberg, and N.F. Ramsey, Phys. Rev. **126**, 603 (1962); D. Kleppner, H.C. Berg, S.B. Crampton, N.F. Ramsey, R.F.C. Vessot, H.E. Peters, and J. Vanier, Phys. Rev. **138**, A972 (1965).
16. H.G. Andresen, Z. Physik, **210**, 113-141 (1968).
17. M.A. Humphrey, D.F. Phillips, and R.L. Walsworth, Phys. Rev. A **62**, 063405 (2000).

A MODERN MICHELSON-MORLEY EXPERIMENT USING ACTIVELY ROTATED OPTICAL RESONATORS

S. HERRMANN, A. SENGER, K. MÖHLE, E.V. KOVALCHUK, A. PETERS

Institut für Physik, Humboldt-Universität zu Berlin
Hausvogteiplatz 5-7, 10117 Berlin
E-mail: sherrma@stanford.edu, achim.peters@physik.hu-berlin.de

We present a new setup of a Michelson-Morley test of the isotropy of the speed of light that achieves an order of magnitude improvement in sensitivity as compared to previous measurements. The experiment compares the resonance frequencies of two orthogonal cavities, implemented in a single block of fused silica and rotated on an air bearing turntable once every 45 s. A preliminary analysis of first data already provides limits on combinations of SME parameters at the level of 10^{-17}.

1. Introduction

The experiment of A. A. Michelson and E. W. Morley,[1] testing the isotropy of the speed of light, has served as a sensitive test of special relativity and Lorentz invariance for more than a century now. Numerous repetitions have been performed so far, increasing the sensitivity by more than six orders of magnitude. During recent years results of several modern versions of this experiment have been used to provide upper limits on test parameters of the photonic sector of the Standard-Model Extension[8] (SME) down to few parts in 10^{16}. These measurements compared the resonance frequencies of rotating optical[2,3,4] or microwave[5,6,7] cavities, either using active rotation on a turntable or relying solely on Earth's rotation. An anisotropy of the speed of light, as a consequence of broken Lorentz invariance, is currently restricted to a level of $\Delta c/c < 10^{-16}$ by these measurements.

Here, we present a new setup aiming to improve this limit by at least one order of magnitude. The basic scheme of the experiment is depicted to the left in Fig. 1. In vacuum the frequency of the TEM_{00} mode of a linear optical Fabry-Pérot cavity is given by an integer multiple of $c/2L$, where c is the speed of light and L is the cavity length. This frequency is read out by stabilizing a laser of frequency ν_1 to the cavity resonance using an electronic feedback loop. A second laser of frequency ν_2 is stabilized to a similar orthogonal cavity and a beat frequency measurement of $\Delta\nu = \nu_1 - \nu_2$ is performed while the whole setup rotates continuously. An anisotropy of c would then lead to a modulation of the beat frequency at twice the rotation rate ω_{rot}:

$$\frac{\Delta c}{c} \sim \frac{\Delta\nu}{\nu_0} = \frac{\nu_1 - \nu_2}{\nu_0} = B \sin 2\omega_{\mathrm{rot}} t + C \cos 2\omega_{\mathrm{rot}} t, \quad (1)$$

where $\nu_1, \nu_2 \approx \nu_0 = 282\,\mathrm{THz}$. If we take into account Earth's rotation at $\omega_\oplus$ and consider the case of $\omega_{\mathrm{rot}} \gg \omega_\oplus$, this leads to a modulation of the amplitudes B and C as described by

$$\begin{aligned} B = {} & B_0 + B_{\mathrm{s}1} \sin \omega_\oplus t + B_{\mathrm{c}1} \cos \omega_\oplus t \\ & B_{\mathrm{s}2} \sin 2\omega_\oplus t + B_{\mathrm{c}2} \cos 2\omega_\oplus t. \end{aligned} \quad (2)$$

and

$$\begin{aligned} C = {} & C_0 + C_{\mathrm{s}1} \sin \omega_\oplus t + C_{\mathrm{c}1} \cos \omega_\oplus t \\ & C_{\mathrm{s}2} \sin 2\omega_\oplus t + C_{\mathrm{c}2} \cos 2\omega_\oplus t, \end{aligned} \quad (3)$$

where the daily modulation amplitudes B_k and C_k depend on the geographical latitude χ of the laboratory and may carry an annual phase shift due to Earth's orbital motion.

In what follows we describe the actual realization of the basic setup of Fig. 1 in more detail. We then present first results on the signal amplitudes B and C of Eq. (1) which we consider to be a null result and we will give results from an evaluation of the experiment in terms of the SME.

2. The experiment

The experiment employs a pair of orthogonal cavities implemented in a single block of fused silica as shown in Fig. 1. The spacer is a 55 mm × 55 mm × 35 mm cuboid with centered perpendicular bore holes of 10 mm diameter along each axis. Four mirror substrates coated with a high-reflectivity dielectric coating at 1064 nm are optically contacted to either side, forming two crossed optical cavities the length of which is matched to better than $2\,\mu m$. The finesse is 380000 resulting in a linewidth of 7 kHz for each cavity.

The cavities are set up in a custom-made vacuum chamber featuring several stages of thermal insulation, combined with high mechanical rigidity. The complete chamber is pumped by an ion pump to a pressure of $< 10^{-5}$ mbar. The innermost stage is designed as an inner vacuum chamber, with the option to be separately pumped to Ultra High Vacuum. To isolate the cavity from ambient vibrations, the vacuum chamber is placed

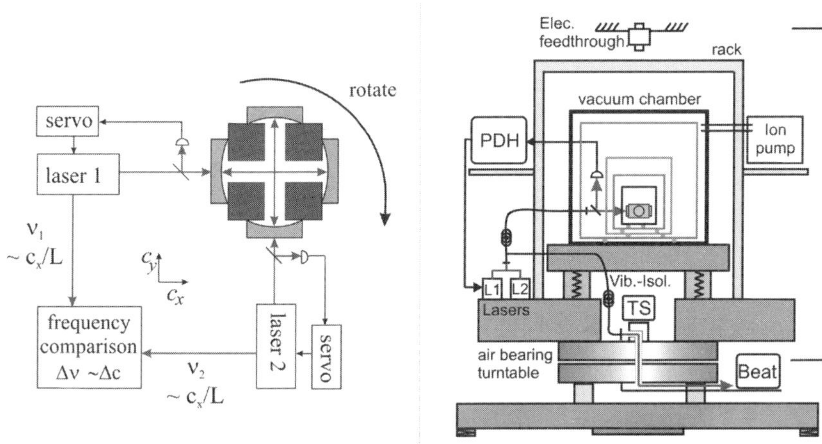

Figure 1. Basic principle of the experiment (left) and schematic of the complete rotating setup (right). TS = Tilt sensor, PDH = Pound-Drever-Hall laser stabilization electronics.

on an active vibration isolation system (HWL Scientific 350M), with a resonance frequency of 1.2 Hz and an isolation approaching 40 dB for frequencies well above resonance.

Two Nd:YAG lasers at 1064 nm are used to read out the cavity frequencies. The light of both lasers is overlapped in a polarization-maintaining single-mode fiber (using identical polarizations), which enters the chamber via a vacuum feedthrough. The chamber contains miniaturized optics to couple the light into the cavities, and detectors to detect the reflected light. The laser frequencies are stabilized to the cavities using feedback loops that apply frequency discrimination according to a modified Pound-Drever-Hall method, demodulating the signal at three times the modulation frequency. The lasers are equipped with piezoelectric actuators, which allow tuning and modulation of the laser frequency. The beat frequency measurement between the two stabilized lasers is accomplished by overlapping a fraction of the laser power on a fast photodiode outside the vacuum chamber. The beat frequency on the order of up to half a free spectral range of the cavities (2.7 GHz) is converted down to below 100 MHz and counted at a rate of 1 data point per second. The observed relative frequency stability reaches a optimum close to 1.5×10^{-15} at an integration time of about 20 s (see Fig. 2). The long term relative frequency drift is on the order of 10 mHz/s. Comparison to a third frequency reference based on an independent cavity shows that this relative drift is reduced by a factor of 100 as compared to the absolute frequency drift of each cavity. This reduction is due to common mode suppression of relative length changes of the spacer block.

The vacuum chamber on the vibration isolation platform, the optical setup, and the laser stabilization electronics are mounted on an optical breadboard which is placed on a precision air bearing turntable (Kugler RTV600) that can be rotated continuously. Electrical connections are made via a 15 channel slip ring feedthrough. The beat measurement and data acquisition is most conveniently done in the non-rotating laboratory frame. Thus, the laser beams that are split off for the beat measurement are overlapped at the center of the turntable, aligned with the downward rotation axis and are transferred to a stationary platform below the table through a center borehole. After passing two quarter-wave plates on axis, one rotating and one stationary, the light maintains a constant linear polarization at the beat detector. Active rotation of the setup on a turntable gives rise to systematic effects that compromise a possible anisotropy signal. The most obvious effects occur due to modulated forces acting on the cavities, such as gravitational bending or centrifugal forces. To minimize these systematics, we use a precision turntable specified for axial and planar true run of better than $1\,\mu$rad and $0.1\,\mu$m respectively. Slowly varying tilt of the rotation axis against the vertical is reduced to less than $1\,\mu$rad by active stabilization[9] and does not contribute to more than 0.1 Hz systematic beat frequency variation. Varying centrifugal forces are also reduced to below

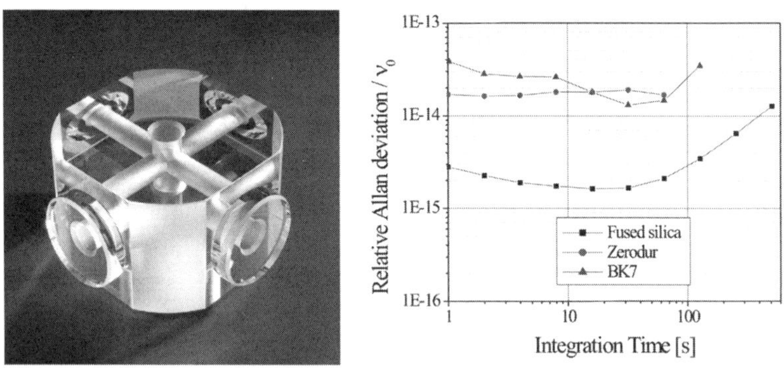

Figure 2. Picture of the crossed resonators (left) and relative frequency stability (right). Also shown are the frequency stability obtained from comparison of two crossed Zerodur resonators of identical design as the fused silica ones and the stability obtained with another set of cavities using BK7 mirror substrates within a previous setup. The achieved ten-fold improvement is attributed mostly to the lower level of thermal noise for fused silica mirror substrates as compared to substrates made from Zerodur or BK7.

0.1 Hz systematic beat frequency variation by actively stabilizing the rotation rate. Further measures to reduce systematic frequency variations include balancing the center of mass of the table (estimated to be better than 1 mm offset from the rotation axis) and shielding the lasers and optics outside the vacuum chamber against air currents, temperature gradients etc. The rotation rate of the turntable was set to 45 s. Even faster rotation would have allowed to enhance data integration, but also resulted in an increase of residual systematic effects compromising the measurement. At the chosen rotation rate the observed residual systematic frequency variations occur mainly at $\omega_{\rm rot}$ and are below 1×10^{-15}, while the residual systematic effects at the relevant Fourier component $2\omega_{\rm rot}$ are even less pronounced. Moreover, as long as these systematic effects are reasonably stationary in the laboratory frame, they will be averaged out by Earth's rotation in the following analysis looking for sidereal effects.

3. Preliminary results

Measurements have been performed almost continuously during a time spanning from April 23d to June 4th 2007. The total data includes 926 hours of measurement corresponding to $\sim$ 74000 turntable rotations. For the analysis, the data sets have been divided into samples of ten table rotations each. This corresponds to a time span of 450 s $\ll \omega_\oplus$ such that a variation of B and C due to Earth's rotation can be neglected within a sin-

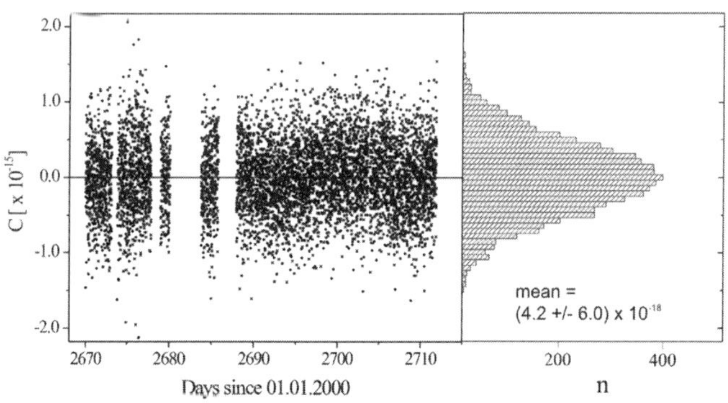

Figure 3. Cosine amplitudes C of frequency variation at twice the turntable rotation. Each point is determined from a sample of 450 s of data comprising 10 table rotations).

gle sample. Each sample is fitted with Eq. (1) including an offset, a linear drift and signals at $\sin\omega_{\rm rot}t$ and $\cos\omega_{\rm rot}t$ to account for the small residual effects at the turntable rotation frequency. The phase of this fit has been set by choosing $t = 0$ when one of the cavities is oriented North-South. The amplitudes $B^{(i)}$ and $C^{(i)}$ obtained from the ith sample are assigned with the center time t_i of this sample on a time axis based on the modified Julian date. Every 24 h of data then yield 192 values of $B^{(i)}$ and $C^{(i)}$ equally distributed over t_i as shown in Fig. 3 for the C amplitudes of the complete measurement span. The distribution is consistent with zero and features a one sigma error bar $\leq 1 \times 10^{-17}$ for the mean of the distribution (assuming uncorrelated noise, which is consistent with the more detailed SME analysis presented below).

For a further analysis of these results we apply the SME in a similar way as in the analysis of a previous experiment described in Refs. 3, 9. The starting point is the photonic sector of the SME, taking into account the Lorentz violating extension as given by[10]

$$\mathcal{L} = -\frac{1}{4}F_{\mu\nu}F^{\mu\nu} - \frac{1}{4}(k_F)_{\mu\nu\kappa\lambda}F^{\mu\nu}F^{\kappa\lambda}. \tag{4}$$

The tensor k_F possesses the symmetries of the Riemann tensor and thus is left with 19 independent components. Ten of these describe polarization dependent effects and have been restricted to values below 2×10^{-32} by astrophysical observations.[10,11] The remaining 9 parameters are only accessible in laboratory experiments. They can be grouped into one scalar $\kappa_{\rm tr}$ and two traceless 3x3 matrices, κ_{e-} (symmetric, parity even) and κ_{o+} (antisymmetric, parity odd). The effect of the Lorentz violating extension in (4) on the resonance frequency of an optical Fabry-Pérot resonator has been elaborated by Kostelecký and Mewes.[10] Based on those results we have derived expressions for the signal amplitudes B_k and C_k of Eqs. (2) and (3) that apply to the specific geometry of our experiment.[9] Fitting the resulting expressions to our data following the same procedure as described previously[3,9] yields estimates on the photonic SME parameters as given in Tab. 1. These estimates are generally consistent with zero within one sigma at a level of 1×10^{-17} for the κ_{e-} parameters and 10^{-13} for the κ_{o+} parameters. This is about an order of magnitude more stringent as compared to the previous best limits.[7] Note that due to the limited measurement span of one month we have to assume that no cancelation occurs between κ_{e-} and κ_{o+} terms. In order to reduce this correlation data spanning more than one year would be required.

Table 1. Preliminary limits on SME parameters assuming no cancelation between κ_{e-} and κ_{o+} parameters. All values are $\times 10^{-17}$. $\beta = v/c = 10^{-4}$ accounts for Earth's orbital boost.

κ_{e-}^{XY}	κ_{e-}^{YZ}	κ_{e-}^{XZ}	$\kappa_{e-}^{XX} - \kappa_{e-}^{YY}$	κ_{e-}^{ZZ}
-0.1 ± 0.6	-0.3 ± 1.4	-2.0 ± 0.9	-2.0 ± 1.7	-0.2 ± 3.1
$\beta\kappa_{o+}^{XY}$	$\beta\kappa_{o+}^{YZ}$	$\beta\kappa_{o+}^{XZ}$		
-2.5 ± 2.5	-1.0 ± 1.5	1.5 ± 1.7		

4. Outlook

To conclude, we have set up a Michelson-Morley experiment which after one month of data integration has already improved current limits on a possible violation of Lorentz invariance by one order of magnitude. These limits exclude an anisotropy of the speed of light at a level of $\Delta c/c < 1 \times 10^{-17}$. The experiment applies cavities specially designed for the use in a Michelson-Morley experiment and benefits from a suppression of systematic effects down to a level of few parts in $\Delta\nu/\nu \sim 10^{-17}$. Further improved limits, as well as a completely independent determination of the κ_{e-} and κ_{o+} terms, should be readily obtained by extending the measurement time span to more than one year. In the longer term, it should be possible to overcome the current sensitivity limitations due to mirror thermal noise by combing cryogenic optical resonators[2] and extremely well controlled active rotation in a single experimental setup. Together with a reasonable improvement in the suppression of systematic effects, this would then allow one to test for potential violations of Lorentz invariance in the $\Delta c/c \sim 10^{-20}$ regime.

References

1. A.A. Michelson and E.W. Morley, Am. J. Sci. **34**, 333 (1887).
2. H. Müller *et al.*, Phys. Rev. Lett. **91**, 020401 (2003).
3. S. Herrmann *et al.*, Phys. Rev. Lett. **95**, 150401 (2005).
4. P. Antonini *et al.*, Phys. Rev. A **71**, 050101 (2005).
5. P. Wolf *et al.*, Phys. Rev. D **70**, 051902 (2004).
6. P. Stanwix *et al.*, Phys. Rev. Lett. **95**, 040404 (2005).
7. P. Stanwix *et al.*, Phys. Rev. D **74**, 081101 (2006).
8. D. Colladay and V.A. Kostelecký, Phys. Rev. D **58**, 116002 (1998).
9. S. Herrmann *et al.*, Lect. Notes Phys. **702**, 385 (2006).
10. V.A. Kostelecký and M. Mewes, Phys. Rev. D **66**, 056005 (2002).
11. B. Altschul, Phys. Rev. Lett. **96**, 201101 (2006).

ROTATING EXPERIMENTS TO TEST LORENTZ INVARIANCE IN THE PHOTON SECTOR

M.E. TOBAR, P.L. STANWIX, M.M. MIAO, E.N. IVANOV, A.C. FOWLER, J.G. HARTNETT, J.-M. LE FLOCH

University of Western Australia, School of Physics M013
35 Stirling Hwy., Crawley 6009 WA, Australia
E-mail: mike@physics.uwa.edu.au

P. WOLF

LNE-SYRTE, Observatoire de Paris
61 Av. de l'Observatoire, 75014 Paris, France

We present results from two complementary tests of Lorentz invariance in electrodynamics. The first test is an even parity test, which compares two orthogonal cryogenic sapphire microwave oscillators rotating in the lab. We have now acquired over 1 year of data, allowing us to place independent limits on all eight $\tilde{\kappa}_{e-}$ and $\tilde{\kappa}_{o+}$ parameters. The second test is a rotating odd parity test, which compares the phase shift of two one-way propagating waves that experience different permeability over the length of propagation. A sensitive carrier suppression microwave interferometer is used to obtain highly sensitive phase comparison using a Mach-Zehnder configuration. This experiment is sensitive to the isotropic Lorentz violating parameter $\tilde{\kappa}_{tr}$, and is the first rotating experiment of this type with the first operation of this experiment seting a limit order 10^{-6} in a few weeks.

1. Rotating Cryogenic Sapphire Oscillator Experiment

Our experiment consists of two cylindrical sapphire resonators of 3 cm diameter and height supported by spindles within super-conducting niobium cavities.[1] The sapphire loaded cavities are situated one above the other, oriented with their cylindrical axes orthogonal to each other in the horizontal plane. The experiment is rotated with a period of 18 seconds around its vertical axis. Whispering gallery modes[2] are excited in each near 10 GHz, with a difference frequency between the two of 226 kHz. The difference frequency along with various experimental parameters are logged by a stationary data acquisition system as a function of the experiment's orientation. A detailed description of the experiment is given in Ref. 3.

To test for Lorentz violations we derive the perturbation of the difference frequency with respect to an alternative test theory. In the photon sector of the Standard-Model Extension (SME) this may be calculated to first order as the integral over the non-perturbed fields (Eq. (34) of Ref. 4; see Refs. 3, 5 for an application to our case). The change in orientation of the fields due to the rotation of the experiment in the lab and Earth's orbital and sidereal motion induces a time varying modulation of the difference frequency, which is searched for in the experiment. In the photon sector of the SME,[4] Lorentz violating terms are parameterized by 19 independent components, which are in general grouped into three traceless and symmetric 3×3 matrices ($\tilde{\kappa}_{e+}$, $\tilde{\kappa}_{o-}$, and $\tilde{\kappa}_{e-}$), one antisymmetric matrix ($\tilde{\kappa}_{o+}$) and one additional scalar, which all vanish when LLI is satisfied. The 10 independent components of $\tilde{\kappa}_{e+}$ and $\tilde{\kappa}_{o-}$ have been constrained[4,6] by astronomical measurements to $< 2 \times 10^{-32}$. Recently two combinations of these parameters have been further constrained[7] to less than parts in 10^{-37}. The scalar $\tilde{\kappa}_{tr}$ component has been constrained[8,9] to $< 2.2 \times 10^{-7}$ through the re-analysis of a previous Ives-Stilwell experiment.[8,10] The interferometric technique presented later in this paper has the potential to improve on this result by four orders of magnitude.[8] Seven components of $\tilde{\kappa}_{e-}$ and $\tilde{\kappa}_{o+}$ have been independently constrained in stationary optical and microwave cavity experiments[11,12,13] at the 10^{-15} and 10^{-11} level respectively. The last remaining component $\tilde{\kappa}_{e-}^{ZZ}$ was only recently constrained for the first time by a group of cavity experiments[14,15,16,17,18] designed to both improve on the results of Refs. 11, 12, 13 and, more importantly, be sensitive to $\tilde{\kappa}_{e-}^{ZZ}$ through the use of active rotation in the laboratory.

However, the most stringent independent limits on the isotropy ($\tilde{\kappa}_{e-}$) and boost terms ($\tilde{\kappa}_{o+}$) can only be achieved with 1 year of data. This is because the maximum boost with respect to the Sun Centered Equatorial Celestial Frame (SCECF) is due to the Earth's annual motion. Thus, over 1 year of data is required to decorrelate the parameters. Previous analysis,[13,14,15] which contained significantly less than one year of data, constrained the $\tilde{\kappa}_{e-}$ and $\tilde{\kappa}_{o+}$ parameters by assuming no cancellation occurred in the case of a non-zero Lorentz violating effect. We have now acquired sufficient data to remove this assumption, producing independent limits on all of the eight components of $\tilde{\kappa}_{e-}$ and $\tilde{\kappa}_{o+}$.

In our previous analysis[14] the amplitude and phase of a Lorentz violating signal was determined by fitting the parameters of Eq. (1) to the data, with the phase of the fit adjusted according to the test theory used:

$$\frac{\Delta\nu_0}{\nu_0} = A + Bt + \sum_i C_i \cos(\omega_i t + \varphi_i) + S_i \sin(\omega_i t + \varphi_i). \qquad (1)$$

Here ν_0 is the average unperturbed frequency of the two sapphire resonators, and $\Delta\nu_0$ is the perturbation of the 226 kHz difference frequency. A and B determine the frequency offset and drift, and C_i and S_i are the amplitudes of a cosine and sine at frequency ω_i respectively. In the final analysis we fit 15 frequencies to the data, $\omega_i = (2\omega_R, 2\omega_R \pm \Omega_\oplus, 2\omega_R \pm \omega_\oplus, 2\omega_R \pm \omega_\oplus \pm \Omega_\oplus, 2\omega_R \pm 2\omega_\oplus, 2\omega_R \pm 2\omega_\oplus \pm \Omega_\oplus)$, where ω_R is the rotation frequency of the experiment in the lab and $\omega_\oplus$ and $\Omega_\oplus$ are the sidereal and annual frequencies of the Earth's rotational and orbital motion respectively. Since the residuals of the fit exhibit a significantly non-white behavior, the optimal regression method is weighted least squares (WLS).[12] WLS involves pre-multiplying both the experimental data and the model matrix by a whitening matrix determined by the noise type of the residuals of an ordinary least squares analysis. However, this method of analysis proved to be computationally intensive due to the large amount of data we have now acquired. For this reason, an alternative approach used by Refs. 15, 16 was adopted. Using this technique we reduce the size of the data set by demodulating it in quadrature with respect to $2\omega_R$ in blocks of 40 periods of rotation. The number of periods was chosen to minimize the net effect of narrow band noise (due to instabilities in the systematic at $2\omega_R$) and broad band noise (due to oscillator frequency noise), which is similar to an optimal filter. By fitting the expression of Eq. (2) to each block of data using an ordinary least squares regression technique we determine the coefficients $S(t)$ and $C(t)$, which can be considered linear combinations of the sidereal, semi-sidereal, and annual modulations and combinations thereof. The relationship between $S(t)$ and $C(t)$ and the various modulation frequencies is expressed in Eqs. (3) and (4), where $\omega_i = (\Omega_\oplus, \omega_\oplus, \omega_\oplus \pm \Omega_\oplus, 2\omega_\oplus, 2\omega_\oplus \pm \Omega_\oplus)$, with the coefficients calculated in Ref. 19:

$$\frac{\Delta\nu_0}{\nu_0} = A + Bt + S(t)\sin(2\omega_R t + \varphi) + C(t)\cos(2\omega_R t + \varphi), \qquad (2)$$

$$S(t) = S_0 + \sum_i S_{s,i} \sin(\omega_i t + \varphi_i) + S_{c,i} \cos(\omega_i t + \varphi_i), \qquad (3)$$

$$C(t) = C_0 + \sum_i C_{s,i} \sin(\omega_i t + \varphi_i) + C_{c,i} \cos(\omega_i t + \varphi_i). \qquad (4)$$

A comparison was made between the two techniques by performing a complete analysis of 30 data sets (3 data sets were later excluded from the analysis due to overly large and varying systematic signals at $2\omega_R$). Both techniques produced consistent results, with the uncertainties associated with the demodulated technique being lower than the WLS technique by no more than 15 percent. The difference between the two techniques is most likely due to the efficiency with which the data analysis could be optimized for the noise type present in the data. WLS only takes into account the broad band noise (spectral density) whereas the optimization used in the demodulated technique takes into account the extra noise source of instability of the systematic at $2\omega_R$. Hence, the latter approach was adopted in further investigations of the data.

Table 1. Results for the SME Lorentz violation parameters determined independently in this work. Also shown for comparison is the previous best independent constraints of seven parameters[12] and more recent short term results that assume no cancellation between the $\tilde{\kappa}_{e-}$ and $\tilde{\kappa}_{o+}$ terms, other than[15,14] $\tilde{\kappa}_{e-}^{ZZ}$ ($\tilde{\kappa}_{e-}$ in 10^{-16}, $\tilde{\kappa}_{o+}$ in 10^{-12}). The P_{MM} parameter from the RMS framework is also listed (in 10^{-11}).

Parameter	This work	Previous analysis[12]	Recent short analysis[15,14]
$\tilde{\kappa}_{e-}^{XY}$	2.9(2.3)	-57(23)	-3.1(2.5)
$\tilde{\kappa}_{e-}^{XZ}$	-6.9(2.2)	-32(13)	1.9(3.7)
$\tilde{\kappa}_{e-}^{YZ}$	2.1(2.1)	-5(13)	-4.5(3.7)
$(\tilde{\kappa}_{e-}^{XX} - \tilde{\kappa}_{e-}^{YY})$	-5.0(4.7)	-32(46)	5.4(4.8)
$\tilde{\kappa}_{e-}^{ZZ}$	143(179)	-	-19.4(51.8)
$\tilde{\kappa}_{o+}^{XY}$	-0.9(2.6)	-18(15)	2.0(2.1)
$\tilde{\kappa}_{o+}^{XZ}$	-4.4(2.5)	-14(23)	-3.6(2.7)
$\tilde{\kappa}_{o+}^{YZ}$	-3.2(2.3)	27(22)	2.9(2.8)
P_{MM}	9.4(8.1)	120(220)[5]	-21(19)

The data used in this analysis span a period from December 2004 to January 2006. It consists of 27 sets of data totalling approximately 121 days (see Ref. 19 for details). An offset and drift has been removed from the coefficients derived from each data set. As described earlier, these data are then used to determine the amplitudes of the frequencies of interest. In Ref. 14 we describe how systematic effects dominate the data at $2\omega_R$, limiting our ability to constrain test theory parameters associated with this frequency (a detailed discussion of the systematics and their effect is thus left out here). Also, we do not consider the nearby annual offsets

$(2\omega_R \pm \Omega_\oplus)$ for two reasons. Firstly, the strong systematic signal at $2\omega_R$ has been shown to have a significant effect on nearby sidebands due to leakage,[14] and secondly, by subtracting a linear drift from the individual data sets after being demodulated it is possible that a signal at the annual frequency may be suppressed so is not included in the analysis. However, all other frequencies of interest are close to the sidereal or semi-sidereal frequencies, so will be unaffected by the removal of an offset and drift from each data set.

In the SME, all $\tilde{\kappa}_{e-}$ and $\tilde{\kappa}_{o+}$ parameters other than $\tilde{\kappa}_{e-}^{ZZ}$ can be constrained from the sidereal and semi-sidereal frequencies and their annual frequency offsets as outlined in Ref. 19. $\tilde{\kappa}_{e-}^{ZZ}$ only appears in the coefficient $C_{C,0}$ so to determine a limit we consider the $C_{C,0}$ coefficients for each data set to be independent and treat them statistically. The systematic at $2\omega_R$ has been shown to be primarily due to tilt variations. It remains relatively constant in phase within a data set but varies between data sets. The mean and standard error is used to calculate $\tilde{\kappa}_{e-}^{ZZ}$.[19] The results for the SME analysis are given in Table 1. We note that the results for $\tilde{\kappa}_{e-}^{XZ}$ and $\tilde{\kappa}_{o+}^{XZ}$ are significant at approximately the 3σ and 2σ level respectively. However, we do not believe this to be an indication of a Lorentz violating effect for reasons similar to those given in Ref. 12, which also used data taken over more than one year. Our result for $\tilde{\kappa}_{e-}^{XZ}$ is inconsistent with other recent measurements shown in Table 1. Also, an examination of the corresponding sideband coefficients from an analysis of the individual data sets (not shown here) shows no coherence in the phase of the signal, which would be expected in the presence of a genuine Lorentz violating effect.

In conclusion, by collecting over one year of data we have been able to set the first independent limits on 8 parameters in the photon sector of the SME, without assuming that no cancellation occurs between the isotropy and boost terms. The results do not indicate any Lorentz violating effects, and compared to previous experiments we see a slight improvement in the constraints on these parameters. We improve on our previous determination of $\tilde{\kappa}_{e-}^{ZZ}$ by more than a factor of three. However, due to the systematic disturbances present at twice the rotation frequency we are unable to measure this parameter with the precision of Ref. 15, which has developed a tilt control system which avoids the major rotation induced systematic.

To improve on these results we intend to replace the resonators with higher quality sapphire loaded cavities, which have a frequency instability approximately 40 times lower than the current experiment.[20] Considerable effort will need to be invested to improve the rotation system and reduce environmental disturbances for this improvement to be realized.

2. Rotating Magnetic Interferometer Experiment

This experiment was originally proposed in Ref. 8 and is a new way to measure the isotropic Lorentz violating parameter $\tilde{\kappa}_{tr}$ from the SME. The experiment uses a magnetically asymmetric MZ interferometer, which provides a null experiment sensitive to the same SME parameters as an Ives-Stillwell experiment. The experiment relies on microwave interferometer technology that we have already developed for low noise phase detection and oscillators,[21,22] and here we briefly report on the first operation of the experiment.

The first operation began in August 2006, with the interferometer consisting of a magnetic ferrite wave guide of 12 cm of length in one arm, with a balancing arm consisting of a variable attenuator and phase shifter. The interferometer was placed in a vacuum tight stainless steel chamber temperature controlled to a temperature above ambient through a heater and a commercial temperature control system. The interferometer was balanced and left rotating for a period of a few days at a frequency of 0.17 Hz. The biggest frequency component of phase shift detected was at the diurnal frequency, of about 10^{-3} radians. This effect was eliminated by introducing voltage controlled devices, and controlling the Dark Port (DP) of the interferometer to a null, with a control loop of 0.4 Hz bandwidth. Since the bandwidth of the cancellation is slightly larger than the rotation frequency the effect of the noise cancellation at 0.17 Hz must be taken into account when calibrating the system. However, the diurnal effect is greatly reduced by four orders of magnitude, and no longer destabilized the interferometer. The calibration from voltage at the output of the readout mixer to phase detected by the interferometer was measured to be 6.5 volts per radian at the 0.17 Hz rotation frequency (which includes the effect of the filter control loop).

The next series of experiments revealed a large rotation systematic of order 10^{-4} radians. It turned out it was due to magnetic field interacting with the rotating wave guide. Thus, we manufactured a Mu-Metal shield for the wave guide, which significantly reduced the amplitude to parts in 10^{-7} radians. This value was about the same if the wave guide was substituted with a non-magnetic piece of cable, implying the effects of magnetic field on the wave guide were eliminated.

By May 2007 we started to take data for analysis. For the first three days, the standard error (calculated using least squares fit) was of order 10^{-9} radians at the frequency components of interest[8] (4 frequencies in total when we apply the small data set approximation). Taking into account the boost factor dependence and the amplitude of the expected signals a limit of order 10^{-6} on $\tilde{\kappa}_{tr}$ has been achieved. To improve on this recycling can be implemented as discussed in Ref. 8, and will be investigated. Detailed analysis is not presented here, as we are still optimizing the experiment, and most likely in a few month a proper analysis and limit will be given.

References

1. A.J. Giles *et al.*, Physica B **165,** 145, (1990).
2. M.E. Tobar and A.G. Mann, IEEE Trans. Microw. Theory Tech. **39 (12),** 2077 (1991).
3. M.E. Tobar *et al.*, in J. Ehlers and C. Lämmerzahl, eds., *Special Relativity: Will it Survive the Next 101 Years?* Lect. Notes Phys. **702**, 416 (2006); arXiv:hep-ph/0506200 (2005).
4. V.A. Kostelecký, M. Mewes, Phys. Rev. D **66**, 056005 (2002).
5. P. Wolf, *et al.*, Gen. Rel. and Grav., **36**, 10, 2351 (2004).
6. V.A. Kostelecký, M. Mewes, Phys. Rev. Lett. **87**, 251304 (2001).
7. V.A. Kostelecký, M. Mewes, arXiv:hep-ph/0607084 (2006).
8. M.E. Tobar *et al.*, Phys. Rev. D **71**, 025004 (2005).
9. M. Hohensee *et al.*, Phys. Rev. D **75**, 049902(E) (2007).
10. G. Saathoff, *et al.*, Phys. Rev. Lett. **91**, 190403 (2003).
11. H. Müller H. *et al.*, Phys. Rev. Lett. **91**, 2 020401, (2003).
12. P. Wolf *et al.*, Phys. Rev. D **70**, 051902(R) (2004).
13. J.A. Lipa *et al.*, Phys. Rev. Lett. **90**, 060403 (2003).
14. P.L. Stanwix *et al.*, Phys. Rev. Lett. **95**, 040404 (2005).
15. S. Herrmann *et al.*, Phys. Rev. Lett. **95**, 150401 (2005).
16. P. Antonini *et al.*, Phys. Rev. A **71**, 050101(R) (2005).
17. M.E. Tobar *et al.*, Phys. Rev. A, **72**, 066101 (2005).
18. P. Antonini *et al.*, Phys. Rev. A, **72**, 066102 (2005).
19. P.L. Stanwix *et al.*, Phys. Rev. D **74**, 081101 (2006).
20. J.G. Hartnett *et al.*, Appl. Phys. Lett., **36**, 203513 (2006).
21. E.N. Ivanov, M.E. Tobar and R.A. Woode, IEEE Trans. Ultrason. Ferroelect. Freq. Contr. **45**, 1526 (1998).
22. E.N. Ivanov and M.E. Tobar, IEEE Trans. Ultrason. Ferroelect. Freq. Contr. **49**, 1160 (2003).

LORENTZ VIOLATION, ELECTRODYNAMICS, AND THE COSMIC MICROWAVE BACKGROUND

MATTHEW MEWES
Physics Department, Marquette University
Milwaukee, WI 53201, U.S.A.

Vacuum birefringence is a signature of Lorentz-symmetry violation. Here we report on a recent search for birefringence in the cosmic microwave background. Polarization data is used to place constraints on certain forms of Lorentz violation.

1. Introduction

The properties of light have proved to be a valuable testing ground for special relativity for more than a century. Contemporary experiments are motivated in part by a possible breakdown of special relativity with origins in Planck-scale physics.[1,2,3] These experiments include modern versions of the classic Michelson-Morley and Kennedy-Thorndike experiments that use highly stable resonant cavities to search for violations of rotation and boost symmetries.[4] However, the highest sensitivities to relativity violations in electrodynamics are found in searches for vacuum birefringence in light from very distant sources.[5,6,7,8] Birefringence studies take advantage of the extremely long baselines that allow the miniscule effects of a Lorentz violation to accumulate to (potentially) detectable levels over the billions of years it takes for the light to reach Earth. The cosmic microwave background (CMB) is the oldest light available to observation and therefore provides an excellent source for birefringence searches. Here we summarize a recent search for signals of Lorentz violation using CMB polarimetry.[8]

General Lorentz violation is described by a framework known as the Standard-Model Extension (SME).[3] The SME provides the theoretical backbone for studies in a number of areas,[2] including photons.[4,6,7,8,9] Most tests of Lorentz violation focus on the minimal SME, which assumes usual gauge invariance and energy-momentum conservation and restricts attention to superficially renormalizable operators. Operators of dimension $d \leq 4$ are of renormalizable dimension and are included in the minimal SME. Two types of Lorentz-violating operators appear in the minimal SME, CPT-odd operators with coefficients $(k_{AF})_\kappa$ and CPT-even operators with coefficients $(k_F)^{\kappa\lambda\mu\nu}$.

In this work, we also consider non-minimal higher-dimensional operators in the photon sector with $d > 4$. In general there are an infinite number of possible operators that emerge when we relax the renormalizable condition. These operators are phenomenologically and theoretically relevant in that they help provide a connection to the underlying Planck-scale physics. They also add a number of new and interesting signals for Lorentz violation that may be tested experimentally.

2. Theory

General Lorentz-violating electrodynamics is given by a lagrangian that takes the same basic form as the minimal-SME photon sector:[8]

$$\mathcal{L} = -\frac{1}{4}F_{\mu\nu}F^{\mu\nu} + \frac{1}{2}\epsilon^{\kappa\lambda\mu\nu}A_\lambda(\hat{k}_{AF})_\kappa F^{\mu\nu} - \frac{1}{4}F_{\kappa\lambda}(\hat{k}_F)^{\kappa\lambda\mu\nu}F_{\mu\nu}\,. \quad (1)$$

We assume a linear theory and impose the usual U(1) gauge invariance. The key difference between this theory and the minimal-SME photon sector is that here the $\hat{k}_{AF}$ and $\hat{k}_F$ coefficients are differential operators. The effects of these operators mimic the effects of a permeable medium whose activity depends on the photon energy and momentum. This introduces a plethora of new effects not found in either the conventional Lorentz-conserving case or the minimal SME. These include drastically different frequency dependences and direction-dependent propagation of light.

Expanding the $\hat{k}_{AF}$ and $\hat{k}_F$ operators in the 4-momentum $p_\mu = i\partial_\mu$ leads to the expressions

$$(\hat{k}_{AF})_\kappa = \sum (k_{AF}^{(d)})_\kappa{}^{\alpha_1\ldots\alpha_{(d-3)}}\partial_{\alpha_1}\ldots\partial_{\alpha_{(d-3)}}\,, \quad (2)$$

$$(\hat{k}_F)^{\kappa\lambda\mu\nu} = \sum (k_F^{(d)})^{\kappa\lambda\mu\nu\alpha_1\ldots\alpha_{(d-4)}}\partial_{\alpha_1}\ldots\partial_{\alpha_{(d-4)}}\,. \quad (3)$$

The coefficients for Lorentz violation associated with the dimension d operators are now given by $(k_{AF}^{(d)})_\kappa{}^{\alpha_1\ldots\alpha_{(d-3)}}$ and $(k_F^{(d)})^{\kappa\lambda\mu\nu\alpha_1\ldots\alpha_{(d-4)}}$. The $\hat{k}_{AF}$ expression contains all CPT-breaking effects, and the sum is restricted to odd-dimensional operators, d = odd. The $\hat{k}_F$ coefficients control all CPT-even violations and have d = even. Imposing gauge invariance places various constraints on these coefficients. For $k_{AF}^{(d)}$ coefficients, any trace that involves the first index vanishes identically. For $k_F^{(d)}$, the antisymmetrization on any three indices vanishes. Standard group theory techniques allow a counting of the independent coefficients for Lorentz violations.[10] For a given dimension d, we find $\frac{1}{2}(d+1)(d-1)(d-2)$ independent $k_{AF}^{(d)}$ coefficients in the CPT-odd case and $(d+1)d(d-3)$ independent $k_F^{(d)}$ coefficients in the CPT-even case.

For studies of Lorentz-violation induced birefringence, certain linear combinations of these general coefficients are important. They result from a spherical-harmonic expansion of plane-waves propagating in the vacuum. This plane-wave expansion is best characterized using the language of Stokes parameters. We begin by defining a Stokes vector $\boldsymbol{s} = (s^1, s^2, s^3)^T$. The direction in which this vector points in the abstract 3-dimensional Stokes space uniquely characterizes the polarization of the radiation. Stokes vectors lying in the s^1-s^2 plane correspond to all possible linear polarizations, while Stokes vectors parallel and antiparallel to the s^3 axis give the two circular polarizations. General right-handed elliptical polarizations point in the upper-half Stokes space, $s^3 > 0$, while left-handed are given by the lower half, $s^3 < 0$.

This formalism provides an intuitive picture of birefringence. It can be shown that birefringence causes a rotation of the Stokes vector $\boldsymbol{s}$ about some axis ς. This occurs whenever the usual degeneracy between the various polarizations in broken. Formally, we solve the modified equations of motion. We find that some types of violations lead to two propagating eigenmodes that have slightly different velocities. They also differ in polarization, and light of an arbitrary polarization is a superposition of the two eigenmodes. This superposition is altered as the eigenmodes propagate at different velocities, causing an oscillatory effect that reveals itself as a rotation of the Stokes vector. The rotation takes the form

$$d\boldsymbol{s}/dt = 2\omega\boldsymbol{\varsigma} \times \boldsymbol{s} \ , \tag{4}$$

where ω is the wave frequency, and the rotation axis ς corresponds to the Stokes vector of the faster eigenmode. In general, ς may depend on both the direction of propagation and the frequency.

The basic idea of a birefringence test is to examine light from a distant polarized source for the above rotation. To do this we need to express the rotation axis ς in terms of the coefficients for Lorentz violation. The general result is rather complicated, but can be written in a relatively simple form in terms of a set of "vacuum" coefficients, which are linear combinations of the general coefficients. The calculation involves decomposing ς into spin-weighted spherical harmonics. The result takes the form

$$\varsigma^1 \mp i\varsigma^2 = \sum_{dlm} \omega^{d-4} (k^{(d)}_{(E)lm} \pm i k^{(d)}_{(B)lm}) \, {}_{\pm 2}Y_{lm}(\hat{\boldsymbol{n}}) \, , \tag{5}$$

$$\varsigma^3 = \sum_{dlm} \omega^{d-4} k^{(d)}_{(V)lm} \, {}_0Y_{lm}(\hat{\boldsymbol{n}}) \, , \tag{6}$$

where ${}_sY_{lm}$ is a spin-weighted spherical harmonic with spin-weight s, and $\hat{\boldsymbol{n}}$ is the radial unit vector pointing toward the source on the sky. The vacuum coefficients $k^{(d)}_{(V)lm}$, $k^{(d)}_{(E)lm}$, and $k^{(d)}_{(B)lm}$ represent the minimal combinations of coefficients for Lorentz violation that cause birefringence and affect polarization. The designations E and B refer to the parity of the coefficient and is borrowed from radiation theory. In the next sections, we describe a search for these effects in existing CMB polarization data.

3. CMB

The CMB is conventionally parameterized by a spin-weighted spherical-harmonic expansion similar to the expansion of ς given above.[11,12] The complete characterization of radiation from a given point on the sky includes the temperature T, the linear polarization, given by Stokes parameters s^1 and s^2, and the circular polarization, given by s^3. The global description is given by the expansion

$$T = \sum a_{(T)lm} \, {}_0Y_{lm}(\hat{\boldsymbol{n}}) \, , \qquad s^3 = \sum a_{(V)lm} \, {}_0Y_{lm}(\hat{\boldsymbol{n}}) \, ,$$
$$s^1 \mp i s^2 = \sum (a_{(E)lm} \pm i a_{(B)lm}) \, {}_{\pm 2}Y_{lm}(\hat{\boldsymbol{n}}) \, . \tag{7}$$

One then constructs various power spectra,

$$C_l^{X_1 X_2} = \tfrac{1}{2l+1} \sum_m \langle a^*_{(X_1)lm} a_{(X_2)lm} \rangle \, , \tag{8}$$

where $X_1, X_2 = T, E, B, V$. These spectra quantify the angular size variations in each mode and any correlation between different modes. Smaller l correlates to larger angular size on the sky.

Within conventional physics, we expect a nearly isotropic ($l = 0$) temperature distribution. However, tiny fluctuations in temperature during recombination not only introduce higher-order multipole moments ($l > 0$) but also provide the necessary anisotropies to produce a net polarization. Only linear polarizations are expected since no circular polarization is produced in Thomson scattering. Furthermore, E-type polarization is expected to dominate and be correlated with the temperature. No correlation is expected between the much smaller B polarization and temperature. This general picture agrees with observation to the extent to which CMB radiation has been measured.[13,14]

A breakdown of Lorentz symmetry may alter these basic features. Some of the new effects can be readily understood as consequences of the Stokes rotations. For example, the CPT-odd coefficients $k^{(d)}_{(V)lm}$ lead to a Stokes rotation axis that points along the s^3 direction. The resulting local rotations in polarization leave the circularly polarized component unchanged. However, it does lead to a rotation in the linear components, causing a simple change in the polarization angle at each point on the sky. Globally this causes a mixing between the E and B polarization. This could introduce an unusually large B component, which gives a potential signal of CPT and Lorentz violation. Similar mixing can arise from the $k^{(d)}_{(E)lm}$ and $k^{(d)}_{(B)lm}$ coefficients. However, since these give a rotation axis that lies in the s^1-s^2 plane, the rotations in this case also introduce circular polarization. So a large circularly polarized component in the CMB might indicate a CPT-even violation of Lorentz invariance.

All except the $d = 3$ coefficients result in frequency-dependent rotations. Also, only the $l = 0$ coefficients cause isotropic rotations that are uniform across the sky. As a result, the coefficient $k^{(3)}_{(V)00}$ provides a simple isotropic frequency-independent special case. A calculation shows that this case causes a straightforward rotation between C_l^{EE}, C_l^{BB}, and C_l^{EB}, as well as between C_l^{TE} and C_l^{TB}.[15] In contrast, more general anisotropic and frequency-dependent cases cause very complicated mixing between the various $C_l^{X_1X_2}$ and require numerical integration of the rotation (4) over the sky and frequency range.

4. Results

To illustrate the kinds of sensitivities that are possible in CMB searches for birefringence, we next examine the results of the BOOMERANG experiment.[14] This balloon-based experiment made polarization measurements in a narrow band of frequencies at approximately 145 GHz. This relatively high frequency implies that BOOMERANG is well suited to birefringence tests since for all violations, except those with $d = 3$, higher photon energy implies a larger rotation in polarization. The small frequency range is also helpful since we can approximate all frequencies as ~ 145 GHz.

In our calculation, we assume conventional polarization is produced during recombination and numerically determine the rotated polarization for points across the sky. The resulting $C_l^{X_1X_2}$ for various values of coef-

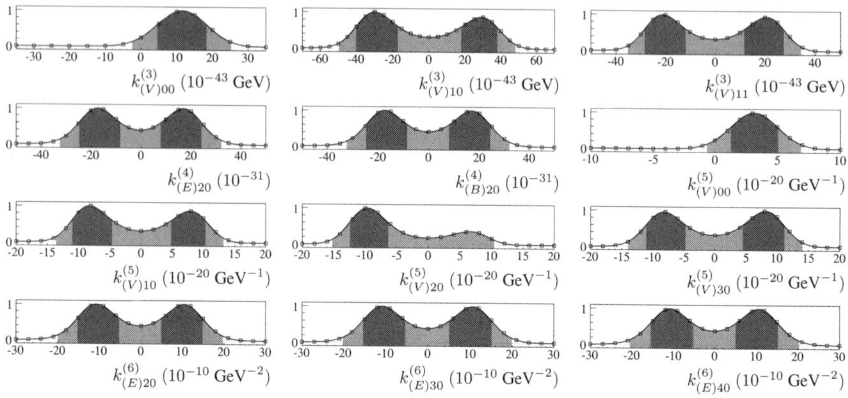

Figure 1. Relative likelihood versus coefficients for Lorentz violation. Boxes indicate numerically calculated values, and the curve is the smooth extrapolation of these points. The dark-gray shaded region indicates the 68% confidence level, and the light-gray shows the 95% level.

ficients for Lorentz violation are determined and compared to published BOOMERANG results. Figure 1 shows the calculated relative likelihood for a sample of 12 coefficients for Lorentz violation. In each case, we vary the value of one coefficient, setting all other coefficients to zero. The 1σ and 2σ regions are shown.

Some generic features are seen in our survey. In each case, the coefficient is nonzero at the 1σ level, hinting at possible Lorentz violation. However, since this occurs in every case, it is likely that this indicates some systemic feature of the BOOMERANG data or our analysis. We also see that each case is consistent with no Lorentz violation at the 2σ level, giving conservative upper bounds on the 12 coefficients in Fig. 1.

These results demonstrate the potential of the CMB for testing Lorentz invariance. Due to the long propagation times, the sensitivities to $d = 3$ coefficients afforded by the CMB are near the limit of what can be expected in birefringence tests. However, for $d \geq 4$, better sensitivities might be obtained using high-frequency sources like gamma-ray bursts.[7] Regardless, because of its all-sky nature, the CMB provides a useful probe that can simultaneously probe large portions of coefficient space, which is difficult in searches involving a handful of point sources that access a limited number of propagation directions.

Acknowledgments

This work was supported in part by the Wisconsin Space Grant Consortium.

References

1. V.A. Kostelecký and S. Samuel, Phys. Rev. D **39**, 683 (1989); V.A. Kostelecký and R. Potting, Nucl. Phys. B **359**, 545 (1991).
2. For a recent review of various experimental and theoretical approaches to Lorentz violation, see, for example, V.A. Kostelecký, ed., *CPT and Lorentz Symmetry III,* World Scientific, Singapore, 2005.
3. D. Colladay and V.A. Kostelecký, Phys. Rev. D **55**, 6760 (1997); **58**, 116002 (1998); V.A. Kostelecký, Phys. Rev. D **69**, 105009 (2004).
4. J. Lipa *et al.*, Phys. Rev. Lett. **90**, 060403 (2003); H. Müller *et al.*, Phys. Rev. Lett. **91**, 020401 (2003); P. Antonini *et al.*, Phys. Rev. A **71**, 050101 (2005); S. Herrmann *et al.*, Phys. Rev. Lett. **95**, 150401 (2005); P.L. Stanwix *et al.*, Phys. Rev. D **74**, 081101 (2006).
5. S.M. Carroll, G.B. Field, and R. Jackiw, Phys. Rev. D **41**, 1231 (1990).
6. V.A. Kostelecký and M. Mewes, Phys. Rev. Lett. **87**, 251304 (2001); Phys. Rev. D **66**, 056005 (2002).
7. V.A. Kostelecký and M. Mewes, Phys. Rev. Lett. **97**, 140401 (2006).
8. V.A. Kostelecký and M. Mewes, Phys. Rev. Lett. **99**, 011601 (2007).
9. H. Müller *et al.*, Phys. Rev. D **67**, 056006 (2003); R. Lehnert and R. Potting, Phys. Rev. Lett. **93**, 110402 (2004); Q.G. Bailey and V.A. Kostelecký, Phys. Rev. D **70**, 076006 (2004); A. Petroff and M. Mewes, Phys. Rev. D **75**, 056002 (2007); A.J. Hariton and R. Lehnert, Phys. Lett. A **367**, 11 (2007); B. Altschul, Phys. Rev. D **75**, 105003 (2007); A. Kobakhidze and B.H.J. McKellar, arXiv:0707.0343.
10. V.A. Kostelecký and M. Mewes, in preparation.
11. For review of current CMB theory and experiment, see, for example, Particle Data Group (http://pdg.lbl.gov), S. Eidelman et al., Phys. Lett. B 592, 1 (2004).
12. For a review of CMB polarimetry, see, for example, W. Hu and M. White, New Astron. **2**, 323 (1997).
13. G.F. Smoot *et al.*, Ap. J. **396**, L1 (1992); J. Kovac *et al.*, Nature **420** 772 (2002); A.C.S. Readnead *et al.*, Science **306**, 836 (2004); D. Barkats *et al.*, Ap. J. **619**, L127 (2005); E.M. Leitch *et al.*, Ap. J. **624**, 10 (2005); W.C. Jones *et al.*, Ap. J. **647**, 823 (2006); G. Hinshaw *et al.*, Ap. J. Supp. **170**, 288 (2007); L. Page *et al.*, Ap. J. Suppl. **170**, 335 (2007).
14. T.E. Montroy *et al.*, Ap. J. **647**, 813 (2006); F. Piacentini *et al.*, Ap. J. **647**, 833 (2006).
15. B. Feng, *et al.*, Phys. Rev. Lett. **96**, 221302 (2006); P. Cabella, *et al.*, arXiv:0705.0810.

HIGH ENERGY ASTROPHYSICAL TESTS OF LORENTZ INVARIANCE

B. ALTSCHUL

Department of Physics and Astronomy
University of South Carolina
Columbia, SC 29208, U.S.A.
E-mail: baltschu@physics.sc.edu

Certain forms of Lorentz violation can give rise to maximum electron velocities less than the speed of light or to maximum energies for subluminal electrons. Either of these effects can depend on the direction of an electron's motion. Data on high-energy astrophysical sources are very important for constraining these possible effects; their synchrotron and inverse Compton spectra can be used to place complementary bounds. Using data from numerous sources spread across the sky, it is possible to place bounds on the dimensionless Lorentz-violating parameters involved at 10^{-14}–10^{-20} levels.

There is growing interest in the possibility that Lorentz symmetry may not be truly exact in nature. This is not a new suggestion; it has been with us since 1905. However, only in the last fifteen years have people looked systematically at the possible ways that relativity could fail, within the framework of effective field theory.[1,2,3] After the grounbreaking theoretical work of our host V.A. Kostelecký and D. Colladay, it was possible to place bounds on many of the newly-introduced Standard-Model Extension (SME) coefficients using existing data or new experiments using the same design paradigms.

However, while some extremely strong experimental bounds rapidly appeared in the literature, some other sectors remained rather poorly constrained. Sometimes, the reasons for this were obvious. There are few bounds in the neutrino sector simply because neutrinos are very hard to measure. On the other hand, some of the most natural parameters in the electron sector were bounded only at the 10^{-2} level by direct laboratory measurements of the Doppler shift.[4,5] The coefficients in question were among the electron c and d coefficients, which are described by the

Lagrange density

$$\mathcal{L} = \bar{\psi}(i\Gamma^{\mu}\partial_{\mu} - m)\psi, \tag{1}$$

where

$$\Gamma^{\mu} = \gamma^{\mu} + c^{\nu\mu}\gamma_{\nu} - d^{\nu\mu}\gamma_{\nu}\gamma_{5}, \tag{2}$$

and which are expected to represent the most important forms of Lorentz violation at large electron energies.

If c is nonzero, then the maximum achievable velocity (MAV) for electrons will be different from the speed of light (which is 1 in our units) and generally also direction dependent. (d has a similar effect, but the MAV is helicity dependent. Because of this complication, we shall not discuss the d terms further, but the bounds on these coefficients are comparable to the bounds on the c.) The MAV for an electron moving in a direction $\hat{e}$ is $1 - c_{jk}\hat{e}_j\hat{e}_k - c_{0j}\hat{e}_j$ to leading order in the c coefficients.[6]

If we want to place bounds on the electron's MAV, we should look at the very highest energy electrons possible. The most energetic electrons are found in astrophysical x-ray sources—pulsar wind nebulae, active galactic nuclei, and the like. Lorentz-violating effects that become more salient at high energies will be easiest to pick out by looking carefully at these emitters. Many of the best bounds on Lorentz-violating coefficients have been derived from high-energy astrophysical data.[7,8,9,10,11,12]

In the Lorentz-violating field theory with c and d, once a charge's trajectory is known, the radiation it emits can be calculated by standard means. As is true conventionally, the electrostatic potential $\Phi = A^0$ is coupled simply to the charge density $e\psi^{\dagger}\psi$, and the vector potential $\vec{A}$ couples to $e\psi^{\dagger}\dot{\vec{x}}\psi$. Radiation from an ultrarelativistic charge is beamed into a narrow pencil of angles centered around the direction of the charge's velocity. So the radiation we observe from a distant source comes from electrons that are moving in the source-to-Earth direction. The most important radiation processes for extremely energetic electrons are synchrotron radiation resulting from the interaction with a magnetic field and inverse Compton (IC), which arises from electron-photon collisions.

If a body's synchrotron spectrum indicates the presence of electrons with Lorentz factors up to at least some value $\gamma_{\max}$, then the Lorentz-violating coefficients for the electron must be bounded according to

$$c_{jk}\hat{e}_j\hat{e}_k + c_{0j}\hat{e}_j < \frac{1}{2\gamma_{\max}^2}, \tag{3}$$

where $\hat{e}$ is in the source-to-Earth direction just mentioned. [By the Lorentz factor γ, we mean precisely the quantity $\gamma = (1 - \vec{v}^2)^{-1/2}$.] This is a one-sided limit, so only coefficients with particular signs are restricted. In particular, we can only find upper limits on the diagonal components of c this way; no negative values of c_{jk} for $j = k$ can be excluded. So a complementary measurement is needed if the Lorentz-violating coefficients are to be restricted to a bounded region of the parameter space.

Fortunately, complementary bounds are possible. Just as a positive $c_{jk}\hat{e}_j\hat{e}_k + c_{0j}\hat{e}_j$ leads to a maximum electron speed that is less than one, a negative value for $c_{jk}\hat{e}_j\hat{e}_k + c_{0j}\hat{e}_j$ can indicate a maximum electron speed that is greater than one. However, for electrons that are coupled conventionally to the electromagnetic field, new effects would come into play if the speeds actually became superluminal. The validity of the model is questionable under these circumstances, so we cannot make use of the maximum velocity directly. On the other hand, there is instead a maximal energy attainable by subluminal electrons, and we can use this to set further bounds on c.

In IC scattering, an ultrarelativistic electron scatters off a comparatively low-energy photon, transferring a substantial fraction of its energy to the photon. For kinematical reasons, the emitted radiation is beamed into a narrow pencil of angles and, when observed, can be used to bound the Lorentz violation. The key is that the initial electron must have a greater energy than the observed IC γ-ray. If the emission from a source is well modeled by known processes, we may infer that superluminal electrons (which would produce vacuum Cerenkov radiation and emit synchrotron radiation at a rate which diverges in the approximation of no back reaction) are not the source of the emission. Then the energies of the most energetic IC photons represent lower bounds on the energies of subluminal electrons; and if there are such electrons up to some energy E_{max}, then the c coefficients must satisfy

$$-c_{jk}\hat{e}_j\hat{e}_k - c_{0j}\hat{e}_j < \frac{1}{2(E_{\text{max}}/m)^2}, \tag{4}$$

since otherwise, an electron with energy E_{max} moving in the Earthward direction would be superluminal.

The similarity between equations (3) and (4) is not coincidental. In each case, the effect we want to measure is a product of how the Lorentz violation affects the relationship between velocity and energy. In the Lorentz-invariant theory, $\gamma = E/m$, and by measuring E and γ separately, we can get bounds on any deviations from Lorentz invariance.

The sources used must be well understood; there must be clearly identifiable synchrotron and/or IC portions of their spectra. If the spectra are anomalous in any way, they cannot be used to place bounds, since what Lorentz violation would predict in this context is precisely an unconventional emission spectrum. Values of $\gamma_{\max}$ must be extracted from models of the sources involved. However, this is quite straightforward. The most important feature of a synchrotron spectrum is the existence of a cutoff frequency ν_c, above which the emission falls off precipitously. A lower limit on ν_c can be easily derived from the data. Then the value of $\gamma_{\max}$ is proportional to $\sqrt{\nu_c/B_\perp}$, where $B_\perp$ is the magnetic field normal to the line of sight. $B_\perp$ is generally inferred to be the minimum-energy field which is sufficient to generate the low-frequency part of the spectrum. Its value is fairly insensitive to ν_c and any other fit parameters. Ultimately, $\gamma_{\max}$ depends on at most the seventh root of any fit parameter (except ν_c) that might be in error, and so its value is quite robust.

The $E_{\max}$ limits are even less model dependent. IC models of γ-ray sources typically require maximum electron energies that are several times larger than the highest observed photon energies (because an IC scattering event does not transfer all of the electron's energy to the photon). So the identification of the highest observed γ-ray energy from a source as a lower bound on $E_{\max}$ is fairly conservative. This choice of $E_{\max}$ ensures that Lorentz-violating distortions of the energy-momentum relation at higher than observed energies (which could invalidate the model results) are not a problem. The only input from a model is that the source's γ-ray emission is well described by the IC process. In the last few years, the excellent data from the HESS γ-ray telescope in Namibia has tremendously improved the available data on $E_{\max}$.

The best bounds on the electron c coefficients come from looking at the Crab nebula. The Crab's synchrotron spectrum indicates the presence of electrons with energies up to approximately 1500 TeV radiating conventionally. This corresponds to a bound on c at the 10^{-20} level. For other sources, farther away and less exhaustively studied, the bounds are more typically at the 10^{-15}–10^{-16} level. Intrinsically, many of these sources are probably not so different from the Crab nebula; only the quality of our observational data has thus far prevented us from resolving the contributions of the small population of very highest energy electrons. So better spectra and models for these sources could probably improve the bounds by several orders of magnitude.

There are nine components of c that can be bounded with the synchrotron and IC data—the three c_{0j} and the six-component symmetric part of c_{jk}. (The same components of d can also be bounded.) If each of these coefficients is to be bounded on both sides, we must obtain at least ten inequalities of the forms (3) or (4), corresponding to emissions from at least nine separate sources. Each of these inequalities generally couples all nine of the coefficients in a nontrivial way, but they may be translated into separate limits on individual coefficients by means of linear programming.

To get the best possible bounds, we can also include in the linear program the comparable bounds on several of the electron c coefficients that come from resonant cavity experiments.[13,14] These experiments are typically used to bound the k_F coefficients analogous to c in the electromagnetic sector, but because of condensed matter effects in the cavity material, Lorentz violations in the electron sector could also make contributions to what is observed experimentally. The cavity-derived bounds are complementary in structure to the astrophysical ones and improve the linear program significantly.

The ultimate bounds on the separate c (and d) coefficients are all at the 10^{-14} level or better, and most are at the 10^{-15}–10^{-16} level typical of the electron energy scales we have observed. Several of the bounds derived this way are much better than any previous constraints. The Doppler shift bounds on the electron c_{Tj} coefficients were at the 10^{-2} level, and so some of the improvements were by more than fourteen orders of magnitude. Moreover, unlike in terrestrial experiments, violations of boost invariance and rotation invariance are constrained at comparable levels.

This method for placing bounds on Lorentz violation could not be used directly to identify actual Lorentz violation. The signature of Lorentz violation in high-energy astrophysical sources would be emission spectra that cannot be modeled by conventional radiation mechanisms. If the calculated bounds on certain Lorentz-violating coefficients were significantly weaker than others, that would be an indication that those coefficients might actually be nonzero. However, because all of the bounds we can derive are comparable in magnitude, there are no indications that any particular components of $c_{\mu\nu}$ are more likely to be nonzero than others.

Better observational data and further modeling of well understood sources would allow all these bounds to be improved. The higher the energy of radiation that can be observed from a source, the better the bounds on the electron MAV in the source-to-Earth direction are. Also, careful models of the synchrotron emission from multiple sources would be quite useful. No other nearby source has been modeled in the same detail as the

Crab nebula. If other pulsar wind nebulae were treated with comparable exhaustiveness, so that the contributions to the synchrotron and IC spectra of the very highest energy electrons could be picked out, the limits on all the electron c coefficients might be pushed down to the 10^{-18}–10^{-20} level, several orders of magnitude's improvement.

References

1. D. Colladay, V.A. Kostelecký, Phys. Rev. D **55**, 6760 (1997).
2. D. Colladay, V.A. Kostelecký, Phys. Rev. D **58**, 116002 (1998).
3. V.A. Kostelecký, Phys. Rev. D **69**, 105009 (2004).
4. G. Saathoff, S. Karpuk, U. Eisenbarth, G. Huber, S. Krohn, R. Muñoz Horta, S. Reinhardt, D. Schwalm, A. Wolf, G. Gwinner, Phys. Rev. Lett. **91**, 190403 (2003).
5. C.D. Lane, Phys. Rev. D **72**, 016005 (2005).
6. B. Altschul, D. Colladay, Phys. Rev. D **71**, 125015 (2005).
7. B. Altschul, Phys. Rev. Lett. **96**, 201101 (2006).
8. B. Altschul, Phys. Rev. D **74**, 083003 (2006).
9. B. Altschul, Phys. Rev. D **75**, 041301 (R) (2007)
10. F.W. Stecker, S.L. Glashow, Astropart. Phys. **16**, 97 (2001).
11. T. Jacobson, S. Liberati, D. Mattingly, Nature **424**, 1019 (2003).
12. T. Jacobson, S. Liberati, D. Mattingly, F.W. Stecker, Phys. Rev. Lett. **93**, 021101 (2004).
13. H. Müller, Phys. Rev. D **71**, 045004 (2005).
14. H. Müller *et al.*, Phys. Rev. Lett. **99**, 050401 (2007).

FUNDAMENTAL PHYSICS EXPERIMENTS IN SPACE (WITHIN ESA)

T.J. SUMNER
Blackett Laboratory, Imperial College London
London, SW7 2AZ, U.K.
E-mail: t.sumner@imperial.ac.uk

As a discipline Fundamental Physics is a relatively new domain within the European Space Agency for carrying out dedicated space experiments. However the availability of new technologies making a very quiet environment in space accessible, together with a powerful scientific motivation have created a community of scientists who are promoting a number of experiments. The motivations, technologies and some specific mission concepts will be reviewed.

1. Motivation

The low-noise environment of space is ideal for weak-force experiments using spacecraft which can be put into precise gravitationally determined orbits. This makes their use particularly appropriate for studying gravity and any other similar strength forces which have yet to be discovered. The past several years, including the most recent ESA 'Cosmic Visions' call,[1] have seen responses to mission calls covering a wide gamut of science goals:

- Precision tests of general relativity
- Pointers to quantum gravity
- Pointers to unification
- Tests of alternative theories of gravity
- Tests of Lorentz invariance
- Searches for new forces
- Speculative new parameter space searches
- [A new astronomy]

This article will not attempt to cover everything comprehensively (but superficially) but will rather focus in more detail on a subset.

2. Enabling Technologies

There are three identifiable essential technologies which have recently matured to a level allowing new missions with significant scientific capabilities.
Gravitational Reference Sensors: Many missions require a spacecraft in a purely gravitationally determined trajectory. This is always compromised by non-gravitational disturbances, due to atmospheric drag, solar radiation pressure, solar wind particles and cosmic rays. The effect of spurious external forces can, to some extent, be negated using a proof-mass contained within the spacecraft and isolated from it. If the relative motion between spacecraft and proof-mass is measured then corrective thrust can be applied to the spacecraft to 'follow' the proof-mass; a technique usually referred to as 'drag-free', first demonstrated in 1972 by DISCOS.[2] More recent examples are GP-B[3] and GRACE,[4] with GOCE[5] due for launch next year.
μN Thrusters: An integral part of a 'drag-free' satellite is a low-noise propulsion system for applying the small corrective thrusts to the spacecraft. The thrusters must have sufficient dynamic range to cope with the magnitude of the spurious forces but must also have sufficient low-noise performance to keep residual disturbances down; ideally the thruster noise should not constrain the drag-free closed loop gain.
Precision Measurement Technology: In many missions the science data themselves are the precision measurements of proof-mass position or motion. Quite often the proof-mass concerned is also the gravitational reference sensor. For macroscopic (conventional) proof-masses sensing can be capacitive, magnetic and/or optical. For experiments proposing microscopic proof-masses matter interferometry is a very promising new technology.

3. Opportunities

Some opportunities within ESA for fundamental physics missions are:
LISA Pathfinder[6] *- Technology demonstrator for LISA*
LISA[7] *- A gravitational wave observatory*
MICROSCOPE[8] *- Equivalence Principle measurement*
Cosmic Visions - What are the fundamental laws of the Universe?
- –GAUGE[9] - GrAnd Unification and Gravity Explorer
- –LATOR[10] - Laser Astrometric Test Of Relativity
- –ASTROD[11] - Astrodynamical Space Test of Relativity using Optical Devices
- –ODYSSEY[12] - Long-range gravity - Pioneer Anomaly
- –MWXG[13] - Matter Wave eXplorer Gravity

At the time of writing the mission concepts listed under the Cosmic Vision heading are those known to have submitted Letters of Intent.[1]

The fundamental physics capabilities of LISA Pathfinder (LISAPF), LISA, and GAUGE will now be described in more detail together with some of the key information concerning their mission/experiment concepts.

3.1. *LISA Pathfinder*

3.1.1. *Technological Concept*

The main goal is demonstration of geodesic motion to better than two orders of magnitude beyond current capabilities. 10^{-12} m resolution laser interferometry will track two proof-masses nominally in free-fall.

The two macroscopic proof-masses are 4.6 cm Au/Pt cubes separated by 30 cm. One will be used as a gravitational reference sensor for drag-free operations. The other will be used to assess relative parasitic accelerations in the frequency range 1mHz to 0.1Hz. A combination of laser interferometry between proof-masses and optical benches, supplemented with capacitive/optical sensing will provide both drag-free signals and science measurements. Field-emission electric propulsion and colloidal thrusters will control the spacecraft once in the final Earth-Sun L1 Lagrange point halo orbit.

3.1.2. *Scientific Objectives*

Along with the high-precision demonstration of geodesic motion will automatically come a drag-free performance similarly better than anything achieved before. Hence a number of secondary science objectives have been considered[14] including: a measurement of the gravitational constant, G, short and long-range inverse-square law departures, and tests of alternative gravity scenarios, such as MOdified Newtonian Dynamics (MOND).

G has one of the largest uncertainties of any physical constant.[15] All measurements have used pendula/torsion balances in the 1-g Earth environment.[16] A space measurement would use a new technique and be done in lower g. The experiment modulates the position of one proof-mass and measures the response of the other. This measurement requires precision knowledge of mass distributions, spring constants and displacements. Without upgrading the mission requirements $1 : 10^4$ is all that is possible. This is within the spread of measurement values[15] but not competitive with the best of them. Useful short-range inverse-square law tests were possible when LISAPF included an additional pair of US proof-masses. Proof-mass pairs with different separations could have been used for independent measurements of G, each to $1 : 10^4$, looking for systematic variations with distance with two orders of magnitude more sensitivity than current limits. Long-range inverse-square law deviations tracking the satellite in drag-free mode on its transfer orbit from near-Earth to L1[17] could again give two orders of magnitude improvement over this scale-length. Unfortunately drag-free operations will not start until L1 is reached. Once in the L1 halo orbit tracking might nonetheless provide some more modest improvement. Finally, it will be possible to test the proof-mass response to applied forces at, and well below the characteristic MOND acceleration.[18] The forces can be applied electrostatically and optically as well as gravitationally to study whether the inertial response depends on how forces are applied. L1 is not a gravity balance point, so the experiment is unlikely to provide a direct test of either MOND or the related TeVeS formulation,[19] even assuming a reasonable transition formula between MOND and Newtonian gravity.[20]

3.2. *LISA*

3.2.1. *Technological Concept*

LISAPF technology will broadly translate directly into LISA with the main differences being interferometry between proof-masses in separated spacecraft with a path-length of 5×10^6 km, the frequency range extends down to ~ 0.1 mHz and the use of a Sun-orbit at 1AU, some 20° behind the Earth.

3.2.2. *Scientific Objectives*

LISA science objectives are a combination of astrophysics and fundamental physics. As a pioneering astrophysics mission it will open insights into compact stellar binaries, extreme mass-ratio black-hole binaries, massive black-hole mergers and primordial cosmological backgrounds. Some observations will bring with them an opportunity to also get at some fundamental physics such as:

- a direct study of inherent gravitational wave properties
- strong field probes confronting GR
- precision measurement of black-hole properties
- establish the graviton mass/propagation speed
- study of phase transitions/strings (speculative)

3.3. *GAUGE*

GAUGE (GrAnd Unification and Gravity Explorer) is a proposal for a drag-free spacecraft platform onto which is attached a number of modular experiments. The possible complement of experiments is designed to address a number of key issues at the interface between gravity and unification with the other forces of nature. Included are:

- A test of string-dilaton theories using a high precision macroscopic equivalence principle experiment
- A test of the effect of quantum space-time fluctuations in a microscopic equivalence principle experiment
- An inverse-square law test at intermediate ranges
- An axion-like mass-spin coupling search
- Measurement of quantum decoherence from Planck-scale space-time fluctuations

3.3.1. *Technological Concept*

The list of experiments is extensive but it has been selected to benefit from commonalities and heritages. For example the assumed drag-free platform is a LISAPF replica including its gravitational reference sensor, LTP. Each experiment uses some variant of precision proof-mass displacement sensing with prior heritage: magnetic (SQUIDS) - STEP and GP-B, electrostatic - LTP and STEP, optical (Interferometry) - LTP and LISA, cold atoms (Interferometry) - HYPER.[21] More specifically, the macroscopic equivalence principle test is closely derived from STEP (and GP-B), the microscopic equivalence principle and quantum decoherence tests are derived from HYPER, and the inverse-square law and spin-coupling tests derive partly from an early version of STEP[22] (and SSPIN, a specific spin-coupling proposal to ESA for M2) and partly from recent laboratory developments.[23] Variants of the inverse-square law and spin-coupling experiments using atom interferometry would be especially interesting with atom laser technology.[24]

VEGA would provide a launch into a low-Earth sun-synchronous orbit.

3.3.2. *Scientific Objectives*

Macroscopic Equivalence Principle

GAUGE will achieve a sensitivity of $1 : 10^{18}$ using three proof-mass pairs working as differential accelerometers. Materials (Be:Pt/Ir, Be:Nb, Nb:Pt/Ir) are chosen to maximise sensitivity within a string-dilaton scenario[25] and to provide a cyclic closure cross-check. The sensitivity grasp is also within the range required by theories incorporating fine structure time variation[26] and, with STEP, represents the most aggressive on offer.[27]

Microscopic Equivalence Principle

Time domain atom interferometry using cohabiting atomic species (Cs/Rb) offers an equivalence principle test with rather more modest sensitivity,[28] $1 : 10^{15}$, but which nonetheless test theoretical scenarios in which quantum fluctuations induce violations for microscopic particles in this range.[29]

Inverse Square Law

At intermediate ranges (10s cm) the macroscopic LTP proof-masses can be used in conjunction with the macroscopic equivalence principle masses and one additional differential accelerometer aligned along the spacecraft axis. This enables an intermediate range test two orders of magnitude more sensitive than existing limits. In the very short-range regime atom interferometry using a atom laser source looks as though it could well be more competitive than the already very interesting ground-based limits.[30]

Spin-coupling

At intermediate ranges (10s cm) a specialised spin-source within the cryogenic macroscopic equivalence principle enclosure can be used in conjunction with the macroscopic equivalence principle masses to look for coupling between quantum mechanical spin and matter.[31] This would achieve an improvement of a few orders of magnitude over existing mass-spin coupling limits.[23] Again in the very short-range regime atom interferometry using a atom laser source looks like it could be much more sensitive even though it would still probably struggle to reach axion search sensitivity.

Quantum Decoherence

Understanding fluctuations in the fabric of space time may lead the way to quantum-gravity. If they exist their effect may well be first seen as a limitation in matter wave interferometry.[32] Demonstrating this requires proper understanding and control of all other limiting disturbances and what could be better than doing this in the quiet environment of space.

References

1. http://sci.esa.int/cv2015/
2. B. Lange, Class. Quant. Grav. **18**, 4153-4158 (2001).
3. C. W. F. Everitt, *The Stanford Relativity Gyroscope Experiment (A): History and Overview* in Near Zero: New Frontiers of Physics, eds. J.D. Fairbank, B.S. Deaver Jr., C.W.F. Everitt, P.F. Michelson (Freeman - New York), 1988, pp 587-639.

4. D. Adam, Nature **416**, 10-11 (2002).
5. M.R. Drinkwater *et al.*, *Proc. 3rd Int. GOCE User Workshop*, 6-8 November, 2006, Frascati, Italy, **ESA SP-627**, 2007.
6. S. Vitale *et al.*, Nucl. Phys. B Proc. Suppl. **110**, 209-216 (2002); R. Dolesi *et al.*, Class. Quant. Grav. **20**, S99-S108 (2003); S. Anza *et al.*, Class. Quant. Grav. **22**, S125-S138 (2005); P.W. McNamara AIP Conf. Ser. **873**, 49-58 (2006).
7. K. Danzmann and LISA Science Team, Adv. Space Res. **32**, 1233-1242 (2003); T.J. Sumner and D.N.A. Shaul, Mod. Phys. Lett. A **19**, 785-800 (2004).
8. P. Touboul and M. Rodrigues, Class. Quant. Grav. **18**, 2487-2498 (2001).
9. T.J. Sumner *et al.*, Exp. Astr. in press (2007).
10. S.G. Turyshev, M. Shao and K. Nordtvedt, Int. J. Mod. Phys. **13** 2035-2063 (2004); S.G. Turyshev *et al.*, Nucl. Phys. B Proc. Suppl. **134**, 171 (2004)
11. W.-T. Ni Int. J. Mod. Phys. D **11**, 947 (2002); W.-T. Ni *et al.*, J. Phys. Conf. Ser. **32**, 161-166 (2006).
12. B. Christophe *et al.*, Exp. Astr. in press (2007).
13. W. Ertmer *et al.*, Exp. Astr. in press (2007).
14. T.J. Sumner, *Gravitational Physics with LTP*, ESA Ref: S2-ICL-TN-3007 (2005); P.W. McNamara AIP Conf. Ser. **873**, 49-58 (2006).
15. CODATA 2002: http://physics.nist.gov/cgi-bin/cuu/Value?bg
16. J. Gundlach and S. Merkowitz, Phys. Rev. Lett. **85**, 2869-2872 (2000).
17. C.C. Speake, private communication, (2005).
18. M. Milgrom, Astrophys. J. **270**, 365 (1983).
19. J. Bekenstein and J. Magueijo, Phys. Rev. D **73** 103513 (2006).
20. M.B. Gerrard and T.J. Sumner, arXiv:gr-qc/0605080, (2006).
21. C. Jentsch *et al.*, Gen. Rel. Grav. **36**, 2197-2221 (2004).
22. C.C. Speake *et al.*, *STEP spin-coupling experiment*, in Perspectives on Neutrinos, Atomic Physics and Gravitation, *Proc. XXVIIIth Rencontre de Moriond*, Eds. J. Tran Thanh Van, T. Damour, E. Hinds and J. Wilkerson. pp445-451 (1993); C.C. Speake *et al.*, Proc. STEP Symposium, PISA. **ESA WPP-115**, 133-147 (1993).
23. G.D. Hammond *et al.*, Phys. Rev. Lett. **98**, 081101 (2007); C.C. Speake *et al..*, J. Meas. Sci. Tech. **10**, 508-513 (1999).
24. W. Guerin *et al.*, Phys. Rev. Lett. **97** 200402 (2006).
25. T. Damour and A.M. Polyakov, Nucl. Phys. B **423** 532 (1994).
26. H.B. Sandvik, J.D. Barrow and J. Magueijo, Phys. Rev. Lett. **88**, 031302 (2002).
27. T.J. Sumner, Gen. Rel. Grav. **36**, 2331-2339 (2004).
28. S. Dimopoulos *et al.*, Phys. Rev. Lett. **98**, 111102 (2007).
29. C. Lämmerzahl and E. Goklu, private communication, (2007).
30. D.J. Kapner *et al.*, Phys. Rev. Lett. **98**, (2007); E.G. Adelberger *et al..*, Phys. Rev. Lett. **98**, 131104 (2007); S. Dimopoulos and A.A. Geraci, Phys. Rev. D **68**, 124021 (2003).
31. J.E. Moody and F. Wilcek, Phys. Rev. D **30**, 130 (1984).
32. C. Wang, R. Bingham and T. Mendonca, Class. Quant. Grav. **23**, L59 (2006).

THE EXPERIMENTAL FOUNDATIONS OF THE DIRAC EQUATION

C. LÄMMERZAHL

Center for Applied Space Technology and Microgravity (ZARM)
University of Bremen, Am Fallturm, 28359 Bremen, Germany
E-mail: laemmerzahl@zarm.uni-bremen.de

The Dirac equation can be derived through an operational approach. Each requirement, (i) well-posedness of the Cauchy problem, (ii) superposition principle, (iii) finite propagation speed, (iv) conservation of probability, and (v) Local Lorentz Invariance can directly be related to experiments. The consequences of these requirements are outlined and corresponding experiments are described.

1. The Dirac equation

The Dirac equation is one of the most fundamental and important equations in physics, in one line with the Schrödinger, Maxwell and Einstein equations. It describes the dynamics of spin-$\frac{1}{2}$ particles and, thus, of most of the particles in our world. In most textbooks this equation is derived from a particular square root of the Klein–Gordon equation where one enlarges the number of components of the wave function so that the square root can be taken in a matrix space. Without enlarging the number of components one has to take the square root in terms of pseudo–differential operators which also leads to a meaningful quantum equation[1] which is, however, non–local and does not describe particles with spin. In any case, the requirement to take the square root is an *ad hoc* procedure which is not based on an operational procedure.

However, it has been shown that there is a scheme by which through a series of requirements the Dirac equation can be derived where each step can be tested in principle. These steps are:[2]

(1) The well–posedness of the Cauchy problem
(2) The superposition principle
(3) Finite propagation speed
(4) Conservation of probability
(5) Local Lorentz Invariance (LLI)

In the following we describe the main steps to derive the Dirac equation from these requirements and also describe experiments which can or have been carried through to test the validity of these assumptions.

2. The generalized Dirac equation

The standard Dirac equation is the result of certain requirements posed upon the dynamics of a complex multicomponent field $\psi \in \mathbb{C}^s$. We first require unique evolution, that is a well posed Cauchy–problem: there is a certain 3+1–splitting of the manifold so that for given Ψ_{t_0} the Ψ_t is uniquely determined. That means that there exists a mapping U so that $\Psi_t = U(t, t_0, \Psi_{t_0})$. The superposition principle then gives that $U(t, t_0, \cdot)$ is a linear mapping. Differentiation of this mapping leads to an abstract Cauchy problem

$$\frac{d\Psi_t}{dt} = G_t \Psi_t \qquad \text{with linear} \quad G_t\,. \tag{1}$$

The further requirement of a finite propagation speed can be recast in the form $\mathrm{supp}(G_t\Psi_t) \subset \mathrm{supp}\Psi_t$ so that G_t is a linear and local operator. Such operators have to be a differential operator

$$(G_t\Psi_t)(x) = \sum_{r=0}^{s} \alpha^{i_1 i_2 \dots i_r} \partial_{i_1}\partial_{i_2} \cdots \partial_{i_r} \Psi_t(x)\,. \tag{2}$$

The finite propagation speed implies further $s = 1$, that is

$$\frac{d\Psi_t(x)}{dt} = \alpha^i \partial_i \Psi_t(x) + \alpha \Psi_t(x) \qquad \Leftrightarrow \qquad 0 = i\gamma^\mu \partial_\mu \psi - M\psi\,. \tag{3}$$

The final requirement of probability conservation $\int_{\Sigma_t} \psi^+(t,x)\psi(t,x) d^3x =$ constant gives for the coefficients that there exists a hermitizing matrix β

$$(\beta\gamma^\mu)^+ = \beta\gamma^\mu\,, \quad (\beta M)^+ = \beta M. \tag{4}$$

Therefore we have real characteristics and mass shells.

The final result is a generalized Dirac equation[3,2]

$$0 = i\gamma^\mu \partial_\mu \psi - M\psi\,, \qquad \gamma^\mu\gamma^\nu + \gamma^\nu\gamma^\mu = 2g^{\mu\nu} + X^{\mu\nu}\,, \quad M = m + \mu, \tag{5}$$

which is a hyperbolic system of partial differential equations giving in general a deformed light cone with birefringence and anisotropy, and a deformed mass shell with Stern–Gerlach effects and also anisotropy. These effects violate LLI and also the Universality of Free Fall (UFF). See also Ref. 4 for an early discussion of a generalized Dirac equation with LLI and UFF violation.

2.1. *General geometrical structures*

In a quasiclassical limit given by the wavefunction $\psi = e^{iS}a$ with $||\gamma^\mu \partial_\mu a|| \ll ||\gamma^\mu p_\mu a||$ we obtain

$$(\gamma^\mu p_\mu - M)\, a = 0 \qquad \text{and} \qquad \gamma^\mu \partial_\mu a = 0\,. \tag{6}$$

The solvability condition of the first one defines the Hamilton function

$$H(x,p) = \det\,(\gamma^\mu p_\mu - M) = 0\,. \tag{7}$$

From this we can also derive a consistency relation between the γ^μ and the traces $g^{\mu_1\cdots\mu_i} = \mathrm{tr}\,(\gamma^{\mu_1}\cdots\gamma^{\mu_i})$

$$0 = \sum_{i=0}^{s} \gamma^{(\mu_1}\gamma^{\mu_2}\cdots\gamma^{\mu_i} g^{\mu_{i+1}\cdots\mu_s)}\,, \tag{8}$$

which plays the role of a generalized Clifford algebra. LLI is violated if $g^{\mu_1\cdots\mu_i} \neq 0$ for $i > 2$.

2.2. *Non–relativistic limit*

The non–relativistic limit of Eq. (5) gives a generalized Pauli equation[5]

$$\begin{aligned} i\frac{\partial\psi}{\partial t} = & -\frac{1}{2m}\left(\delta^{ij} + \frac{\delta m_{\mathrm{i}}^{ij}}{m} + \frac{\delta m_{\mathrm{i}}^{ij}{}_k}{m}\sigma^k\right)\partial_i\partial_j\psi + i\left(\frac{1}{m}\lambda^i_j + \Lambda^i_j\right)\sigma^j\partial_i\psi \\ & + \left(T_i\sigma^i + mB_i\sigma^i\right) + \left(m\left(1 + C_i\sigma^i\right)\delta_{ij} + \delta m_{\mathrm{g}ij}\right)U^{ij}\psi \end{aligned} \tag{9}$$

with

$$U^{ij}(\boldsymbol{x}) = \int \frac{\rho(\boldsymbol{x}')(x-x')^i(x-x')^j}{|\boldsymbol{x}-\boldsymbol{x}'|^3} d^3x'\,.$$

Here a violation of Lorentz invariance[6] is related to $\delta m_{\mathrm{i}}^{ij}$, $\delta m_{\mathrm{i}}^{ij}{}_k$, λ^i_j, T_i, and B_i, and a direct violation of the UFF to C_i, $\delta m_{\mathrm{g}ij}$. According to Schiff's conjecture, a violation of LLI also leads to a violation of the UFF. If one of these anomalous parameters is non–vanishing, then $X^{\mu\nu} \neq 0$ or $\mu \neq 0$.

3. Tests of LLI and UFF

First we can perform interference in spin space. A beam of polarized atoms will be split coherently. The spin of one beam is flipped after the beam splitter and before the recombiner. In the case of an anomalous spin–coupling

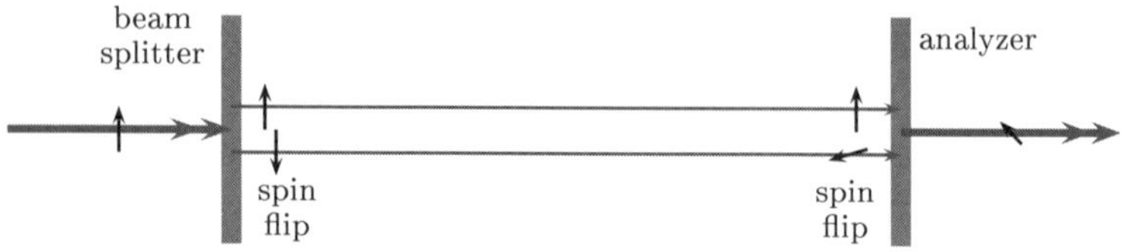

Figure 1. Interference in spin space

the different spin states accumulate different energies. The corresponding phase shift is[7,5]

$$\begin{aligned}\delta\phi &= (H(\boldsymbol{x},\boldsymbol{p},\boldsymbol{S}) - H(\boldsymbol{x},\boldsymbol{p},-\boldsymbol{S}))\,\delta t \qquad (10)\\ &= \left(\frac{\delta m_{\mathrm{i}}^{ij}{}_k}{m} p_i \delta_{jk} l^k - 2\left(\lambda^i_j + m\Lambda^i_j\right)\delta_{ik} l^k + 2mB_j\delta t + 2C_j m U\delta t + T_j\delta t\right) S^j .\end{aligned}$$

If such an experiment is being performed one may reach estimates of the order

$$\left|\frac{\delta m_k^{ij}}{m}\right| \leq 5\cdot 10^{-15}, \quad |\Lambda^i_j| \leq 8\cdot 10^{-24}, \quad |\lambda^i_j| \leq 5\cdot 10^{-6}\mathrm{m}^{-1} \qquad (11)$$

$$|B_k| \leq 3\cdot 10^{-30}, \quad |T_k| \leq 2\cdot 10^{-9}\mathrm{m}^{-1}, \quad |C_k| \leq 3\cdot 10^{-23}. \qquad (12)$$

From Eq. (9) we can derive the classical acceleration for polarized matter

$$a^i = -\left(\delta^{ij} + \frac{\delta m_{\mathrm{i}}^{ij}}{m} + 2\left(\frac{\delta m_{\mathrm{i}}^{ij}{}_k}{m} + \delta^{ij} C_k\right) S^k\right)\partial_j U - \delta^{ij}\frac{\delta_{gkl}}{m}\partial_j U^{kl}. \qquad (13)$$

For an atomic interferometry experiment in a gravitational field with polarized matter the phase shift is then given by

$$\delta\phi = -k_i a^i T^2 . \qquad (14)$$

For polarized atoms and an expected null result one may reach the estimates

$$\left|\frac{\delta \bar{m}_{\mathrm{i}k}^{ij}}{m}\right| \leq 10^{-8}, \qquad \left|\frac{\delta \bar{m}_{\mathrm{i}k}^{ij}}{m}\right| \leq 10^{-10}, \qquad |C_i| \leq 10^{-10}. \qquad (15)$$

4. Test of basic principles

Now we use interferometry and also other experiments in order to get information about the quality of the assumptions leading to the generalized Dirac equation.

4.1. *Well posedness of Cauchy problem*

The most simple violation of the Cauchy problem is to assume that there is a small term with a second time derivative in the Dirac equation. In 3+1 form and with a minimally coupled electromagnetic field this reads[8]

$$i\frac{\partial\psi}{\partial t} = -i\boldsymbol{\alpha}\cdot(\boldsymbol{\nabla} - ie\boldsymbol{A})\,\psi + \beta m\psi + e\phi\psi - \epsilon\frac{1}{m}\left(\frac{\partial}{\partial t} - ie\phi\right)^2\psi, \tag{16}$$

where ϵ is a small dimensionless parameter.

From this equation one determines the plane wave solution, the Hamilton function and the group velocity which depends on the energy different from ordinary plane waves of massive particles. Comparison with astrophysical time–of–flight measurements one arrives at an estimate[8] $\epsilon \leq 10^{-35}$.

Also quantum experiments are sensitive to the term under consideration. For that we first derive the non–relativistic version of (16)

$$i\frac{\partial\psi}{\partial t} = -\frac{1}{2m}(\boldsymbol{\nabla} - ie\boldsymbol{A})^2\psi + c^*\phi\psi + \frac{e}{m}\boldsymbol{\sigma}\cdot\boldsymbol{B}\psi - \epsilon\frac{1}{m^*}\left(\frac{\partial}{\partial t} - ie\phi\right)^2\psi\,, \tag{17}$$

where $m^* := (1+2\epsilon(1+\epsilon))m$, $e^* = e/(1+2\epsilon(1+\epsilon))$. From this equation we can exactly determine the energy values of the energy levels.[8] For simplicity we mention here just the first order modifications

$$\Delta E_{n'n} = Ry\left(\frac{1}{n'^2} - \frac{1}{n^2}\right) + 3\epsilon\frac{Ry}{mc^2}\left(\frac{1}{n'^4} - \frac{1}{n^4}\right), \tag{18}$$

where Ry is the Rydberg constant. An $\epsilon \neq 0$ modifies the structure of atomic series and of ionization energies. A comparison with best measured hydrogen spectrum gives an estimate[8] $\epsilon \leq 7 \cdot 10^{-10}$.

For this issue interferometry seems not to be very appropriate. From a minimal coupling of a Newtonian potential U we obtain to first order in ϵ and U the Hamilton function

$$H = \frac{p^2}{2m} + mU - \epsilon\frac{p^4}{m^3c^2}\,. \tag{19}$$

The corresponding acceleration ($g = -\nabla U$)

$$a = -g\left(1 - 3\epsilon\frac{v_0^2}{c^2}\right) \tag{20}$$

gives a phase shift $\delta\phi = -k_i a^i T^2$ in atomic interferometry. Here v_0 is the velocity of the atoms at the first laser pulse. The modifications are too small to give a reasonable estimate for ϵ. Further experiments for this model have been proposed in Ref. 9.

4.2. *Locality*

Another issue is the test of the locality of fundamental field equations. In the case of the Dirac equations this means that there might be a small term with a second order spatial derivative. Higher spatial than time derivatives imply an infinite propagation speed.

The most simple model thus is

$$i\partial_t\psi = -i\alpha^i\partial_i\psi + \alpha^{ij}\partial_i\partial_j\psi + \beta m\psi\,, \tag{21}$$

where a non–vanishing of the parameter α^{ij} under consideration also implies a violation of LI. Note that α^{ij} in general is an element of the Dirac algebra.

The non–relativistic limit of this equation is

$$i\partial_t\psi = -\frac{1}{2m}\Delta\psi + \frac{1}{m^2}\left(A^{ijk} + B_l^{ijk}\sigma^l\right)\partial_i\partial_j\partial_k\psi, \tag{22}$$

where the parameters are traces of the coefficient α^{ij}. With an atomic spin flip experiment as described above we obtain a phase shift

$$\Delta\phi = \frac{1}{\hbar}\left(H(S) - H(-S)\right)\Delta t = \frac{2}{\hbar c}B_m^{jki}S^m m v_i v_j l_k\,, \tag{23}$$

which for an assumed null–result for current interferometers yield an estimate of $B_m^{(ijk)} \leq 10^{-10}$. The calculation of modifications of atomic spectra is under way.[10] A further question which has to be addressed is whether one can conclude from a null results in these experiments a vanishing of all components of the spin–matrix α^{ij}. Further issues are estimates for higher order spatial derivatives. We also like to mention that a non–locality not necessarily implies a violation of LI. This can be seen from the square root of the Klein–Gordon equation.[1]

4.3. *The superposition principle*

Here we use as our starting point a general non–linear Schödinger equation

$$i\frac{\partial\psi}{\partial t} = -\frac{1}{2m}\Delta\psi + F(\psi^*\psi)\psi\,. \tag{24}$$

The requirement of the separability of quantum systems then specifies[11] $F(\psi^*\psi) = a\ln(b\psi^*\psi)$.

Based on this non–linear equation Shimony[12] proposed an interference experiment which is uniquely sensitive to the non–linearity. The main idea is to place an attenuator in one arm of a Mach–Zehnder setup, see Fig. 2. A change of the position of the attenuator yields a phase shift

$$\delta\phi = \frac{d}{\hbar}\sqrt{\frac{m}{2E}}\left(F(|\varphi|^2) - F(\alpha^2|\varphi|^2)\right)\,, \tag{25}$$

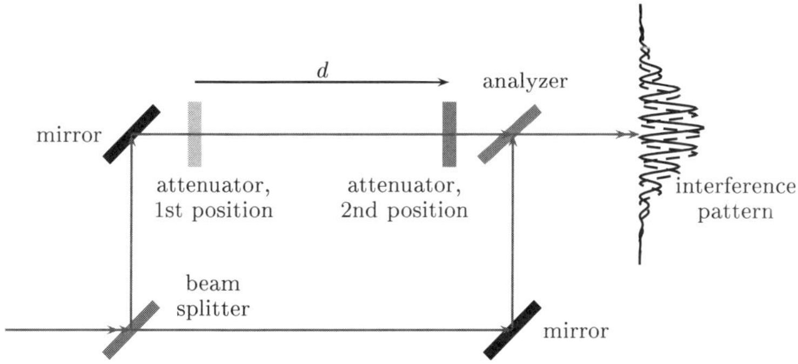

Figure 2. Interference experiment to test the non-linearity of the Schrödinger equation. Moving the attenuator from the 1st to the 2nd position shifts the interference fringes (dashed line → solid line).

where α is the attenuation factor. This proposal has subsequently been implemented with neutron interferometry[13] with the result $a \leq 3.4 \cdot 10^{-13}$ eV. It is obvious that atomic interferometry has the potential to improve this by many orders of magnitude. Atomic spectroscopy gives weaker estimates.

Acknowledgments

I would like to thank the German Aerospace Center DLR for financial support.

References

1. C. Lämmerzahl, J. Math. Phys. **34**, 3918 (1993).
2. J. Audretsch and C. Lämmerzahl, in U. Majer and H.-J. Schmidt, eds., *Semantical Aspects of Space-Time Geometry*, BI Verlag, Mannheim, 1993, p.21.
3. C. Lämmerzahl, in J. Audretsch and V. DeSabbata, eds., *Quantum Mechanics in Curved Space-Time*, NATO ASI Series B, Physics Vol. 230, Plenum, New York, 1990, p.23.
4. D.-E. Liebscher, Ann. Phys. (Leipzig) **42**, 35 (1985).
5. C. Lämmerzahl, Class. Quantum Grav. **14**, 13 (1998).
6. V.A. Kostelecký and C.D. Lane, Phys. Rev. D **60**, 116010 (1999).
7. J. Audretsch, U. Bleyer, and C. Lämmerzahl, Phys. Rev. A **47**, 4632 (1993).
8. C. Lämmerzahl and Ch.J. Bordé, in C. Lämmerzahl, C.W.F. Everitt, and F.W. Hehl, eds., *Gyros, Clocks, and Interferometers: Testing Relativistic Gravity in Space*, Springer–Verlag, Berlin, 2001, p.464.
9. A. Camacho and A. Macias, Phys. Lett. B **582**, 229 (2004).
10. C. Lämmerzahl and A. Macias, in preparation.
11. I. Bialynicki–Birula and J. Mycielski, Ann. Phys. (NY) **100**, 62 (1976).
12. A. Shimony, Phys. Rev. A **20**, 394 (1979).
13. C.G. Shull *et al.*, Phys. Rev. Lett. **44**, 765 (1980).

PERSPECTIVES ON LORENTZ AND CPT VIOLATION

V. ALAN KOSTELECKÝ

Physics Department, Indiana University
Bloomington, IN 47405, U.S.A.

This talk offers some comments and perspectives on Lorentz and CPT violation.

1. Introduction

Lorentz symmetry is the invariance of physical laws under rotations and boosts. As a global symmetry over Minkowski spacetime, it underlies the theory of Special Relativity and the Standard Model of particle physics, where it is intimately tied to CPT invariance. As a local symmetry of freely falling frames, it is an essential component of General Relativity. Nonetheless, the possibility exists that nature may exhibit tiny violations of Lorentz symmetry. This talk presents some perspectives on the basic motivations and reasoning in this subject.

Since Lorentz symmetry has been verified in many experiments, as has CPT invariance, it is reasonable to ask why relativity violations are worth considering. A sufficient theoretical motivation is the need for a consistent description of Lorentz and CPT violation to offer guidance for experimental tests. However, a stronger motivation is the prospect that Lorentz and CPT violation can serve as a sensitive potential signal for physics at the Planck scale. In fact, the present interest in the subject was triggered by the realization that natural mechanisms for Lorentz and CPT violation exist in unified theories at the Planck scale.[1] The large range of existing phenomenological and experimental activities stems from the application of effective field theory[2] and the construction of the Standard-Model Extension (SME)[3,4] to catalogue and predict observable effects.

2. Approaches and origins

The study of Lorentz and CPT violation can be approached on three distinct levels. First, at the level of fundamental theory at the unification scale, one can investigate possible mechanisms and determine their features and implications. Second, at the level of theory at accessible scales, one can seek a description of the resulting effects that is quantitative, general, and compatible with the established physics of the Standard Model and General Relativity. Finally, at the level of observation and experiment, one can study and perform both high sensitivity tests and broad searches, preferably ones that are feasible with existing or near-future technology.

The first point to establish is whether Lorentz violation can indeed occur in a fundamental theory. Without at least one viable mechanism, the interest of the idea would be much reduced. A plausible origin for Lorentz violation has been identified in string field theory,[1] which has interactions with a generic structure that could in principle trigger spontaneous violation of Lorentz symmetry and generate vacuum expectation values for Lorentz tensor fields. More recently, numerous other mechanisms for Lorentz violation at the fundamental level have been proposed including, for example, ones involving noncommutative field theories,[5] spacetime-varying fields,[6] quantum gravity,[7] random-dynamics models,[8] multiverses,[9] brane-world scenarios,[10] supersymmetry,[11] and massive gravity.[12]

Given that mechanisms exist for Lorentz and CPT violation in an underlying theory, it is natural to ask about the consequences for physics at observable scales. In particular, the question arises as to the best method to develop a description of the possible effects.

3. Describing Lorentz violation

Some key criteria offer valuable guidance in the search for a suitable theoretical framework for describing Lorentz violation at attainable scales. The first is *coordinate independence.* It has been accepted since long before 1905 that the physics of a system should be independent of a change of observer coordinates. This holds whether a coordinate change is implemented via a Lorentz transformation or in any other way.

The second is *realism.* Since 1905, when virtually no fundamental particles were known and quantum physics was at its dawn, thousands of people have invested millions of person-hours and billions of dollars in establishing the Standard Model of particle physics and General Relativity as an accurate description of nature. To be of real interest nowadays, any proposed theoretical framework for Lorentz violation must incorporate this well-established physics.

The third is *generality.* No compelling evidence for Lorentz violation exists at present. Physics is therefore currently in the position of searching for a violation, as opposed to attempting to understand an observed effect. In the searching phase, it is desirable to have the most general possible formulation so that no region is left unexplored. This is in strong contrast to the modeling phase, where considerations such as simplicity are important in attempts to understand a known effect.

Armed with these criteria, we can follow the basic reasoning that leads to the application of effective field theory and the construction of the SME.

3.1. *Modified Lorentz transformations*

Since the essential content of Special Relativity is the idea that physics is invariant under Lorentz transformations, the most obvious approach to describing relativity violations is to investigate modifications of the Lorentz transformations. In fact, the literature since 1905 abounds with various *ad hoc* proposals of this type. However, independent of any specific proposal, this approach has some serious disadvantages.

One is that a textbook Lorentz transformation acts on the observer and therefore corresponds merely to a change of coordinates, i.e., a change of reference frame. However, according to the above criterion of coordinate independence, a frame change cannot have physical implications by itself. The key feature of Special Relativity is really the requirement that the equations for the system being observed must be covariant under a Lorentz transformation, which intrinsically assumes that Lorentz symmetry is exact. This approach is therefore problematic for investigating violations.

It is of course possible to construct special models imposing form covariance of the system under some *ad hoc* alternative transformation. However, any specific such proposal runs counter to the criterion of generality. Moreover, some kinds of violations are difficult and perhaps even impossible to countenance via this approach. For example, Lorentz violation in nature might well be particle-species dependent, but it is very challenging to formulate a description of this flavor dependence based on modified Lorentz transformations of the observer. The criterion of realism presents a further substantial obstacle, since it is awkward at best to implement such models in the context of the Standard Model and General Relativity.

3.2. *Modified dispersion laws*

The above discussion suggests that a general and realistic investigation of Lorentz violation is most naturally performed directly in terms of the properties of a system rather than via modifications of the Lorentz transformations. A simple implementation of this is to study modifications of particle dispersion laws. However, this also suffers from serious drawbacks.

One issue involves the criterion of generality. Modifications of dispersion laws can only describe changes in the free propagation of particles and perhaps also partially account for interaction kinematics. However, physics is far more than free propagation, and this approach therefore disregards a large range of interesting Lorentz-violating effects involving interactions.

There are also various issues associated with the choice of modifications to the dispersion law. Not all choices are compatible with desirable features such as originating from an action. Also, meaningful physical measurements must necessarily compare two quantities, so some choices may be unphysical. In particular, calculations with a modified dispersion law that yield apparent changes of properties are insufficient by themselves to demonstrate physical Lorentz violation. A simple example of a modified dispersion law with no observable consequences in Minkowski spacetime is[3] $p^\mu p_\mu = m^2 + a_\mu p^\mu$, where a_μ is a prescribed set of four numbers in a given frame. Direct calculations with this dispersion law appear to give Lorentz-violating properties that depend on the preferred vector a_μ, but in fact they are unobservable because a_μ can be eliminated via a physically irrelevant redefinition of the energy and momentum. The observability of modifications to a dispersion law can be challenging to demonstrate.

3.3. *Effective field theory and the SME*

We see that the desiderata for a satisfactory description of Lorentz violation include a comprehensive treatment of free and interacting effects in all particle species. Remarkably, a model-independent and general approach of this type exists.

The key is to take advantage of the idea that Lorentz violation at attainable energies can be described using effective field theory, independent of the underlying mechanism.[2,13] Starting from the Standard Model coupled to General Relativity, we can add to the action all possible scalar terms formed by contracting operators for Lorentz violation with coefficients that control the size of the effects. The operators are naturally ordered according to their mass dimension. The resulting realistic effective field theory is the SME.[3,4] Since CPT violation in realistic field theories comes with Lorentz violation,[14] the SME also incorporates general CPT violation.

By virtue of its construction, the SME satisfies the three guiding criteria of coordinate invariance, realism, and generality. Moreover, it handles simultaneously all particle species, including both propagation and interaction properties, so its equations of motion contain all action-compatible modifications of realistic dispersion laws. The coordinate invariance implies that physics is unaffected by observer frame changes, including Lorentz and other transformations, while particle transformations can produce observable effects of Lorentz violation.

The primary disadvantage of the SME approach is its comparative complexity and the investment required to become proficient with its use. However, this is outweighed by its advantages as a realistic, general, and calculable framework for describing Lorentz violation. Judicious choices establishing relations among the SME coefficients for Lorentz violation yield elegant and simple models that can serve as a theorist's playground, while the general case offers guidance for broad-based experimental searches.

4. Gravity and Lorentz violation

The SME allows for both global[3] and local[4] Lorentz violation, and interesting effects arise from local Lorentz violation in the gravitational context. In general, local Lorentz violation can be understood as arising when a nonzero coefficient $t_{abc\ldots}$ for Lorentz violation exists in local freely falling frames.[4] The cofficient $t_{abc\ldots}$ can be converted to a coefficient $t_{\lambda\mu\nu\ldots}$ on the spacetime manifold using the vierbein ${e_\mu}^a$.

One result is that spontaneous violation of local Lorentz symmetry is always accompanied by spontaneous diffeomorphism violation, and vice versa.[15] A nonzero coefficient $t_{abc\ldots}$ is the vacuum value of a local Lorentz tensor field, and it implies Lorentz violation because it is invariant instead of transforming like a tensor under particle transformations. The vierbein ensures that there is a corresponding spacetime tensor field with vacuum value $t_{\lambda\mu\nu\ldots}$ on the spacetime manifold, which in turn implies spontaneous diffeomorphism breaking because it is invariant instead of transforming like a tensor under particle diffeomorphisms.

A more surprising result is that explicit Lorentz violation is generically incompatible with Riemann geometry.[4] Explicit violation occurs when the SME coefficients are externally prescribed, but the ensuing equations of motion turn out to be inconsistent with the Bianchi identities. This result also holds in Riemann-Cartan spacetime. However, spontaneous violation evades the difficulty because it generates the SME coefficients dynamically, thereby ensuring compatibility with the underlying spacetime geometry.

Spontaneous local Lorentz violation is accompanied by up to 10 Nambu-Goldstone (NG) modes.[15] With a suitable choice of gauge, these modes can be identified with components of the vierbein normally associated with local Lorentz and diffeomorphism gauge freedoms. The physical role of the NG modes varies, but in general they represent long-range forces that can be problematic for phenomenology. However, in certain models the NG modes can be interpreted as photons, thus offering the intriguing prospect that the existence of light could be a consequence of Lorentz violation instead of local U(1) gauge invariance, with concomitant observable signals.[15,16] A similar interpretation is possible for the graviton.[17] Other potential experimental signals arise from NG modes in the gravity[18] and matter[19] sectors, and from torsion.[20] The spectrum of vacuum excitations typically also includes massive modes that may lead to additional observable effects.[21]

5. The search for signals

The SME predicts some unique signals, such as rotational, sidereal, and annual variations. The effects are likely to be heavily suppressed, perhaps as some power of the ratio of an accessible scale to the underlying scale, but they could be detected using sensitive tools such as interferometry. For example, meson interferometry offers the potential to identify flavor- and direction-dependent energy shifts of mesons relative to antimesons,[22] while exquisite interferometric sensitivity to polarization-dependent effects of photons is attained using cosmological birefringence.[23] Conceivably, SME effects might even be reflected in existing data, such as those for flavor oscillations of neutrinos.[24] Overall, an impressive range of sensitivities in the matter, gauge, and gravitational sectors of the SME has been achieved.[25]

Despite a decade of intense activity, most of the SME coefficient space is still unexplored by experiments, and many basic theoretical issues are unaddressed. The study of relativity violations remains fascinating, with the enticing prospect of identifying a signal from the Planck scale.

Acknowledgments

This work was supported in part by DoE grant DE-FG02-91ER40661 and NASA grant NAG3-2914.

References

1. V.A. Kostelecký and S. Samuel, Phys. Rev. D **39**, 683 (1989); V.A. Kostelecký and R. Potting, Nucl. Phys. B **359**, 545 (1991).

2. V.A. Kostelecký and R. Potting, Phys. Rev. D **51**, 3923 (1995).
3. D. Colladay and V.A. Kostelecký, Phys. Rev. D **55**, 6760 (1997); Phys. Rev. D **58**, 116002 (1998).
4. V.A. Kostelecký, Phys. Rev. D **69**, 105009 (2004).
5. See, for example, I. Mocioiu *et al.*, Phys. Lett. B **489**, 390 (2000); S.M. Carroll *et al.*, Phys. Rev. Lett. **87**, 141601 (2001).
6. V.A. Kostelecký *et al.*, Phys. Rev. D **68**, 123511 (2003).
7. See, for example, G. Amelino-Camelia *et al.*, AIP Conf. Proc. **758**, 30 (2005); N.E. Mavromatos, Lect. Notes Phys. **669**, 245 (2005); Y. Bonder and D. Sudarsky, arXiv:0709.0551.
8. C.D. Froggatt and H.B. Nielsen, hep-ph/0211106.
9. J.D. Bjorken, Phys. Rev. D **67**, 043508 (2003).
10. See, for example, C.P. Burgess *et al.*, JHEP **0203**, 043 (2002).
11. M. Berger and V.A. Kostelecký, Phys. Rev. D **65**, 091701(R) (2002); P.A. Bolokhov *et al.*, Phys. Rev. D **72**, 015013 (2005).
12. See, for example, G. Dvali *et al.*, Phys. Rev. D **76**, 044028 (2007); D.S. Gorbunov and S.M. Sibiryakov, JHEP **0509**, 082 (2005); M.V. Libanov and V.A. Rubakov, JHEP **0508**, 001 (2005); N. Arkani-Hamed *et al.*, JHEP **0507**, 029 (2005); V.A. Kostelecký and S. Samuel, Phys. Rev. D **42**, 1289 (1990); Phys. Rev. Lett. **66**, 1811 (1991).
13. V.A. Kostelecký and R. Lehnert, Phys. Rev. D **63**, 065008 (2001).
14. O.W. Greenberg, Phys. Rev. Lett. **89**, 231602 (2002).
15. R. Bluhm and V.A. Kostelecký, Phys. Rev. D **71**, 0065008 (2005).
16. B. Altschul and V.A. Kostelecký, Phys. Lett. B **628**, 106 (2005).
17. V.A. Kostelecký and R. Potting, Gen. Rel. Grav. **37**, 1675 (2005).
18. Q.G. Bailey and V.A. Kostelecký, Phys. Rev. D **74**, 045001 (2006).
19. V.A. Kostelecký and J. Tasson, in preparation.
20. V.A. Kostelecký and N. Russell, in preparation.
21. R. Bluhm *et al.*, in preparation; V.A. Kostelecký and S. Samuel, Phys. Rev. D **40**, 1886 (1989); Phys. Rev. Lett. **63**, 224 (1989).
22. H. Nguyen (KTeV), hep-ex/0112046; A. Di Domenico *et al.* (KLOE), these proceedings; B. Aubert *et al.* (BaBar), hep-ex/0607103; arXiv:0711.2713; D.P. Stoker (BaBar), these proceedings; J.M. Link *et al.* (FOCUS), Phys. Lett. B **556**, 7 (2003); V.A. Kostelecký, Phys. Rev. Lett. **80**, 1818 (1998); Phys. Rev. D **61**, 016002 (2000); Phys. Rev. D **64**, 076001 (2001).
23. V.A. Kostelecký and M. Mewes, Phys. Rev. Lett. **87**, 251304 (2001); Phys. Rev. D **66**, 056005 (2002); Phys. Rev. Lett. **97**, 140401 (2006); Phys. Rev. Lett. **99**, 011601 (2007).
24. L.B. Auerbach *et al.*, Phys. Rev. D **72**, 076004 (2005); B.J. Rebel and S.F. Mufson, these proceedings; V.A. Kostelecký and M. Mewes, Phys. Rev. D **69**, 016005 (2004); Phys. Rev. D **70**, 031902(R) (2004); Phys. Rev. D **70**, 076002 (2004); T. Katori *et al.*, Phys. Rev. D **74**, 105009 (2006); V. Barger *et al.*, Phys. Lett. B **653**, 267 (2007); K. Whisnant, these proceedings.
25. Results are tabulated in V.A. Kostelecký and N. Russell, these proceedings.

SEARCH FOR LORENTZ AND CPT VIOLATION EFFECTS IN MUON SPIN PRECESSION

B. LEE ROBERTS

ON BEHALF OF THE MUON $(g-2)$ COLLABORATION*

Department of Physics, Boston University
590 Commonwealth Ave., Boston, MA 02215, U.S.A.
E-mail: roberts@bu.edu

Two signals of CPT/Lorentz violation have been searched for in data collected in the muon $(g-2)$ experiment E821 at the Brookhaven Lab AGS. Muons of 3.1 GeV/c were stored in a magnetic storage ring, and their spin precession frequency ω_a relative to their momentum was measured. The difference in the average precession frequencies for μ^+ and μ^- restricts the Standard-Model Extension parameter b_Z. The non-observation of a sidereal variation in the muon frequencies over a several-month data collection period restricts the parameters $\check{b}_{\perp}^{\mu^{\pm}}$.

1. Introduction

A charged particle with spin $\vec{s}$ has a magnetic moment $\vec{\mu}_s = g_s(e/2m)\vec{s}$ and an anomaly $a \equiv (g_s - 2)/2$ with $\mu_s = (1 + a)e\hbar/2m$, where g_s is the

*Muon $(g-2)$ Collaboration: G.W.Bennett,[2] B.Bousquet,[10] H.N.Brown,[2] G.Bunce,[2] R.M. Carey,[1] P. Cushman,[10] G. T. Danby,[2] P.T.Debevec,[8] M.Deile,[13] H.Deng,[13] W. Deninger,[8] S. K. Dhawan,[13] V. P. Druzhinin,[3] L.Duong,[10] E.Efstathiadis,[1] F.J.M.Farley,[13] G.V.Fedotovich,[3] S.Giron,[10] F.E.Gray,[8] D.Grigoriev,[3] M.Grosse-Perdekamp,[13] A.Grossmann,[7] M.F.Hare,[1] D.W.Hertzog,[8] X.Huang,[1] V.W.Hughes,[13] M.Iwasaki,[12] K.Jungmann,[6,7] D.Kawall,[13] M.Kawamura,[12] B.I.Khazin,[3] J.Kindem,[10] F. Krienen,[1] I.Kronkvist,[10] A.Lam,[1] R.Larsen,[2] Y.Y.Lee,[2] I.Logashenko,[1,3] R.McNabb,[10,8] W. Meng,[2] J. Mi,[2] J. P. Miller,[1] Y.Mizumachi,[9,11] W.M.Morse,[2] D.Nikas,[2] C.J.G.Onderwater,[8,6] Y.Orlov,[4] C.S.Özben,[2,8] J.M.Paley,[1] Q.Peng,[1] C.C.Polly,[8] J.Pretz,[13] R.Prigl,[2] G.zu Putlitz,[7] T.Qian,[10] S.I.Redin,[3,13] O.Rind,[1] B.L.Roberts,[1] N.Ryskulov,[3] S.Sedykh,[8] Y.K.Semertzidis,[2] P.Shagin,[10] Yu.M. Shatunov,[3] E.P. Sichtermann,[13] E. Solodov,[3] M. Sossong,[8] A.Steinmetz,[13] L.R.Sulak,[1] C.Timmermans,[10] A.Trofimov,[1] D.Urner,[8] P.von Walter,[7] D.Warburton,[2] D.Winn,[5] A.Yamamoto,[9] D.Zimmerman[10]

[1]Boston U., [2]BNL, [3]Budker Institute, [4]Cornell U., [5]Fairfield U., [6]KVI, U Groningen, [7]U. Heidelberg, [8]U. Illinois Urbana-Champaign, [9]KEK, [10]U. Minnesota, [11]Science U. Tokyo, [12]Tokyo Inst. of Tech., [13]Yale U.

spin g−factor. This latter expression is what one finds in the Particle Data Tables.[1] The g-value is exactly 2 for a point-like fermion described by the Dirac equation. Substructure can cause a large anomalous magnetic moment, e.g., the proton anomaly is 1.79. For the leptons, e, μ and τ, radiative corrections give rise to an anomaly, the largest being the "Schwinger" contribution of $\alpha/2\pi \simeq 10^{-3}$.

The interaction of an external magnetic field with the magnetic dipole moment gives rise to a torque, causing the muon spin to precess with the frequency ω_S. If the muon is moving, the momentum vector will turn with the cyclotron frequency ω_C, and the spin will precess relative to the momentum with the difference frequency

$$\omega_a = \omega_S - \omega_C = \frac{(g-2)}{2}\frac{eB}{m} = a\frac{eB}{m}. \tag{1}$$

With both electric and magnetic fields present, the spin difference frequency is given by

$$\vec{\omega}_a = -\frac{e}{m}\left[a_\mu \vec{B} - \left(a_\mu - \frac{1}{\gamma^2 - 1}\right)\vec{\beta}\times\vec{E}\right], \tag{2}$$

under the assumption that $\vec{\beta}\cdot\vec{B} = 0$. At $\gamma_{\mathrm{magic}} = 29.3$ ($p_{\mathrm{magic}} = 3.09$ GeV/c) the second term vanishes, and the electric field does not contribute to the spin motion relative to the momentum.

In the Standard-Model Extension, the muon spin rotation can be affected by Lorentz and CPT violating terms in the Lagrangian:[2]

$$\begin{aligned}\mathcal{L}' = &- a_\kappa \bar{\psi}\gamma^\kappa\psi - b_\kappa\bar{\psi}\gamma_5\gamma^\kappa\psi - \frac{1}{2}H_{\kappa\lambda}\bar{\psi}\sigma^{\kappa\lambda}\psi \\ &+ \frac{1}{2}ic_{\kappa\lambda}\bar{\psi}\gamma^\kappa \overleftrightarrow{D}^\lambda \psi + \frac{1}{2}id_{\kappa\lambda}\bar{\psi}\gamma_5\gamma^\kappa \overleftrightarrow{D}^\lambda \psi,\end{aligned} \tag{3}$$

where $iD_\lambda \equiv i\partial_\lambda - qA_\lambda$, and the small parameters a_κ, b_κ, $H_{\kappa\lambda}$, $c_{\kappa\lambda}$ $d_{\kappa\lambda}$ represent the Lorentz and CPT violation. All terms violate Lorentz invariance with $-a_\kappa\bar{\psi}\gamma^\kappa\psi$ and $-b_\kappa\bar{\psi}\gamma_5\gamma^\kappa\psi$ CPT odd; all the other terms are CPT even. In this model the conventional figure of merit $r_g^\mu \equiv |g_{\mu^+} - g_{\mu^-}|/g_{average}$ is zero at leading order; however effects on the anomalous spin precession frequency ω_a do exist in lowest-order.[2]

To compare results from different experiments, we use the non-rotating standard celestial equatorial frame[3] $\{\hat{X}, \hat{Y}, \hat{Z}\}$. The $\hat{Z}$ axis is along the earth's rotational north pole, with the $\hat{X}$ and $\hat{Y}$ axes lying in the plane of the earth's equator. The 26,000 year precession period of the earth's rotational axis can be ignored. Using this frame, the correction to the

(Standard Model) muon anomalous precession frequency $\omega_a^{\mu^\pm}$ in Eq. (2) is

$$\delta\omega_a^{\mu^\pm} \approx 2\check{b}_Z^{\mu^\pm} \cos\chi + 2(\check{b}_X^{\mu^\pm} \cos\Omega t + \check{b}_Y^{\mu^\pm} \sin\Omega t)\sin\chi \qquad (4)$$

with

$$\check{b}_J^{\mu^\pm} \equiv \pm\frac{b_J}{\gamma} + m_\mu d_{J0} + \frac{1}{2}\varepsilon_{JKL}H_{KL} \qquad (J = X, Y, Z), \qquad (5)$$

where χ is the geographic colatitude (= 90°− latitude) of the experiment location. For E821, $\chi_{BNL} = 49.1°$. The sidereal angular frequency is $\Omega = 2\pi/T_s$, where $T_s \approx$ 23 hours 56 minutes and $\gamma = 1/\sqrt{1-(v/c)^2}$ is the Lorentz factor. Equation (4) predicts two signatures of Lorentz and CPT violation: a difference between the time averages of ω_{μ^+} and ω_{μ^-}, and oscillations in the values of $\omega_a^{\mu^+}$ and $\omega_a^{\mu^-}$ at the sidereal angular frequency.

2. Experiment E821 at the BNL AGS

In E821 at the Brookhaven AGS, muons with the magic momentum were stored in a magnetic storage ring that used electrostatic quadrupoles to provide vertical focusing.[4] The parity violating decay of the muon,[5] $\mu^- \to e^-\nu_\mu\bar{\nu}_e$, was used to determine the muon spin precession frequency, ω_a. As the muon spin turns relative to the momentum, the number of high-energy decay electrons (or positrons for μ^+ decays) is modulated by the frequency ω_a. The simplest functional form for the arrival-time spectrum contains five parameters:

$$f(t) = N_0 e^{-\frac{t}{\gamma\tau}}[1 + A\cos(\omega_a t + \phi)]. \qquad (6)$$

While the the analysis of the data to determine a_μ required more sophisticated fitting functions,[4] this simple parameterization was adequate for the studies[6] to search for a sidereal variation in ω_a.

A determination of a_μ requires knowledge of the magnetic field (see Eq. (2)), which is measured by nuclear magnetic resonance techniques, and calibrated to the Larmor precession of the free proton,[4] ω_p. We use the frequency ratio $\mathcal{R} = \omega_a/\omega_p$ and the relationship

$$a_\mu = \frac{\omega_a/\omega_p}{\lambda - \omega_a/\omega_p}, \quad \text{where} \quad \lambda = \frac{\mu_\mu}{\mu_p}, \qquad (7)$$

to determine a_μ. The quantity λ comes from measurements with the μ^+, so to obtain a_{μ^-} one needs to assume CPT invariance. The values obtained for the anomaly[7,8,9,10,11] are displayed graphically along with the Standard-Model value.[5] The difference between the Standard-Model and experimental values is $\Delta a_\mu^{(\text{today})} = a_\mu^{(\text{Exp})} - a_\mu^{(\text{SM})} = (295 \pm 88) \times 10^{-11}$, a 3.4 standard deviation difference, that has generated significant interest in the theory community.[12,13,14]

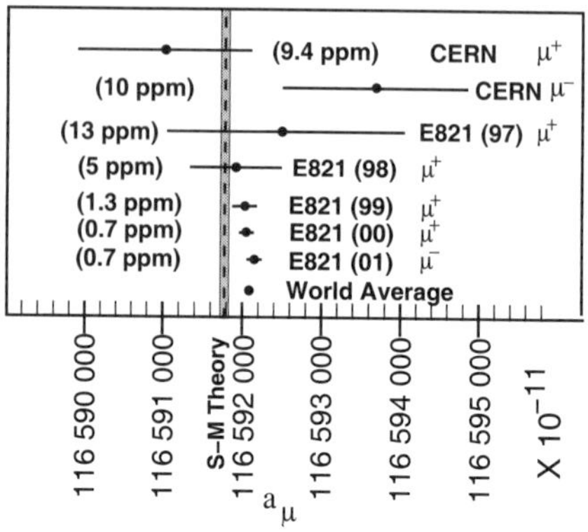

Figure 1. Measurements of the muon anomaly. To obtain the value of a_{μ^-}, as well as the world average, CPT invariance is assumed.

The extraction of a_μ for both μ^+ and μ^- requires ω_a, ω_p along with the value of λ. However, a direct comparison of $\mathcal{R}$ is equivalent to a comparison of ω_a, since any sidereal dependence of ω_p is much smaller than the sensitivity of our measurement.[6]

A comparison of the time averages over many sidereal days of $\mathcal{R}$ for μ^+ and μ^- (Eq. (4)) gives

$$\Delta\omega_a \equiv \langle \omega_a^{\mu^+} \rangle - \langle \omega_a^{\mu^-} \rangle = \frac{4b_Z}{\gamma} \cos\chi. \tag{8}$$

For measurements made at different χ and/or ω_p,

$$\begin{aligned}\Delta\mathcal{R} =& \frac{2b_Z}{\gamma}\left(\frac{\cos\chi_1}{\omega_{p1}} + \frac{\cos\chi_2}{\omega_{p2}}\right) \\ &+ 2(m_\mu d_{Z0} + H_{XY})\left(\frac{\cos\chi_1}{\omega_{p1}} - \frac{\cos\chi_2}{\omega_{p2}}\right).\end{aligned} \tag{9}$$

The colatitudes for the E821 μ^+ and μ^- measurements were identical, and the slightly different values of ω_p can be neglected.[6]

The E821 result[4] is $\Delta\mathcal{R} = -(3.6 \pm 3.7) \times 10^{-9}$, which corresponds to

$$\begin{aligned} b_Z &= -(8.7 \pm 8.9) \times 10^{-24}\ \text{GeV} \\ r_{\Delta\omega_a} &\equiv \frac{\Delta\omega_a}{m_\mu} = -(1.0 \pm 1.1) \times 10^{-23}\,, \end{aligned} \tag{10}$$

a factor of 22 improvement over the limit that can be obtained from the CERN muon $(g-2)$ experiment.[16]

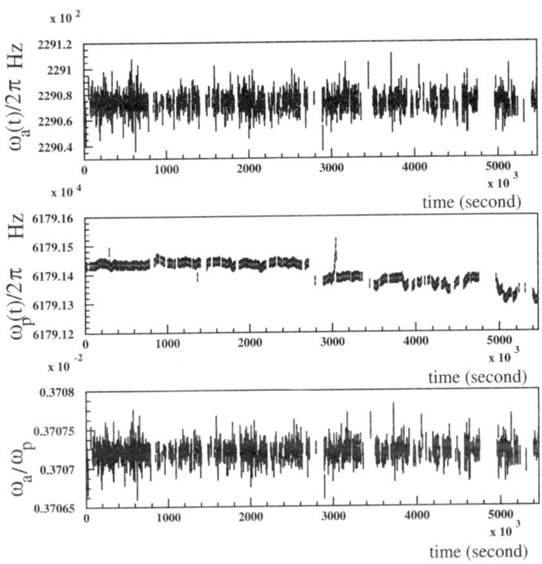

Figure 2. Values of $\omega_a(t)$, $\omega_p(t)$ and $\mathcal{R}(t) \equiv \frac{\omega_a}{\omega_p}(t)$ from the 2001 μ^- run. The uncertainty on each ω_a point is about 20 ppm.

The data were also analyzed for a sidereal variation of ω_a. Data from the 2001 running period are shown in Fig. 2 as a function of time. Because the magnetic field changed slightly with time, $\mathcal{R}$ was used instead of ω_a directly, to search for a variation. A sidereal variation of ω_a can be written as $\omega_a(\omega_p(t_i), t_i) = K\omega_p(t_i) + A_\Omega \cos(\Omega t_i + \phi)$. Dividing by ω_p gives

$$\mathcal{R}(t_i) = K + \frac{A_\Omega}{\omega_p(t_i)} \cos(\Omega t_i + \phi)\,, \tag{11}$$

where $K = \lambda a_\mu/(1 + a_\mu)$ is a constant, and A_Ω is the amplitude of the sidereal variation with the sidereal period $2\pi/\Omega$.

Both a multi-parameter least-square fit, and the Lomb-Scargle test[15] were used. The results of this analysis are given in Table 1. No significant signal was found, and the 95% confidence level limits of

$$\begin{aligned} \check{b}_{\perp}^{\mu^+} &= \sqrt{(\check{b}_X^{\mu^+})^2 + (\check{b}_Y^{\mu^+})^2} < 1.3 \times 10^{-24}\,\text{GeV}, \\ \check{b}_{\perp}^{\mu^-} &= \sqrt{(\check{b}_X^{\mu^-})^2 + (\check{b}_Y^{\mu^-})^2} < 2.5 \times 10^{-24}\,\text{GeV} \end{aligned} \tag{12}$$

were obtained.

Table 1. The signal amplitude in ppm needed for 95% of the simulated data to have larger A_Ω, or $P(\Omega)$, than that of the real data. MPF means multi-parameter fit, L-S stands for Lomb-Scargle.

data set	MPF amplitude (ppm)	L-S amplitude (ppm)
1999 μ^+	5.5	5.2
2000 μ^+	2.2	2.0
1999/2000 μ^+	2.2	2.0
2001 μ^-	4.2	4.2

For the dimensionless figure of merit[2] obtained by dividing by m_μ, we obtain $r_{A\Omega}^{\mu^+} < 2.2 \times 10^{-23}$ and $r_{A\Omega}^{\mu^-} < 3.8 \times 10^{-23}$, which can be compared with the ratio of the muon to Planck mass, $m_\mu/m_P = 8.7 \times 10^{-21}$. Using our result for b_Z in Eq. (10), along with the CERN result[16] for $\Delta\omega$ and Eq. (9), we find $(m_\mu d_{Z0} + H_{XY}) \simeq (1.6 \pm 5.6 \times 10^{-23})$ GeV.

Other experiments involving leptons have also obtained null results. An experiment that searched for sidereal variation in transitions between muonium hyperfine energy levels[17] obtained $r^\mu \leq 5.0 \times 10^{-22}$. Penning trap experiments with a single trapped electron[18] obtained $r^e \leq 1.6 \times 10^{-21}$.

While the anomalous magnetic moment of the muon seems to differ from the Standard-Model value by about 3.4 standard deviations, the $\mathcal{R}$ values obtained for μ^+ and μ^- agree well with each other. We conclude that no significant signal of CPT/Lorentz violation was present in the E821 data, either in a difference of $\mathcal{R}$ for μ^+ and μ^-, or as a sidereal variation in $\mathcal{R}$. Limits on the Standard-Model Extension parameters[2] are of the order 10^{-23} to 10^{-24} GeV, with the dimensionless figures of merit $\sim 10^{-23}$, which represent the best tests of this model for leptons. Both $r_{\Delta\omega_a}^{\mu}$ and $r_{A\Omega}^{\mu}$ are much less than m_μ/m_P, so E821 probes Lorentz and CPT violation signatures beyond the Planck scale. The interested reader is referred to our recent paper[6] for additional details.

Acknowledgments

This work formed part of the Ph.D. dissertation of Xiaobo Huang. Thanks to Mario Deile, Xiaobo Huang, Klaus Jungmann and Jim Miller for comments on this manuscript. Thanks to V.A. Kostelecký for helpful discussions. This project was supported in part by the U.S. Department of Energy, the U.S. National Science Foundation, the German Bundesminister für Bildung und Forschung, the Alexander von Humboldt Foundation, the Russian Ministry of Science, and the U.S.-Japan Agreement in High Energy Physics.

References

1. W.-M. Yao, *et al.*, Particle Data Group, J. Phys. **G 33**, 1 (2006).
2. R. Bluhm, V.A. Kostelecký and C.D. Lane, Phys. Rev. Lett. **84**, 1098 (2000).
3. V.A. Kostelecký and C.D. Lane, Phys. Rev. **D 60** 116010 (1999); J. Math. Phys. **40**, 6245 (1999).
4. G.W. Bennett *et al.* (Muon $(g-2)$ collaboration), Phys. Rev. **D73** 072003 (2006).
5. J.P. Miller, E. de Rafael and B.L. Roberts, hep-ph/0703049, and Rep. Prog. Phys., **70**, (2007) 795-881 (http://stacks.iop.org/0034-4885/70/795).
6. G..W. Bennett *et al.* (Muon $(g-2)$ collaboration), arXiv:0709.4670v2 [hep-ex] and submitted to Phys. Rev. Lett.
7. R.M. Carey *et al.*, Phys. Rev. Lett. **82**, 1632 (1999).
8. H.N. Brown *et al.* (Muon $(g-2)$ Collaboration), Phys. Rev. **D62**, 091101 (2000).
9. H.N. Brown, *et al.* (Muon $(g-2)$ Collaboration), Phys. Rev. Lett. **86** 2227 (2001).
10. G.W. Bennett, *et al.* (Muon $(g-2)$ Collaboration), Phys. Rev. Lett. **89**, 101804 (2002).
11. G.W. Bennett, *et al.* (Muon $(g-2)$ Collaboration), Phys. Rev. Lett. **92**, 161802 (2004).
12. A. Czarnecki and W. J. Marciano, Phys. Rev. D **64**, 013014 (2001).
13. S.P. Martin, J.D. Wells, Phys. Rev. **D 67**, 015002 (2003).
14. D. Stöckinger, hep-ph/0609168, (2006) and J. Phys. **G34**, R45, (2007).
15. N. R. Lomb, Ap. Space Sci., **39**, 447, (1976); J. Scargle, The Ap. J., **263**, 835 (1982). Also see W.H.Press *et al.*, *Numerical Recipes in C*, Cambridge U. Press, 2nd edition, §13.8, p575, (1999).
16. J. Bailey *et al.*, Nucl. Phys. **B 150**, 1 (1979).
17. V.W. Hughes *et al.*, Phys. Rev. Lett. **87** 111804 (2001).
18. R.K. Mittleman, I.I. Ioannou, H.G. Dehmelt and N. Russell, Phys. Rev. Lett. **83** 2116 (1999).

LORENTZ VIOLATION IN A DIFFEOMORPHISM-INVARIANT THEORY

R. JACKIW

Department of Physics
Massachusetts Institute of Technology
Cambridge, MA 02139, U.S.A.

In a diffeomorphism invariant theory, symmetry breaking may be a mask for coordinate choice.

1. Gauge Theory Preliminary

My collaborators and I introduced a Lorentz violating modification of the Maxwell theory many years ago,[1] long before Alan Kostelecký constructed a research program on this topic. We were not systematically studying a reasonable class of Lorentz violating interactions, like in Kostelecký's program. Rather we focused on a particular term, which possesses a fine pedigree in physics and mathematics: we added to the Maxwell Lagrange density the quantity

$$CS(A) = \frac{1}{4}\,\varepsilon^{ijk}\,F_{ij}\,A_k = \frac{1}{2}\,\mathbf{A}\cdot\mathbf{B}, \tag{1}$$

which in mathematics is known as the Chern-Simons term and in physics is recognized as the Gauss linking number (density). Thus we modified the action to read

$$I = \int d^4x\,\left[-\frac{1}{4}\,F^{\mu\nu}F_{\mu\nu} + \frac{m}{2}\,\mathbf{A}\cdot\mathbf{B}\right], \tag{2}$$

where m sets the strength of the modification. The dimensional mismatch between the 4-dimensional Maxwell term and the 3-dimensional Chern-Simons term is responsible for the Lorentz (and CPT) violation. An alternative presentation of the action makes use of a time-like embedding vector $v_\mu = (m, \mathbf{0})$:

$$I = \int d^4x\,\left(-\frac{1}{4}\,F^{\mu\nu}F_{\mu\nu} + \frac{1}{4}\,v_\mu\,\varepsilon^{\mu\alpha\beta\gamma}\,F_{\alpha\beta}\,A_\gamma\right). \tag{3}$$

Now Lorentz non-invariance is attributed to the presence of the external embedding vector v_μ, which picks out a "direction" in space-time. Finally the covariant vector present in Eq. (3)

$$\begin{aligned} K^\mu &\equiv \frac{1}{2}\,\varepsilon^{\mu\alpha\beta\gamma}\,F_{\alpha\beta}\,A_\gamma = \varepsilon^{\mu\alpha\beta\gamma}\,\partial_\alpha\,A_\beta\,A_\gamma \\ &= {}^*F^{\mu\gamma}A_\gamma \qquad (4) \\ \left({}^*F^{\mu\nu}\right. &\equiv \left.\frac{1}{2}\,\varepsilon^{\mu\nu\alpha\beta}\,F_{\alpha\beta}\right) \end{aligned}$$

is recognized as the Chern-Simons current, whose divergence leads to the mathematically/physically significant Chern-Pontryagin density/axial anomaly,

$$\partial_\mu K^\mu = \frac{1}{2}\,{}^*F^{\mu\nu}F_{\mu\nu}. \qquad (5)$$

Consequently another presentation of the modified action uses an externally prescribed function of time $\theta = mt$, so that its gradient is our embedding vector $v_\mu = \partial_\mu\theta = (m, \mathbf{0})$. After partial integration the action (2) or (3) reads

$$\begin{aligned} I &= \int d^4x\,\left(-\frac{1}{4}\,F^{\mu\nu}\,F_{\mu\nu} - \frac{1}{4}\,\theta^*F^{\mu\nu}F_{\mu\nu}\right) \\ &= \int d^4x\,\left(-\frac{1}{4}\,F_{\mu\nu}\,M_{\mu\nu\alpha\beta}\,F^{\alpha\beta}\right), \qquad (6) \\ M_{\mu\nu\alpha\beta} &\equiv g_{\mu\alpha}g_{\nu\beta} + \frac{1}{2}\,\varepsilon_{\mu\nu\alpha\beta}\,\theta, \end{aligned}$$

and Lorentz non-invariance is attributed to the externally prescribed θ.

While there is ample formal evidence for Lorentz violation in the model, it is useful to deduce physical consequences of the modified theory and to recognize Lorentz violation in its physical predictions. The equations of the modified theory coincide with the usual Maxwell's equation, except for Ampère's law, in which the source electromagnetic current $\mathbf{J}$ is modified by the addition of the magnetic field: $\mathbf{J} \to \mathbf{J} + m\mathbf{B}$. (Such an alteration is familiar in plasma physics; here it is offered as a change of fundamental physics.) The equations are easily analyzed, and two important physical consequences emerge: gauge invariance is retained, so the "photon" possesses two polarizations, which propagate in vacuum with velocities that differ from c — the velocity of light — and from each other. The former is a clear indication of Lorentz non-invariance; the latter signals violation of parity. This causes the vacuum to behave as birefringent medium and light undergoes a Faraday-like rotation. The modification has a clear observational signature; it has been looked for and none was found. This brings to a very satisfactory conclusion the analysis of a Chern-Simons modification to Maxwell theory.

Before passing to my discussion of the gravity theory modification, let me record the modified Maxwell equations, for later comparison with the gravitational situation

$$J^\nu = \partial_\mu F^{\mu\nu} + v_\mu {}^*F^{\mu\nu} . \tag{7}$$

The left side is the conserved current arising from gauge invariant couplings to charged sources, the first term on the right is the variation of the usual Maxwell action; the second comes from varying the Chern-Simons contribution. Taking one further divergence causes each term to vanish separately: $\partial_\nu J^\nu$ because the current is conserved; $\partial_\nu \partial_\mu F^{\mu\nu}$ because $F^{\mu\nu}$ is anti-symmetric; $v_\mu \partial_\nu {}^*F^{\mu\nu}$ because $F^{\mu\nu}$ satisfies a Bianchi identity. Thus no further information is contained in the longitudinal component of the equation of motion.

2. Diffeomorphism Invariant Gravity Theory

It is widely appreciated that a diffeomorphism invariant theory contains structures that parallel analogous quantities in a gauge theory, with the Christoffel connection acting like a (non-Abelian) gauge potential.

In particular, there exists in four dimensions a gravitational Chern-Simon current,

$$K^\mu = \varepsilon^{\mu\alpha\beta\gamma} \left(\Gamma^{\sigma}_{\alpha\,\tau}\, \partial_\beta \Gamma^{\tau}_{\gamma\,\sigma} + \frac{2}{3}\, \Gamma^{\sigma}_{\alpha\,\tau}\, \Gamma^{\tau}_{\beta\,\eta}\, \Gamma^{\eta}_{\gamma\,\sigma} \right) , \tag{8}$$

whose divergence is the gravitational Chern-Pontryagin density,

$$\partial_\mu K^\mu = \frac{1}{2}\, {}^*R^\sigma{}_\tau{}^{\mu\nu}\, R^\tau{}_{\sigma\,\mu\nu} \equiv \frac{1}{2}\, {}^*RR$$
$$\left({}^*R^\sigma{}_\tau{}^{\mu\nu} \equiv \frac{1}{2} \varepsilon^{\mu\nu\alpha\beta}\, R^\sigma{}_{\tau\alpha\beta} \right) . \tag{9}$$

Therefore, analogously to the gauge theoretic case, we propose to modify the Einstein-Hilbert action as[2]

$$I = \frac{1}{16\pi G} \int d^4x \left(\sqrt{-g}\, R + \frac{1}{4}\, \theta\, {}^*RR \right)$$
$$= \frac{1}{16\pi G} \int d^4x \left(\sqrt{-q}\, R - \frac{1}{2}\, v_\mu\, K^\mu \right) \tag{10}$$

where $v_\mu = \partial_\mu \theta$ it is taken to be time-like $v_\mu = (\frac{1}{m}, \mathbf{0})$; it is the divergence of $\theta = t/m$. Our anticipation was that here also we shall find physical effects that violate Lorentz invariance. But first let us record the equation of motion for the modified theory

$$-8\pi G T^{\mu\nu} = G^{\mu\nu} + C^{\mu\nu}{}_{\alpha} v^{\alpha}. \tag{11}$$

On the left is the covariantly conserved matter energy momentum tensor, arising from covariantly coupled matter. The first right-hand side term is the conventional Einstein tensor, while the second term is the $g_{\mu\nu}$ variation of the Chern-Simons modification. Next we take the covariant divergence of Eq. (11). Since both $D_\mu T^{\mu\nu}$ and $D_\mu G^{\mu\nu}$ vanish, we are left with

$$D_\mu C^{\mu\nu}{}_{\alpha} v^{\alpha} = 0. \tag{12}$$

But, unlike the electromagnetic analog, the divergence is non-zero; it is identically given by

$$D_\mu C^{\mu\nu}{}_{\alpha} v^{\alpha} = \frac{1}{8\sqrt{-g}} u^{\nu}\, {}^{*}RR. \tag{13}$$

Hence a consistency condition on our theory is that for $u^\nu \neq 0$

$${}^{*}RR = 0. \tag{14}$$

This consistency condition has important consequences. First of all it suppresses the symmetry breaking Chern-Simons term in the action — even though its variation results in a modified equation of motion. Moreover, the following curious observation must be noted. If θ is an external variable (as above) then $\int d^4x \theta\, {}^{*}RR$ apparently violates diffeomorphism invariance and Lorentz invariance. However, let us for a moment consider θ to be a dynamical quantity, which undergoes variations in the various derivations of the dynamics in the theory. Varying the action with respect to $g_{\mu\nu}$ still produces the equation of motion (11). But with dynamical θ, we must also vary that quantity, and we obtain the equation (14). But that equation also emerged with non-dynamical θ, as a subsidiary condition on Eq. (11). So it appears that the same equations hold, whether θ is dynamical or external. It is our purpose to clarify this further.

In order to illuminate the issue of Lorentz violation, we should look to physical consequences, rather than to the confusing formal properties of the theory. Although the non-linear equations are much too complicated to be solved exactly, much has been done by many, at least on the approximate level.

(a) The classical, solar system test of general relativity survive unchanged.[2] The reason is that they are based on a static, radially symmetric geometry for which the Chern-Simons term vanishes. But Smith *et al.* have found effects of the modification on bodies orbiting the earth.[3]

(b) Since ${}^*RR$ must vanish, the Kerr solution, for which ${}^*RR \neq 0$, must be deformed to accomodate rotating black holes. Some investigation of this is due to Konna *et al.*[4]

(c) Both polarizations of gravitational waves propagate with velocity c, but with different intensity.[2] This puts into evidence parity violation, but not Lorentz violation. Parity violating effects are also established in the post Newtonian expansion by Alexander and Yunes,[5] and in cosmological solutions by Lue *et al.* as well as by Alexander.[6]

(d) A useful summary of work, also performed in China, is by Ni.[7]

All this gives ample evidence for physical consequences of the Chern-Simons modification, but no persuasive consequence of Lorentz violation is identified. To illuminate this further, two young researchers at MIT performed a formal analysis of the theory from which a definite conclusion can be drawn.

Let us recall the formal approach to Lorentz and Poincaré invariance in a Lagrangian ($\mathcal{L}$) field theory for a field φ. (φ need not be a scalar field, but its vectorial/fermionic indices are suppressed; in fact in the present application φ stands for the metric tensor $g_{\alpha\beta}$.)

An infinitesimal Poincaré transformation of the coordinates $x^\alpha \to x'^\alpha = x^\alpha - a^\alpha - \omega^\alpha{}_\beta\, x^\beta$ is accompanied by the field transformation

$$\varphi \to \varphi'; \quad \varphi'(x') = \varphi(x) + \frac{1}{2}\,\omega^\alpha{}_\beta\, S^\beta{}_\alpha\, \varphi(x) \tag{15}$$

where $\omega^0{}_0 = 0, \omega^0{}_i = \omega^i{}_0 = 0, \omega^i{}_j = -\omega^j{}_i$ while $S^\alpha{}_\beta$ represents the Lorentz group on the field,

$$S^0{}_0 = 0, \quad S^0{}_i = S^i{}_0 = 0, \quad S^i{}_j = -S^j{}_i. \tag{16}$$

Invariance of the action against translations ($a^\alpha \neq 0, \omega^\alpha{}_\beta = 0$) leads by Noether's theorem to a two-index conserved "tensor" $\theta^\mu{}_\alpha, \partial_\mu\, \theta^\mu{}_\alpha = 0$. When the action is also Lorentz invariant ($\omega^\alpha{}_\beta \neq 0$) one can show that

$$\theta^\mu{}_\alpha = \theta^\mu{}_{B\alpha} + \partial_\nu\, X^{[\nu,\mu]}{}_\alpha \tag{17}$$

where $X^{[\nu,\mu]}{}_\alpha$ is antisymmetric in $[\nu,\mu]$ and $\theta_B^{\mu\nu} \equiv \theta^\mu_{B\alpha}\,\eta^{\alpha\nu}$ is symmetric in (μ,ν). [$\eta^{\alpha\nu}$ is the flat metric tensor.] Due to the antisymmetry of $X^{[\nu,\mu]}{}_\alpha$, conservation of $\theta^{\mu}{}_{\alpha}$ implies conservation of $\theta^{\mu\nu}_{B}$. Thus presence of Poincaré invariance in a field theoretic action ensures the existence of two-index, symmetric and conserved "tensor" — the Belinfante energy-momentum "tensor." Also vice-versa: the existence of a symmetric, conserved two-index object guarantees invariance against Poincaré transformations. (The quotation marks on "tensor" remind us that quantity is not a generally covariant tensor, but only Poincaré convariant. Henceforth the distinction will not be made.)

When this information is brought to gravity theory, various complications arise. First, the Lagrangian involves second derivatives of the field, but Noether's theorem is usually presented for first-derivative Lagrangians. In fact this is not a problem for the Einstein-Hilbert action: because of the identity

$$\int d^4x \sqrt{-g}\, R = \int d^4x \sqrt{-g}\; g^{\mu\nu} \left(\Gamma^{\alpha}_{\mu\beta}\,\Gamma^{\beta}_{\nu\alpha} - \Gamma^{\alpha}_{\mu\nu}\,\Gamma^{\beta}_{\alpha\beta}\right) + \text{surface term} \tag{18}$$

we may use a first-deviative action. But higher derivatives are irremovably present in our modification $^{*}RR$.

To overcome this problem, Noether's theorem can be extended. One finds for translation invariance

$$\theta^{\mu}{}_{\alpha} = \Pi^{\mu}\partial_{\alpha}\varphi + \Pi^{\mu\nu}\,\partial_{\nu}\,\partial_{\alpha}\,\varphi - \partial_{\nu}\,\Pi^{\mu\nu}\,\partial_{\alpha}\,\varphi - \delta^{\mu}_{\alpha}\mathcal{L} \tag{19}$$

and the equation of motion, which shows that the above is conserved, reads

$$\partial_{\mu}\Pi^{\mu} = \frac{\partial\mathcal{L}}{\partial\varphi} + \partial_{\mu}\partial_{\nu}\Pi^{\mu\nu}, \tag{20}$$

where

$$\Pi^{\mu} \equiv \frac{\partial\mathcal{L}}{\partial\partial_{\mu}\varphi}\,, \quad \Pi^{\mu\nu} = \frac{\partial\mathcal{L}}{\partial\partial_{\mu}\partial_{\nu}\varphi}. \tag{21}$$

When this procedure was applied to the Einstein-Hilbert action by Papapetrou and later by Bak *et al.*[8] the result was

$${}^{EH}\theta^{\mu\nu}_{B} = \frac{1}{8\pi G}\;\partial_{\alpha}\partial_{\beta}\;\left[\sqrt{-g}\,\eta^{[\mu}g^{\alpha]\beta} + \eta^{\beta[\alpha}g^{\mu]\nu}\right]. \tag{22}$$

Note that in harmonic gauge, $\partial_{\alpha}(\sqrt{-g}\; g^{\alpha\beta}) = 0$, the above reduces to the simple expression

$${}^{EH}\theta^{\mu\nu}_{B} = \frac{1}{16\pi G}\,\Box\, g^{\mu\nu}.$$

This program can now be repeated for Chern-Simons modified gravity. The tensor arising from translation invariance reads

$$^{CS}\theta^{\mu}{}_{\alpha} = \theta^{\mu}{}_{\alpha} + \frac{1}{2}\, v_{\alpha}\, K^{\mu}. \tag{23}$$

$\theta^{\mu}{}_{\alpha}$ is constructed as in Eqs. (19)-(21), except that the field derivatives in Eqs. (20), (21) are taken from the complete modified Lagrange density. The analog to Eq. (17) applies to the above constructed $\theta^{\mu}{}_{\alpha}$,

$$\theta^{\mu}{}_{\alpha} = \theta^{\mu}{}_{B\alpha} + \partial_{\nu} X^{[\nu,\mu]}{}_{\alpha} + A^{\mu}{}_{\alpha}. \tag{24}$$

The first two terms are as in Eq. (17); the last term depends on $^{*}RR$, and vanishes on shell. Therefore, on shell, Eq. (23) may be presented as

$$^{CS}\theta^{\mu}{}_{\alpha} = \theta^{\mu}{}_{B\alpha} + \partial_{\nu}\, X^{[\nu,\mu]}{}_{\alpha} + \frac{1}{2}\, v_{\alpha}\, K^{\mu}. \tag{25}$$

Conseervation of $^{CS}\theta^{\mu}{}_{\alpha}$ implies that

$$\partial_{\mu}\theta^{\mu}{}_{B\alpha} = -\frac{1}{2}\, v_{\alpha}\partial_{\mu}K^{\mu} = -\frac{1}{2}\, v_{\alpha}{}^{*}RR = 0. \tag{26}$$

Thus the Chern-Simons modified theory admits a symmetric, conserved two-index tensor, leading to the conclusion that Poincaré invariance holds with the Chern-Simons modification in place.

3. Reprise

Can we understand *a priori* why no Lorentz breaking effects are visible in the Chern-Simons modified model? Here is a possible argument. We have previously remarked that the equations of motion are the same, regardless whether θ is a dynamically varying field or the externally prescribed field, $\theta = \frac{1}{mt}$. Also we saw that the equation $^{*}RR = 0$, which is a consistency condition on the equation of motion with θ external, extinguishes the symmetry breaking term in the action.

Let us now view the fully dynamical theory, with θ taken to be a dynamical variable. Complete diffeomorphism invariance holds, with θ transforming as a scalar. But now we can use the diffeomorphism transformation to fix θ at $\frac{1}{mt}$. In other words, what appears to be a symmetry breaking external field, is in fact merely a choice of coordinates (choice of gauge) in a coordinate invariant (gauge invariant) model.

So I leave you with the following caution. Apparent symmetry breaking may in fact be just a choice of gauge!

Acknowledgments

This work was supported in part by funds provided by the U.S. Department of Energy under cooperative research agreement #DF-FC02-94ER40818.

References

1. S.M. Carroll, G.B. Field, and R. Jackiw, Phys. Rev. D **41**, 1231 (1990).
2. R. Jackiw and S.Y. Pi, Phys. Rev. D **68**, 104012 (2003).
3. T.L. Smith *et al.*, arXiv:0708.0001.
4. K. Konno *et al.*, arXiv:0706.3080.
5. S. Alexander and N. Yunes, Phys. Rev. D **75**, 124002 (2007).
6. A. Lue *et al.*, Phys. Rev. Lett. **83**, 1506 (1999); S. Alexander, hep-th/0601034.
7. W.-T. Ni, Int. J. Mod. Phys. D **14**, 901 (2005).
8. A. Papapetrou, Proc. Roy. Irish. Acad. **52A**, 11 (1948); D. Bak *et al.*, Phys. Rev. D **49**, 5173 (1994).
9. D. Guerrera and A.J. Hariton, Phys. Rev. D **76**, 044011 (2007).

STUDIES OF CPT SYMMETRY WITH ASACUSA

R.S. HAYANO

Department of Physics, The University of Tokyo
7-3-1 Hongo, Bunkyo-ku
Tokyo 113-0033, Japan
E-mail: hayano@phys.s.u-tokyo.ac.jp

The ASACUSA collaboration at CERN's antiproton decelerator (AD) has so far used the laser spectroscopy of antiprotonic helium atom, a neutral three-body Coulomb system consisting of an antiproton, a helium nucleus and an electron, to 'weigh' the antiproton, and will measure the ground-state hyperfine splitting of antihydrogen in the future.

1. Introduction

The common physics goal of the three major experiments at CERN's antiproton decelerator (AD), ATRAP, ASACUSA[a] and ALPHA, is to test CPT invariance by comparing particle (matter) vs. antiparticle (antimatter) masses. ATRAP and ALPHA both aim at measuring the $1s - 2s$ transition frequencies of antihydrogen ($\bar{\mathrm{H}}$), as discussed in detail in these proceedings by Madsen.[1] ASACUSA has so far used the laser spectroscopy of antiprotonic helium atom ($\bar{p}\mathrm{He}^+$),[2] a metastable three-body system comprising of an antiproton, an electron and a helium nucleus, and is now preparing for the antihydrogen ground-state hyperfine splitting (GS-HFS) measurement (discussed in these proceedings by Juhász[3]). At the CPT'04 conference, I presented how the spectroscopy of antiprotonic helium atoms can be used to 'weigh' the antiproton.[4] I here report results of our recent improved measurements.

2. Weighing the antiproton - the method

Spectroscopy of exotic atoms offers a powerful tool to study fundamental interactions as well as particle properties. For example, the μ^+ mass[5] as

[a]ASACUSA - Atomic Spectroscopy and Collisions using Slow Antiprotons.

listed in the review of particle physics[6] was determined by the exotic-atom (in this case muonium) spectroscopy, and the CPT theorem was used to assign the same value to the mass of μ^-. The masses of $\pi^{\pm}$[7] and $K^{\pm}$[8] were determined using exotic-atom X-ray transitions, by noting that the transition frequency $\Delta\nu$ from the state (n,ℓ) to (n',ℓ') of the exotic atom (in which a particle X^- is bound to a nucleus) can be written as

$$\Delta\nu_{n,\ell\to n',\ell'} = Rc\frac{m_X^*}{m_e}Z^2\left(\frac{1}{n'^2}-\frac{1}{n^2}\right), \tag{1}$$

wherein R denotes the Rydberg constant, c the speed of light, m_X^*/m_e the reduced 'X^-'-to-electron mass ratio and Z the nuclear charge. For the sake of simplicity we here adopted the Bohr model, and ignored the relativistic and quantum-electrodynamical (QED) corrections, as well as the effect of nuclear-charge screening by the electron(s) which in general can be still present when X-rays are emitted.

The situation is different for the proton, since the proton mass rather than the antiproton mass has been more precisely determined. The 2002 CODATA recommended value[9] for m_p/m_e of 1836.152 672 61(85) (relative standard uncertainty of 4.6×10^{-10}) was based on a measurement carried out by the GSI-Mainz collaboration,[10] who compared the $^{12}C^{5+}$ cyclotron frequency $\omega_c = eB/m_{^{12}C^{5+}}$ and its spin precession frequency $\omega_L = geB/(2m_e)$ in a Penning trap. From the measured ratio ω_c/ω_L and the QED prediction for the bound-electron g factor, the electron's mass in the atomic mass unit was obtained (from which m_p/m_e was derived). The m_p/m_e value thus obtained is much more precise than the antiproton mass value obtained by the antiprotonic-atom X-ray measurements (having relative uncertainties of $\sim 4\times10^{-5}$).[11]

The laser spectroscopy of $\bar{p}He^+$ (see Fig. 1) has already surpassed the precision of antiprotonic-atom X-ray measurements by many orders of magnitude, and is becoming competitive to the Penning-trap measurement of m_p/m_e. In the laser spectroscopy of $\bar{p}He^+$, the antiproton mass is determined from the transition frequencies between (n,ℓ) and (n',ℓ') levels using Eq. (1), like in the case of the X-ray spectroscopy, but with an important difference that Z therein should now read $Z_{\text{eff}}(n,\ell,n',\ell')$. If the system was a two-body antiprotonic-helium ion, the nuclear charge Z is just 2, and hence calculating transition frequencies would be rather trivial. In the case of $\bar{p}He^+$, the effective nuclear charge Z_{eff} is less than 2 due to the charge screening by the remaining electron and depends on the quantum numbers (n,ℓ) of the antiproton orbit, so that state-of-the-art three-body QED cal-

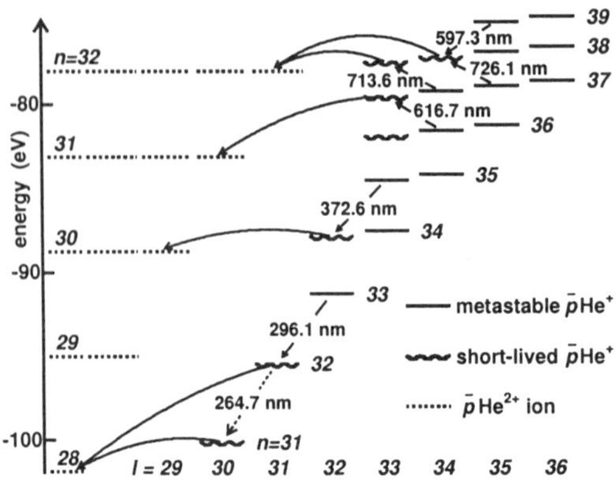

Figure 1. Level structure of $\bar{p}^4He^+$, with E1 transition wavelengths shown in nanometers. The solid lines indicate radiation-dominated metastable states, the wavy lines Auger-dominated short-lived states. The broken lines indicate $\bar{p}He^{2+}$ ionic states formed after Auger emission, the curved arrows Auger transitions with minimum $|\Delta\ell|$. On the left-hand scale the theoretical absolute energy of each state is plotted relative to the three-body breakup threshold.

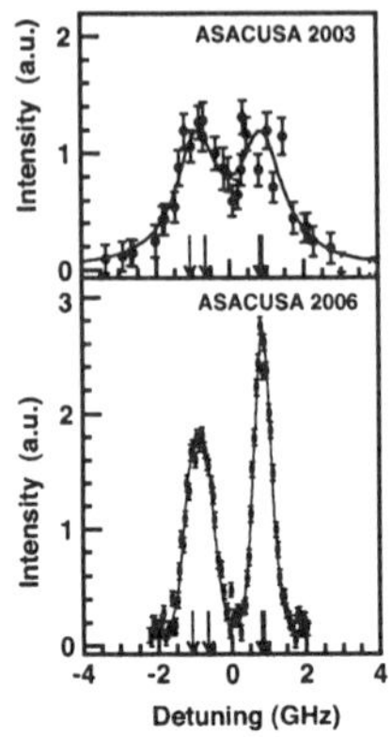

Figure 2. Frequency profiles of the $(n, \ell) = (36, 34) \to (37, 33)$ transition in $\bar{p}^3He$ (analogue of the $(37, 35) \to (38, 34)$ transition at 726.1 nm in $\bar{p}^4He$). Top: with RFQD,[15] bottom: with RFQD pulse-amplified CW laser and frequency comb.[16]

culations are needed in order to obtain theoretical values for the transition frequencies.[12,13]

It has been established[2] that metastable antiprotonic helium atoms can be abundantly produced by stopping low-energy antiprotons in helium. About 97% of the stopped antiprotons annihilate promptly on the helium nucleus, while the remaining 3% form the metastable $\bar{p}He^+$ having a mean lifetime of $\sim 3\mu s$. Near-circular states of excited $(n \sim 35)$ $\bar{p}He^+$ have lifetimes of $\tau \sim 1\mu s$, and deexcite radiatively by the $\Delta n = \Delta\ell = -1$ electric-dipole transitions. Such metastable levels are indicated by the solid lines in Fig. 1. Next to the metastable levels, there are short-lived ($\tau \lesssim 10$ ns) levels, indicated by the wavy lines, which decay via Auger transitions to the circular-states of the antiprotonic helium ion ($\bar{p}He^{2+}$). Due to Stark collisions, the antiprotonic helium ions annihilate almost instantly. We use pulsed laser beam(s) to deexcite a metastable level to an Auger-dominated short-lived state, and detect the resonance by observing a sharp increase in the antiproton annihilation rate.

During the years 2000–2006, we carried out three generations of laser spectroscopy experiments on $\bar{p}He^+$. These increased the fractional precision on the atom's transition frequency ν_{exp}, from $1.3 - 1.6 \times 10^{-7}$, to $\sim 9 \times 10^{-9}$

Table 1. Various sources of experimental error in three measurements of the $\bar{p}He^+$ transition frequencies, carried out using the 5.3-MeV antiproton beam of AD,[14] the 100-keV beam of the RFQD,[15] and using a cw pulse-amplified dye laser stabilized to an optical frequency comb generator.[16]

Error source	AD only[14] (MHz)	RFQD[15] (MHz)	Freq. comb[16] (MHz)
Density extrapolation	20-50	0.1-1	0.1-2
Statistical	20-40	20-200	3-13
Laser freq. calibration	70-140	20-50	< 1
Freq. chirp in dye laser	-	-	2-4
Freq. chirp in SHG/THG[a)]	< 10	< 10	1-2
Magnetic effects	< 1	< 1	< 1
AC Stark effects	< 1	< 1	< 1
HFS[b)] calc. uncertainties	< 1	< 1	< 1
Total error	80-150	30-200	4-15

a) SHG/THG: second and third harmonic generation
b) HFS: hyperfine splitting

(Table 1 and Fig. 2). Most of these experimental efforts improved the following two factors which, as in other types of laser spectroscopy measurements involving normal atoms, limited the precision on $\nu_{\rm exp}$. i) *Spectral purity and accuracy of the spectroscopic laser:* the optical frequencies of the laser beams used here were calibrated and controlled to a precision of 1 part in 10^7–10^8. The linewidth (i.e., the energy spread of the photons) was narrowed to less than 1 part in $10^6 - 10^7$. ii) *"Quality" of the* $\bar{p}He^+$ *target:* if these atoms were to be studied by high-precision laser spectroscopy, they had to be cooled to Kelvin-scale temperatures, and preferably isolated in low-density targets where their atomic energy levels would not be perturbed by collisions with the surrounding atoms. This was not easy to do, because antiprotons can only be produced by GeV-scale collisions, and so a 13–14 orders of magnitude difference in energy scale had to be first bridged to carry out these experiments.

The first generation of experiments[14] used the 5.3-MeV antiproton beam of AD to synthesize the $\bar{p}He^+$ in a target of relatively high density (atomic density $\rho \geq 10^{21}$ cm^{-3}). The spectroscopy was carried out with a type of pulsed dye laser which is widely used in other accelerator-based experiments. The second generation[15] used a radiofrequency quadrupole post-decelerator (RFQD) to first slow down the AD antiprotons from 5.3 MeV to 60 keV, before stopping them in an ultra-low density ($\rho \lesssim 10^{17}$ cm^{-3}) target. This technique reduced the collision rate of $\bar{p}He^+$ with other helium atoms in the target from GHz to a few MHz. The third generation[16] utilized

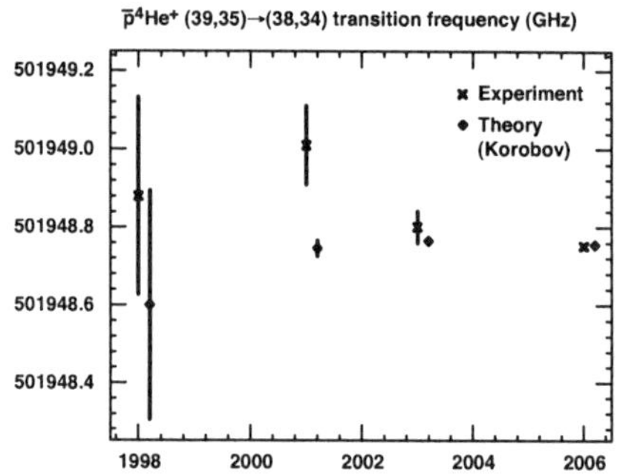

Figure 3. Comparison of measured and calculated frequencies for the $(39, 35) \rightarrow (38, 34)$ transition of $\bar{p}^4\mathrm{He}^+$.

new laser technologies, including an optical frequency comb, to further improve the spectral resolution and precision. Fig. 2 compares the resonance profiles of the $(36, 34) \rightarrow (37, 33)$ transition at $\lambda = 723.9$ nm in $\bar{p}^3\mathrm{He}^+$ of the second-generation experiment (top) and the third-generation experiment (bottom). The narrower line width achieved in the third-generation experiment is evident from the better-resolved hyperfine splitting.[b]

Although the techniques, instrumentation, and sources of systematic errors in the three experiments were quite different, consistent results on ν_{exp} were obtained, as demonstrated in Fig. 3. The figure also shows that the accuracy of theoretical prediction has steadily improved.

3. Weighing the antiproton - results

Figure 4 compares experimental and theoretical values for transition frequencies (seven in $\bar{p}^4\mathrm{He}^+$ and five in $\bar{p}^3\mathrm{He}^+$), where the theoretical values were calculated using the CODATA2002 recommended values for all the fundamental constants (including $m_{\bar{p}}/m_e \equiv m_p/m_e$). The $\bar{p}^4\mathrm{He}^+$ results show satisfactory experiment-theory agreement, while there appears to be a hint of some discrepancies in the case of $\bar{p}^3\mathrm{He}^+$, the source of which is currently unknown, and requires further study.

The antiproton-to-electron mass ratio $m_{\bar{p}}/m_e$ was obtained by mini-

[b]This profile in fact contains eight intense lines corresponding to $E1$ transitions involving no spin-flip between the eight hyperfine substates of states (36, 34) and (37,33) and 12 weak lines wherein one of the constituent particles flips its spin. Only the two peaks separated by 1.8 GHz that arise from the interaction between the orbital angular momentum of the antiproton and electron spin could be clearly resolved, however, due to the 400-MHz Doppler broadening.

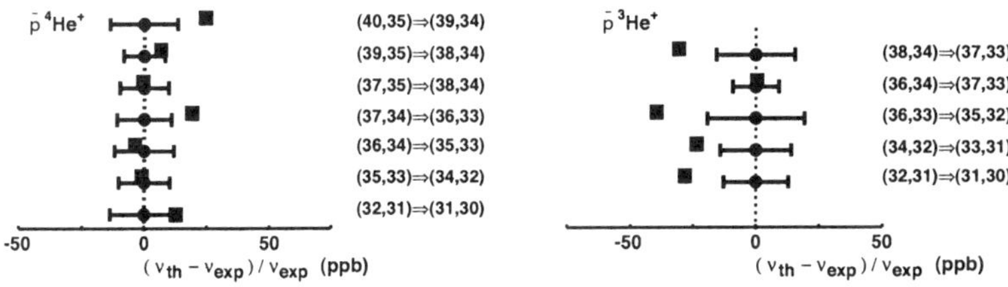

Figure 4. Experimental ν_{exp} (circles with 1σ errors)[16] vs. theoretical ν_{th} (squares)[13] transition frequencies.

mizing chi-squares $\chi^2 = \sum(\nu_{\rm th}(m_{\bar{p}}/m_e) - \nu_{\rm exp})^2/(\sigma^2_{\rm exp} + \sigma^2_{\rm th})$. The result, $m_{\bar{p}}/m_e = 1\ 836.152\ 674(5)$, agrees with the CODATA recommended value of $m_p/m_e = 1\ 836.152\ 672\ 61(85)$. See Fig. 5.

In deriving $m_{\bar{p}}/m_e$ we assumed the equality of proton and antiproton charge magnitude $|Q_{\bar{p}}| = Q_p$. If we let both $\bar{p}$ mass and charge to vary with respect to those of the proton, we obtain a 90%-confidence limit contour as shown in Fig. 6. Naïvely, since the $\bar{p}He^+$ binding energies are roughly proportional to $m_{\bar{p}}Q^2_{\bar{p}}$, we expect that the fractional mass change $\delta m_{\bar{p}}/m_p$ should be proportional to $\approx -2\delta Q_{\bar{p}}/Q_p$. Figure 6 shows this is indeed the case. By combining our contour with the proton-to-antiproton cyclotron-frequency comparison of the TRAP group,[17] a bound on the equality of antiproton mass and charge with those of the proton of 2×10^{-9} (90-% confidence level) is obtained.

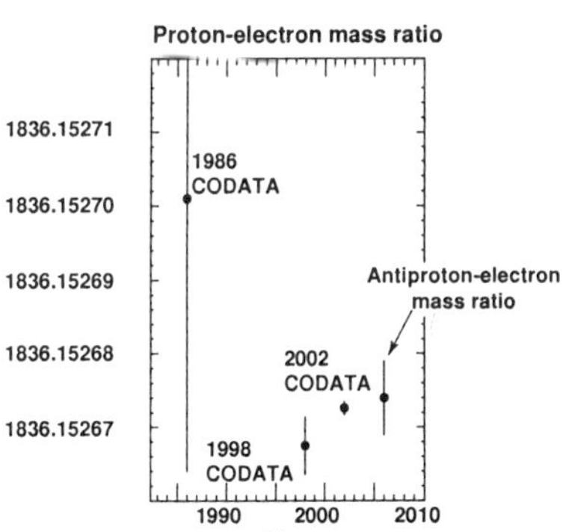

Figure 5. Proton-to-electron[9] and antiproton-to-electron[16] mass ratios.

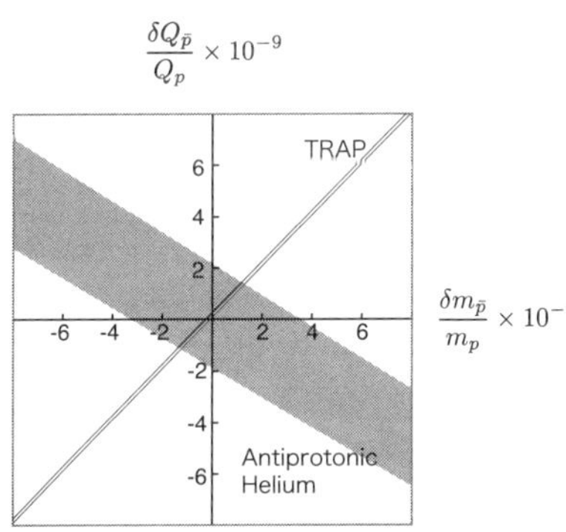

Figure 6. A 90-%-confidence limit contour on the antiproton mass and charge deduced from the $\bar{p}He^+$ laser spectroscopy[16] and the $\bar{p}$–H^- cyclotron frequency comparison[17] is shown.

4. Outlook

ASACUSA's measurement of $m_{\bar{p}}/m_e$ has evolved in three phases, and has now reached a relative precision of 2×10^{-9}. While this is still about a factor 6 less precise than the standard relative uncertainty of m_p/m_e, Table 1 shows that there is still room for improvement. Namely, since we have so far used only single-photon transitions, the spectral lines are Doppler broadened. In our 'phase-4' experiment, we will use counter-propagating two-color two-photon transition such as $(n, \ell) = (36, 34) \rightarrow (34, 32)$, in order to (nearly) cancel the Doppler width, and to achieve higher (sub-MHz) precisions.

Acknowledgments

This work has been carried out working closely with the members of ASACUSA collaboration, in particular, E. Widmann, D. Horváth, and M. Hori. Thanks are also due to V. I. Korobov and Y. Kino for theoretical inputs. This work was supported by the Grant-in-Aid for Specially Promoted Research (15002005) of MEXT Japan, AOARD, the Hungarian Scientific Research Fund (OTKA T046095).

References

1. N. Madsen (for ALPHA), these proceedings.
2. T. Yamazaki *et al.*, Phys. Rep. **366**, 183 (2002).
3. B. Juhász (for ASACUSA), these proceedings.
4. R.S. Hayano, in V.A. Kostelecký, ed., *CPT and Lorentz Symmetry III*, World Scientific, Singapore, 2005.
5. F. Mariam *et al.*, Phys. Rev. Lett. **49**, 993 (1982).
6. Particle Data Group, W.-M. Yao *et al.*, J. Phys. G **33**, 1 (2006).
7. S. Lenz *et al.*, Phys. Lett. B **416**, 50 (1988).
8. K.P. Gall *et al.*, Phys. Rev. Lett. **60**, 186 (1988); A.S. Denisov *et al.*, JETP Lett. **54**, 558 (1991).
9. P.J. Mohr and B.N. Taylor, Rev. Mod. Phys. **77**, 1 (2005).
10. T. Beier *et al.*, Phys. Rev. Lett. **88**, 011603 (2002).
11. J. Eades and F. J. Hartmann, Rev. Mod. Phys. **71**, 373 (1999).
12. Y. Kino *et al.*, Nucl. Instr. Methods B **214**, 84 (2004).
13. V.I. Korobov, In A. Hirtl, J. Marton, E. Widmann, J. Zmeskal, eds., *Exotic Atoms and Related Topics (EXA05)*, Austrian Academy of Sciences, 2005.
14. M. Hori *et al.*, Phys. Rev. Lett. **87**, 093401 (2001).
15. M. Hori *et al.*, Phys. Rev. Lett. **91**, 123401 (2003).
16. M. Hori *et al.*, Phys. Rev. Lett. **96**, 243401 (2006).
17. G. Gabrielse *et al.*, Phys. Rev. Lett. **82**, 3198 (1999).

NEUTRINO OSCILLATIONS AND LORENTZ VIOLATION WITH MINIBOONE

R. TAYLOE AND T. KATORI
FOR THE MINIBOONE COLLABORATION
Physics Department, Indiana University
Bloomington, IN 47405, U.S.A.
E-mail: rtayloe@indiana.edu

The MiniBooNE experiment was designed to test the result for neutrino oscillations as seen by the LSND experiment. If the LSND signal is indeed due to neutrino oscillations the current model that uses three massive neutrinos is inadequate to explain all the evidence for oscillations. We have investigated an alternative model (Tandem Model) based on the Standard-Model Extension which allows for Lorentz and CPT violation and explains all the neutrino oscillation data, including LSND. A prediction was made for the MiniBooNE experiment. The MiniBooNE results show no evidence for oscillations of the standard form above 475 MeV yet do show an anomalous excess at lower energies. This excess is compared to the prediction of the Tandem Model.

1. Neutrino Oscillations, Lorentz Violation, and LSND

The LSND experiment that was performed at Los Alamos National Lab in the 1990s observed an excess of $\bar{\nu}_e$ events in a beam of $\bar{\nu}_\mu$.[1] This observation was consistent with $\bar{\nu}_\mu \to \bar{\nu}_e$ oscillations with Δm^2 in the range 0.1 - 10 eV^2. However, it is inconsistent with other observations using solar and atmospheric neutrinos in a model where the oscillations are occurring within among only three massive neutrinos. If LSND is correct, then there must be additional neutrinos or the 3ν massive model[2] is inadequate.

The Standard-Model Extension (SME)[3] has been developed to include the possible effects of new physics, including those that violate Lorentz and CPT symmetry, in a variety of physical systems. Neutrinos are natural quantum interferometers and so neutrino oscillations are a natural phenomena in which to search for new physics that may violate Lorentz and CPT symmetry.[4]

The LSND data has been analyzed[5] within the SME using a short-baseline approximation.[6] The scale required to generate a neutrino-oscillation signal at the LSND energies $E \simeq 10^{-2}$ GeV was found to be of order 10^{-19} GeV. In addition, an analysis of the sidereal dependence of the oscillation signal was performed and allowed ranges for the SME neutrino parameters were determined. While no distinct signature for Lorentz violation (LV) was found, the LSND data could be reasonably described within the SME. However, a description of the global neutrino oscillation data was not addressed.

2. The Tandem Model

The Tandem Model (TM) for neutrino oscillations[7] was built upon the so-called "bicycle" model[8] that uses the SME in the neutrino sector. The goal for the TM was to provide a description of all the neutrino oscillation data including that from solar, reactor, long-baseline, atmospheric, and LSND experiments. It is an alternative model for neutrino oscillations and has the following features:

(1) only 3 parameters (standard 3 massive neutrino model has 4 parameters);
(2) describes the solar,[9] atmospheric,[10] and KamLAND[11] oscillation data reasonably well;
(3) reproduces the LSND signal and predicts a relatively large signal for the proposed oscSNS experiment;[12]
(4) does not produce large signals for experiments with null results, including KARMEN;[13]
(5) predicts null signals for long baseline ν_e appearance experiments, such as NOvA[14] and T2K;[15] and
(6) predicts a signal for MiniBooNE[16] at low energy.

Figure 1 shows the predictions of the Tandem Model for various short baseline $\nu_e(\bar{\nu}_e)$ appearance experiments. Note the prediction for the MiniBooNE experiment. The oscillation probability for MiniBooNE in the TM is very large at low energy, but rapidly falls off for energies detectable by MiniBooNE (> 300 MeV).

3. The MiniBooNE Experiment

The MiniBooNE experiment[16] was designed to test the LSND result[1] that indicated $\bar{\nu}_\mu \to \bar{\nu}_e$ oscillations by searching for $\nu_\mu \to \nu_e$ oscillations. The experiment is located at Fermilab and is shown in the schematic of Fig. 2.

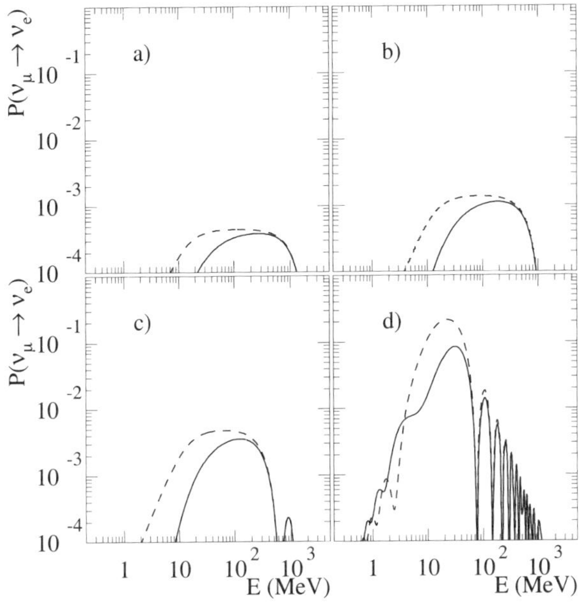

Figure 1. The theoretical predictions of the ν_e (solid line) and $\bar{\nu}_e$ (dashed) oscillation probability for (a) KARMEN, (b) LSND, (c) oscSNS, and (d) MiniBooNE.

The neutrino beam is produced with 8 GeV protons incident on a beryllium target which creates copious numbers of charged pions. The positive pions are focused by a magnetic horn into a 50 m pipe where they decay, producing ν_μ. The resulting beam is 99%-pure ν_μ and of average energy of ≈ 0.7 GeV.

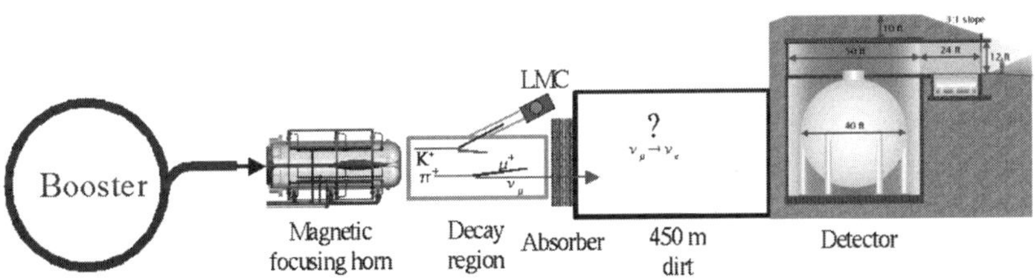

Figure 2. Schematic figure of the MiniBooNE experiment showing the arrangment of the ν-production region and detector.

The detector is located 541 m from the neutrino target and consists of a spherical tank containing 800 tons of mineral oil (CH_2) viewed by 1280 20-cm-diameter photomultiplier tubes for 10% photocathode coverage of the surface area of the tank. In addition, a veto region with 240 photomultiplier tubes provides a signal for particles that enter or exit the main tank region. The time and charge recorded by these tubes in response to the Cerenkov radiation and the small amount of scintillation light that is emitted from charged particles enable the reconstruction of neutrino interactions of various types.

Comparing to LSND, MiniBooNE uses a neutrino beam with approximately 10 times the energy and the detector is approximately 10 times further from the target. This results in a baseline distance to energy ratio (L/E) for MiniBooNE that is about the same as it was for LSND. Therefore, if oscillations are interpreted in a two neutrino framework, with an oscillation probability,

$$P_{\nu_\mu \to \nu_e} = \sin^2 2\theta \sin^2 \left(1.27\Delta m^2 \frac{L}{E}\right), \tag{1}$$

then MiniBooNE will have sensitivity to the same values of neutrino mixing, $\sin^2 2\theta$, and mass scale, Δm^2, as was LSND. It should also be noted that CPT conservation is assumed as MiniBooNE is seeking to test the LSND result for *antineutrinos* using *neutrinos*. A MiniBooNE search with antineutrinos is more challenging as antineutrino production rates and cross sections are substantially lower. However, MiniBooNE is currently running in antineutrino mode and the sensitivity to antineutrino oscillations is being evaluated.

4. Analysis and Results

The data set analyzed for the recently reported MiniBooNE result[16] was collected in the period 2002-2006 and used 5.6×10^{20} protons incident on the neutrino production target. Two separate and blinded oscillation analyses were performed on the data. One of the analyses (TB) used a track-based maximum likelihood approach, the other (BDT) used a boosted-decision tree algorithm. Each analysis was carried through to completion and cuts were tuned on each independently to obtain maximum sensitivity for two-neutrino oscillations. The TB analysis had slightly better sensitivity (determined before data unblinding) and was chosen to be the primary analysis. The BDT analysis was used as a powerful cross check.

With cuts predetermined based on maximum sensitivity to oscillations, the neutrino data set was opened in March, 2007. The reconstructed neutrino energy spectrum is shown in Figure 3. The data are consistent with ν oscillations using a two-neutrino formalism (Eq. (1)). The event excess in the oscillation fit energy region, $475 > E_\nu > 1250$ MeV is not significant, $22 \pm 19 \pm 35$ events, and the oscillation fit in the region yields a χ^2 probability of 93% for the null hypothesis. The 90% exclusion regions resulting from both analyses are shown in Figure 4. The results from the two different analyses are consistent with each other and inconsistent with LSND at 98% CL.

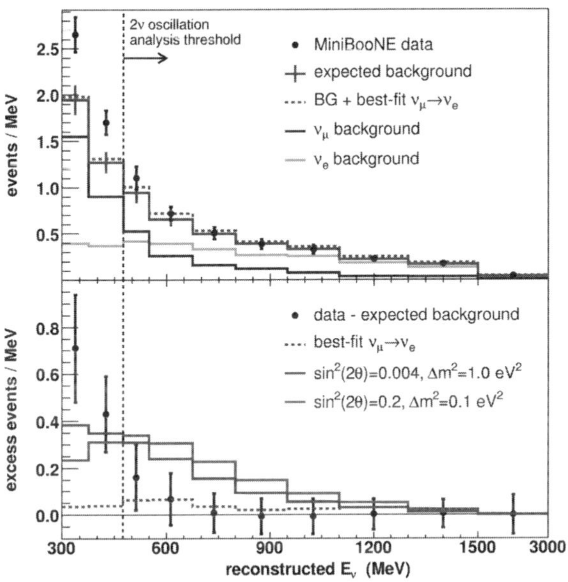

Figure 3. Number (per MeV) of candidate ν_e events as a function of reconstructed neutrino energy. In the top plot, the points show the data with statistical error, while the histogram is the expected background with systematic errors from all sources. The vertical dashed line indicates the threshold used in the two-neutrino oscillation analysis. Also shown are the best-fit oscillation spectrum (dashed histogram) and the background contributions from ν_μ and ν_e events. The bottom plot shows the number of events with the predicted background subtracted as a function of neutrino energy, where the points represent the data with total errors and the two histograms correspond to LSND solutions at high and low Δm^2.

This result sets a stringent limit for two neutrino oscillations with an energy dependence described by Eq. (1) and contradicts the LSND result if the oscillations are interpreted in the two-neutrino framework. However, as

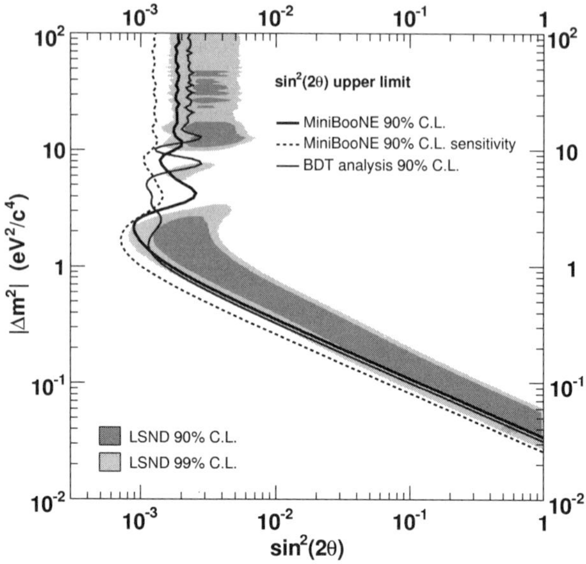

Figure 4. MiniBooNE oscillation result sensitivity. The MiniBooNE 90% confidence limit (thick solid curve) and sensitivity (dashed curve) are shown for events with $475 > E_{/nu} > 3000$ MeV within a two-neutrino oscillation model. Also shown is the limit from the boosted decision tree analysis (thin solid curve) for events with $300 > E_{/nu} > 3000$ MeV.

can be seen in Fig. 3, there is an excess of events ($96 \pm 17 \pm 20$) in the MiniBooNE data set in the energy region below the oscillation analysis window (475 MeV). This may be due to a unaccounted background or it may be due to oscillations of another type (i.e., not described by Eq. (1)). Further background studies are in progress and alternative oscillation hypotheses are under investigation.

One of these alternative oscillation hypotheses is that of the Tandem Model. As shown in Fig. 1, the TM predicts a small yet non-zero signal for neutrino energies above 300 MeV. This model result shown in Fig. 1 was a prediction, published before the MiniBooNE result was known. The TM prediction, as it currently stands, is too small to explain the MiniBooNE excess of events in the $300 > E_\nu > 475$ MeV range. However, the three parameters used in this version of the TM, were set with data not including MiniBooNE. It may be, with further adjustment of parameters, the TM can provide a good description of the data, including MiniBooNE. This is the subject of future work.

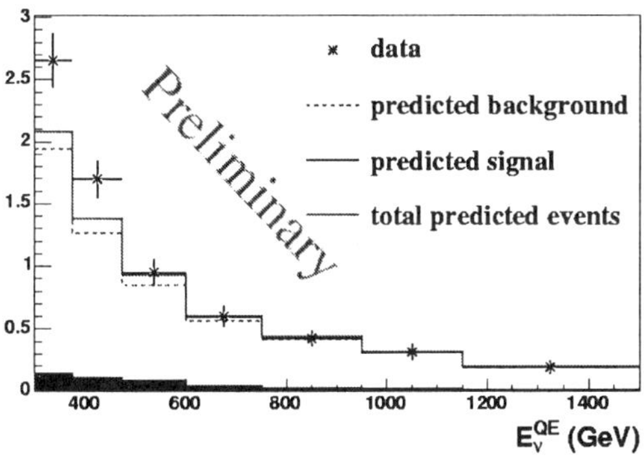

Figure 5. Number per MeV of candidate ν_e events as a function of reconstructed neutrino energy in MiniBooNE. The data are shown (stars) with statistical errors only. The oscillation signal as predicted by the Tandem Model is shown in the solid histogram, the predicted background (dashed), and total signal plus background (solid) is also shown. The systematic errors are not shown on this plot but may be seen in Fig. 3. The Tandem Model prediction is preliminary.

References

1. A. A. Aguilar-Arevalo *et al.* (LSND Collaboration), Phys. Rev. **D64**, 112007 (2001).
2. Particle Data Group, S. Eidelman *et al.*, Phys. Lett. B **592**, 1 (2004).
3. D. Colladay and V.A. Kostelecký, Phys. Rev. D **55**, 6760 (1997); Phys. Rev. D **58**, 116002 (1998).
4. V.A. Kostelecký and M. Mewes, Phys. Rev. D **69**, 016005 (2004).
5. L.B. Auerbach *et al.* (LSND Collaboration), Phys. Rev. D **72**, 076004 (2005).
6. V.A. Kostelecký and M. Mewes, Phys. Rev. D **70**, 076002 (2004).
7. T. Katori, R. Tayloe, and V.A. Kostelecký, Phys. Rev. **D74**, 105009 (2006).
8. V.A. Kostelecký and M. Mewes, Phys. Rev. D **70**, 031902(R) (2004).
9. M.B. Smy *et al.* (Super-Kamiokande Collaboration), Phys. Rev. D **69**, 011104 (2004).
10. Y. Ashie *et al.* (Super-Kamiokande Collaboration), Phys. Rev. Lett. **93**, 101801 (2004).
11. T. Araki *et al.* (KamLAND Collaboration), Phys. Rev. Lett. **94**, 081801 (2005).
12. G. T. Garvey *et al.*, Phys. Rev. **D72**, 092001 (2005).
13. B. Armbruster *et al.* (KARMEN Collaboration), Phys. Rev. **D65**, 112001 (2002).
14. D. S. Ayres *et al.* (NOvA Collaboration), arXiv:hep-ex/0503053.
15. Y. Itow *et al.* (T2K Collaboration), arXiv:hep-ex/0106019.
16. A. A. Aguilar-Arevalo *et al.* (MiniBooNE Collaboration), Phys. Rev. Lett. **98**, 231801 (2007).

TESTING LORENTZ AND CPT INVARIANCE WITH MINOS NEAR DETECTOR NEUTRINOS

B.J. REBEL

Fermi National Accelerator Laboratory
Batavia, Illinois 60510, U.S.A.
E-mail: brebel@fnal.gov

S.L. MUFSON

Astronomy Department
727 E. Third Street
Bloomington, Indiana 47405, U.S.A.
E-mail: mufson@astro.indiana.edu

for the

MINOS COLLABORATION

We present an analysis designed to search for Lorentz and CPT violations as predicted by the SME framework using the charged current neutrino events in the MINOS near detector. In particular we develop methods to identify periodic variations in the normalized number of charged current neutrino events as a function of sidereal phase. To test these methods, we simulated a set of 1,000 experiments without Lorentz and CPT violation signals using the standard MINOS Monte Carlo. We performed an FFT on each of the simulated experiments to find the distribution of powers in the sidereal phase diagram without a signal. We then injected a signal of increasing strength into the sidereal neutrino oscillation probability until we found a 5σ deviation from the mean in the FFT power spectrum. By this method, we can establish upper limits for the Lorentz and CPT violating terms in the SME.

1. Introduction

At experimentally accessible energies, signals for Lorentz and CPT violation can be described by the Standard Model Extension (SME),[1,2] a theory whose foundation is the Standard Model (SM) of particle physics. Since the SM is expected to be the low-energy limit of a more fundamental theory that unifies quantum physics and gravity at the Planck scale, $m_p \simeq 10^{19}$ GeV, the violations of Lorentz and CPT symmetries predicted by the SME provide a link to Planck scale physics. Assuming a quantum-gravity origin for the violations, however, suggests their magnitude in the accessible energy limit is suppressed by a factor of 10^{-17} or more.[3] Despite this huge suppression, these low-energy signatures from new physics at the Planck scale can be probed with current experimental technologies.

The SME framework predicts several unconventional neutrino signals, among which is one that arises from the dependence of the neutrino oscillation probability on the direction of neutrino propagation.[3] For experiments like MINOS[4] with both beam neutrino source and detector fixed on the Earth's surface, the Earth's sidereal rotation causes the direction of neutrino propagation $\hat{p}$ to change with respect to the Sun-centered inertial frame in which the SME is formulated.[5] The theory predicts that this rotation introduces a sidereal variation in the number of neutrinos detected from the beam. In this paper we use simulated beam neutrinos in the MINOS near detector[6] to develop methods to search for this sidereal signal.

We developed our search algorithms by injecting an SME signal into our simulated data set of MINOS near detector ν_μ neutrinos and then developing methods to find it. According to SME, the probability that a ν_μ oscillates from $\nu_\mu \to \nu_\tau$ over a distance L due to Lorentz and CPT violation is given by[3]

$$P_{\nu_\mu \to \nu_\tau} \simeq L^2 \left[(C)_{\nu_\mu \nu_\tau} + (A_c)_{\nu_\mu \nu_\tau} \cos(\omega_\oplus T_\oplus) + (A_s)_{\nu_\mu \nu_\tau} \sin(\omega_\oplus T_\oplus) + (B_c)_{\nu_\mu \nu_\tau} \cos(2\omega_\oplus T_\oplus) + (B_s)_{\mu_\mu \nu_\tau} \sin(2\omega_\oplus T_\oplus)\right]^2, \quad (1)$$

where $\omega_\oplus = 2\pi/(23^h\,56^m\,04.0982^s)$ is the Earth's sidereal frequency and $T_\oplus$ is the Local Sidereal time of the event. In this equation, the expressions for $(A_c)L$, $(A_s)L$, and $(C)L$ include both CPT and Lorentz violating terms; the expressions for $(B_c)L$ and $(B_s)L$ include only Lorentz violating terms. There are no terms that depend on CPT violation alone because CPT violation implies Lorentz violation.[7] In Eq. (1), the CPT violating terms depend only on L and the Lorentz violating terms depend on $L \times E_{\nu_\mu}$, where E_{ν_μ} is the energy of the neutrino. This unconventional behavior is to be compared with the L/E_{ν_μ} dependence of the oscillation probability when oscillations result from neutrino mass.

The sensitivity of the MINOS near detector neutrinos to a sidereal oscillation signal is shown in Fig. 1 based on a figure from Kostelecký and Mewes.[8]

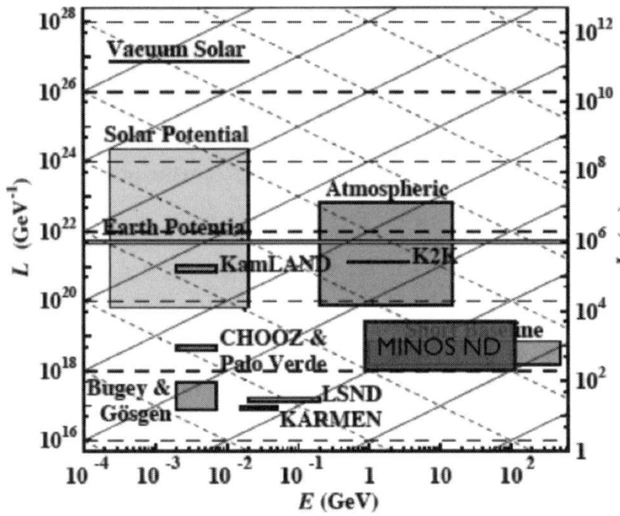

Figure 1. The region in this figure labeled "MINOS ND" shows the MINOS sensitivity to sidereal oscillation signals resulting from SME.

2. Data Sample

The MINOS experiment has been described elsewhere.[4]

The simulated data set of MINOS near detector ν_μ neutrinos we use here was generated by the standard MINOS Monte Carlo.[6] In total 4.0×10^6 beam spills were generated. In each simulated beam spill we identified the charged current (CC) events. These are events with at least one reconstructed muon track, as found by the standard MINOS reconstruction. In addition, these events must pass the following cuts: (a) particle ID cut,[6] (b) a cut requiring the reconstructed vertex position be $>$ 50 cm from the edge of a partial plane or its outline on a full plane, and (c) a cut that requires the event be found in the range $1.73 < z < 4.74$ m within the detector. Both μ^+ and μ^- CC events were accepted.

Using the total data set, we constructed a histogram of the number of CC events/spill. Using this histogram, we simulated 1,000 experiments. We generated each experiment by simulating the total number of spills collected in the MINOS near detector, 2.78×10^6. For each spill in the experiment we picked the number of events found in it from the #CC events/spill distribution and we assigned to each a random sidereal time. We then binned these events into a sidereal phase histogram spanning 0-1 in sidereal phase, $\phi = \omega_\oplus T_\oplus$. In addition, we binned the protons-on-target (POT) in the spill into a second sidereal phase histogram. We then divided the two histograms, with the result being a histogram of #CC events/POT as a function of sidereal phase. Since the #CC events/spill is proportional to POT, this final histogram gives the normalized quantity in which we search for sidereal variations. Figure 2 shows a typical Monte Carlo experiment.

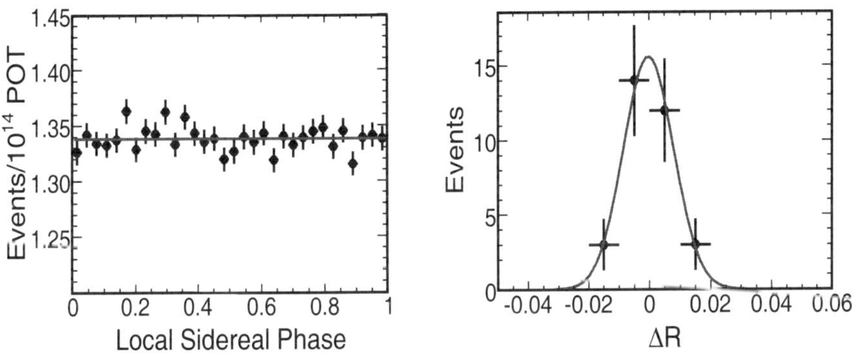

Figure 2. Left panel: a typical MC experiment; the mean rate is superposed on sidereal phase distribution. Right panel: distribution of the fluctuations about the mean rate for this experiment; superposed is a Gaussian fit to the fluctuation distribution.

There are 32 bins in the sidereal phase histograms. Since we are searching for sidereal variations with an FFT[9] and Eq. (1) puts power into frequencies associated with Fourier terms $\omega_{\oplus} \times n$, where n = 1-4, the optimal number of bins is given by 2^N with $N = 5$. For $N = 4$, there are too few Fourier terms kept in the analysis.

For each of the 32 sidereal phase bins in a simulated experiment, we computed the fractional deviations of the rate in phase bin i, R_i, from the mean rate, $\bar{R}$,

$$\Delta R_i = \frac{\bar{R} - R_i}{\bar{R}}. \tag{2}$$

The distribution of these fluctuations ΔR_i for a typical experiment is shown in Fig. 2 with a Gaussian fit superposed. This figure shows that the distribution is consistent with statistical fluctuations and suggests that residual sidereal systematics in the simulated experiments have been minimized.

3. Results

We performed an FFT[9] analysis on each of the 1,000 simulated experiments without a sidereal signal. The distribution of the powers in the even and odd terms out to frequencies $\omega_{\oplus} \times n$, where n = 1-4, are shown on the left and right panels of Fig. 3, respectively. Superposed on these distributions

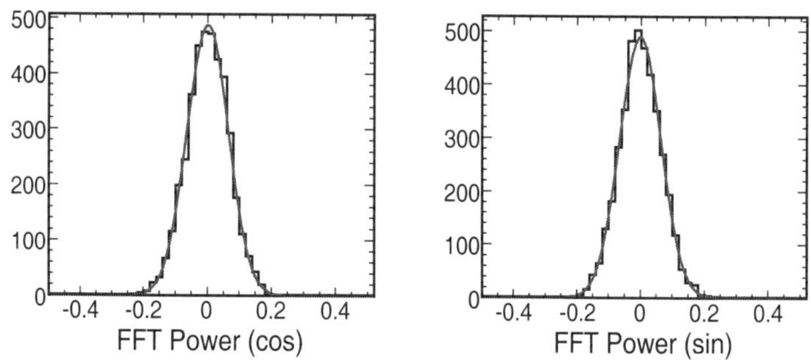

Figure 3. The results of an FFT analysis of the 1,000 simulated experiments without a sidereal signal. The left panel shows the distribution of powers in the even terms and the right panel shows the distribution of the odd terms. Only terms out to $\omega_{\oplus} \times n$, where n = 1-4, are represented in the histogram. Superposed on these distributions are Gaussian fits of widths $\sigma = 6.3 \times 10^{-2}$ for the even terms and $\sigma = 6.6 \times 10^{-2}$ for the odd terms.

are Gaussian fits. We use these distributions as a measure of the signal power required for a detection. In this analysis, a detection was defined as power in a Fourier expansion term that falls 5σ from the mean as defined in Fig. 3.

We can now set limits on the individual Lorentz and CPT violating terms $(a_L)^\mu$ and $(c_L)^{\mu\nu}$ making up the coefficients A_c, A_s, B_c, and B_s in Eq. (1) as follows. We first set all $(a_L)^\mu$ and $(c_L)^{\mu\nu}$ terms equal to zero except one. Then we increase the magnitude of the nonzero term until there is an oscillatory signal with power in a Fourier component that falls 5σ from the mean. Fig. 4 shows the 5σ signal that results from the nonzero Lorentz violating term[3] $(a_L)^X = 1.7 \times 10^{-19}$ in both the sidereal phase diagram and in the power spectrum.

4. Future Work

In the near future we expect to apply these analysis methods to the MINOS near detector CC neutrino event data sample. We then plan to extend

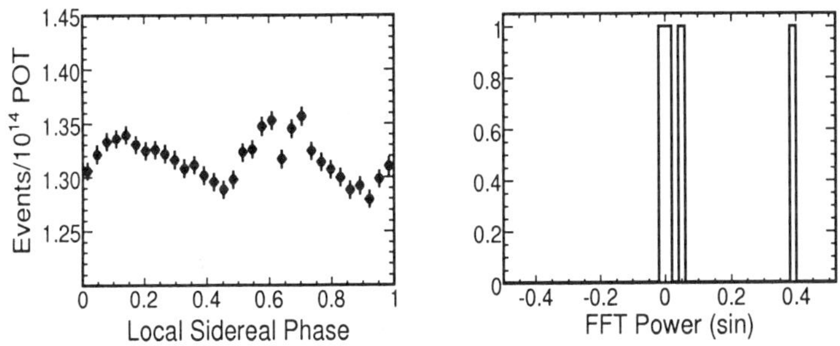

Figure 4. The 5σ oscillatory signal resulting from the nonzero Lorentz violating term $(a_L)^X$. The left panel shows the effect of this signal on the sidereal phase diagram. The right panel shows the 5σ deviation in the power in the $\sin(2\omega_\oplus T_\oplus)$ term.

this analysis to include the neutral current (NC) neutrino events. These additional events increase the statistics in the data sample by approximately 50%, thereby improving the significance of the analysis.

References

1. D. Colladay and V.A. Kostelecký, Phys. Rev. D **55**, 6760 (1997); Phys. Rev. D **58**, 116002 (1998).
2. V.A. Kostelecký, Phys. Rev. D **69**, 105009 (2004).
3. V.A. Kostelecký and M. Mewes, Phys. Rev. D **70**, 076002 (2004).
4. D.G. Michael *et al.*, Nucl. Instrum. Methods Phys. Res. Sect. B (2007), to be submitted.
5. L.B. Auerbach *et al.*, Phys. Rev. D **72**, 076004 (2005).
6. D.G. Michael *et al.*, Phys. Rev. Lett **97**, 191801 (2006).
7. O.W. Greenberg, Phys. Rev. Lett **89**, 231602 (2002); Phys. Lett. B **567**, 179 (2003).
8. V.A. Kostelecký and M. Mewes, Phys. Rev. D **69**, 016005 (2004).
9. W.H. Press *et al.*, *Numerical Recipes in C^{++}*, 2nd ed., Cambridge University Press, Cambridge, 1999.

EINSTEIN-ÆTHER GRAVITY: THEORY AND OBSERVATIONAL CONSTRAINTS

TED JACOBSON

Center for Fundamental Physics
University of Maryland
College Park, MD 20742-4111, U.S.A.
E-mail: jacobson@umd.edu

Einstein-æther theory is general relativity coupled to a dynamical unit timelike vector field. A brief review of current theoretical understanding and observational constraints on the four coupling parameters of the theory is given.

1. Introduction

In general relativity (GR), spacetime structure is determined by a dynamical metric tensor field g_{ab} and nothing else, and the theory is both diffeomorphism invariant and locally Lorentz invariant. Einstein-æther theory is the extension of GR that incorporates a dynamical unit timelike vector field u^a—the "æther"—which breaks the local Lorentz symmetry down to a 3d rotation subgroup. Direct coupling of matter to the æther would violate local Lorentz symmetry yet preserve diffeomorphism invariance. This paper presents a brief overview of the current theoretical and observational status of this theory, assuming that matter does not couple directly to the æther.

The action involving metric and æther is highly constrained. Besides the cosmological constant term, the only independent diffeomorphism invariant local terms containing no more than two derivatives are

$$S = -\frac{1}{16\pi G}\int \sqrt{-g}\ (R + K^{ab}_{mn}\nabla_a u^m \nabla_b u^n)\ d^4x, \tag{1}$$

where R is the Ricci scalar, K^{ab}_{mn} is defined as

$$K^{ab}_{mn} = c_1 g^{ab} g_{mn} + c_2 \delta^a_m \delta^b_n + c_3 \delta^a_n \delta^b_m + c_4 u^a u^b g_{mn} \tag{2}$$

with dimensionless coupling constants c_i, and the unit timelike constraint on the æther is implicit. (The metric signature is $(+---)$ and the speed of light defined by the metric g_{ab} is unity.) Higher derivatives would be suppressed by powers of a (presumably) small length, e.g., the Planck length. It is assumed here that the æther is aligned at large scales with the rest frame of the microwave background radiation.

Einstein-æther theory—"æ-theory" for short—is similar to the vector-tensor gravity theories studied by Will and Nordvedt,[1] but with the crucial difference that the vector field is constrained to have unit norm. This constraint eliminates a wrong-sign kinetic term for the length-stretching mode,[2] hence gives the theory a chance to be viable. An equivalent theory using the tetrad formalism was first studied by Gasperini,[3] and in the above form it was introduced by Jacobson and Mattingly.[4]

2. Newtonian and post-Newtonian limits

In the weak-field, slow-motion limit æ-theory reduces to Newtonian gravity,[5] with a value of Newton's constant $G_{\rm N}$ related to the parameter G in the action (1) by

$$G_{\rm N} = \frac{G}{1 - c_{14}/2}, \tag{3}$$

where $c_{14} \equiv c_1 + c_4$. (Similar notation is used below for other additive combinations of the c_i.) For any choice of the c_i, all parameterized post-Newtonian (PPN) parameters[6] of æ-theory agree with those of GR[7,8] except the preferred frame parameters $\alpha_{1,2}$ which are given by[8]

$$\alpha_1 = \frac{-8(c_3^2 + c_1 c_4)}{2c_1 - c_1^2 + c_3^2} \tag{4}$$

$$\alpha_2 = \frac{\alpha_1}{2} - \frac{(c_1 + 2c_3 - c_4)(2c_1 + 3c_2 + c_3 + c_4)}{c_{123}(2 - c_{14})}. \tag{5}$$

(This particular way of expressing α_2 was given in Ref. 9. The small c_i form of α_2 was first computed in Ref. 10.)

Observations currently impose the strong constraints $\alpha_1 \lesssim 10^{-4}$ and $\alpha_2 \lesssim 4 \times 10^{-7}$.[6] Since æ-theory has four free parameters c_i, we may set $\alpha_{1,2}$ exactly zero by imposing the conditions[8]

$$c_2 = (-2c_1^2 - c_1 c_3 + c_3^2)/3c_1 \tag{6}$$

$$c_4 = -c_3^2/c_1. \tag{7}$$

With (6,7) satisfied, *all* the PPN parameters of æ-theory are equivalent to those of GR. (The parameters $\alpha_{1,2}$ can also be set to zero by imposing $c_{13} = c_{14} = 0$, but this case is pathological, as discussed in Section 8.)

3. Homogeneous isotropic cosmology

Assuming spatial homogeneity and isotropy, u^a necessarily coincides with the 4-velocity of the isotropic observers, and the æther stress tensor is just a certain combination of the Einstein tensor and the stress tensor of a perfect fluid with energy density proportional to the inverse square of the scale factor, like the curvature term in the Friedman equation.[11,5] The latter contribution plays no important cosmological role since the spatial curvature is small, while the former renormalizes the gravitational constant appearing in the Friedman equation, yielding[5]

$$G_{\rm cosmo} = \frac{G}{1 + (c_{13} + 3c_2)/2}. \tag{8}$$

Since $G_{\rm cosmo}$ is not the same as $G_{\rm N}$ the expansion rate of the universe differs from what would have been expected in GR with the same matter content. The ratio is constrained by the observed primordial ^{4}He abundance to satisfy $|G_{\rm cosmo}/G_{\rm N} - 1| < 1/8$.[5] When the PPN parameters $\alpha_{1,2}$ are set to zero by (6,7), it turns out that $G_{\rm cosmo} = G_{\rm N}$, so this nucleosynthesis constraint is automatically satisfied.[8]

4. Linearized wave modes

When linearized about a flat metric and constant æther, æ-theory posesses five massless modes for each wave vector: two spin-2, two spin-1, and one spin-0 mode. The squared speeds of these modes relative to the æther rest frame are given by[12]

$$\text{spin-2} \qquad 1/(1 - c_{13}) \tag{9}$$

$$\text{spin-1} \qquad (c_1 - \tfrac{1}{2}c_1^2 + \tfrac{1}{2}c_3^2)/c_{14}(1 - c_{13}) \tag{10}$$

$$\text{spin-0} \qquad c_{123}(2 - c_{14})/c_{14}(1 - c_{13})(2 + c_{13} + 3c_2). \tag{11}$$

The corresponding polarization tensors were found in one gauge in Ref. 12 and in another gauge in Ref. 9. The energy density of the spin-2 modes is always positive, while for the spin-1 modes it has the sign of $(2c_1 - c_1^2 + c_3^2)/(1 - c_{13})$, and for the spin-0 modes it has the sign of $c_{14}(2 - c_{14})$.[13,9] (These reduce to the results of Ref. 14 in the decoupling limit where gravity is turned off.)

These squared speeds correspond to (frequency/wavenumber)^2, so must be non-negative to avoid imaginary frequency instabilities. They must moreover be greater than unity (super-luminal), to avoid the existence of vacuum Čerenkov radiation by matter.[2] (The strongest constraints arise from the existence of ultra high energy cosmic rays.) And the mode energy densities should be positive, to avoid dynamical instabilities. With the $\alpha_{1,2} = 0$ conditions (6,7) imposed, all of these conditions are met for all of the modes if and only if $c_\pm = c_1 \pm c_3$ are restricted by the inequalities[8]

$$0 \leq c_+ \leq 1 \tag{12}$$

$$0 \leq c_- \leq c_+/3(1 - c_+). \tag{13}$$

Interestingly, if the mode speeds are instead required to be *less* than unity (sub-luminal), then the spin-1 and spin-0 energy densities are negative. Hence not only the Čerenkov constraint, but also energy positivity (together with $\alpha_{1,2} = 0$) requires mode speeds greater than unity.

Note that when (7) holds, we have $c_{14} = 2c_+c_-/(c_+ + c-)$, which satisfies $0 \leq c_{14} < 2$ when the constraints (12), (13) hold. Thus in particular the condition for attractive gravity mentioned in Section 2 need not be separately imposed, and c_{14} is non-negative.

5. Primordial perturbations

Given the same G_N, and assuming the PPN parameters $\alpha_{1,2}$ vanish, the primordial power in cosmological spin-0 and spin-1 perturbations is unchanged relative to GR, while the power in spin-2 perturbations differs from that in GR by the factor $(1 - c_{14}/2)(1 - c_{13})^{1/2}$.[14,15] When the constraints (12), (13) are satisfied this factor is smaller than unity, hence these spin-2 perturbations are even more difficult to detect than in GR. As for the late time evolution of these perturbations, neutrino stresses in the radiation dominated epoch source the spin-1 mode, which leads to modified matter and CMB spectra. The effect is rather small however, and is degenerate with matter-galaxy bias and with neutrino masses.[15]

6. Radiation damping and strong self-field effects

If the fields are weak everywhere (including inside the radiating bodies), and the PPN parameters $\alpha_{1,2}$ vanish, radiation is sourced only by the quadrupole. Waves of spins 0, 1 and 2 are radiated, and the net power is given by $(G_N\mathcal{A}/5)\dddot{Q}_{ij}^2$, where Q_{ij} is the quadrupole moment and $\mathcal{A} = \mathcal{A}[c_i]$ is a function of the coupling parameters c_i that reduces to unity in the case of GR.[9] Agreement with the damping rate of GR (confirmed to $\sim 0.1\%$ in binary pulsar systems[6]) can be achieved by imposing the condition $\mathcal{A}[c_i] = 1$, which is consistent with the constraints (12,13).

Compact sources with strong internal fields such as neutron stars or black holes can be handled[16] using an "effective source" dynamics specified by a worldline action integral

$$S = -m_0 \int d\tau \, [1 + \sigma(v^a u_a - 1) + \sigma'(v^a u_a - 1)^2 + \ldots], \qquad (14)$$

where v^a is the 4-velocity of the body, u_a is the local background value of the æther, and σ and σ' are constants characterizing the body, called a "sensitivity parameters" or just "sensitivities". The sensitivites scale as c_i for small c_i.

The effects of nonzero sensitivities on two-body dynamics and radiation rates lead to a number of phenomena that are constrained by observations, including violations of the strong equivalence principle, modifications of the post-Newtonian dynamics, modifications of quadrupole sourced radiation, and both monopole and dipole sourced radiation. When $\alpha_{1,2} = 0$, all of these constraints are met provided the sensitivities are less than ~ 0.001, which will certainly be the case if $c_i \lesssim 0.01$.[16a] To be more precise would require knowing the actual dependence of the sensitivities on the c_i, which has so far only been determined for σ and only at leading order (where σ vanishes when $\alpha_{1,2} = 0$). (The speed V of the observed binaries with respect to the background æther frame can be neglected in formulating these constraints provided $V \lesssim 10^{-2}$, which is easily satisfied for any known proper motion relative to the rest frame of the microwave background radiation.[16])

7. Spherically symmetric stars and black holes

Unlike GR, æ-theory has a spherically symmetric mode, corresponding to radial tilting of the æther. For each mass, there is a two parameter family of spherically symmetric static vacuum solutions, rather than a unique solution as in GR.[18] Asymptotic flatness reduces this to a one parameter family.[7,18] The solution outside a static star is the unique solution for a given mass in which the æther is aligned with the Killing vector.[18] This "static æther" vacuum solution depends on the c_i only through the combination c_{14}, and was found analytically (up to inversion of a transcendental equation).[18] It is stable to linear perturbations under the same conditions as for stability of flat spacetime, with the exception of the case $c_{123} = 0$.[19]

---

[a]This corrects an error in version 1 of Ref. 16, where σ is said to scale as c_i^2. (Also the a prefactor c_{14} in Eq. (70) should be deleted.) As a result of this correction, the likely constraints on c_i are an order of magnitude stronger, as stated here.[17]

The solution inside a fluid star has been found by numerical integration, both for constant density[18] and for realistic neutron star equations of state.[20] The maximum masses for neutron stars range from about 6 to 15% smaller than in GR when $c_{14} = 1$, depending on the equation of state. The corresponding surface redshifts can be as much as 10% larger than in GR for the same mass. Measurements of high gravitational masses or precise surface redshifts thus have the potential to yield strong joint constraints on c_{14} and the equation of state. The radius of the innermost stable circular orbit (ISCO) differs from the GR value $6G_{\rm N}M$ by a small term of relative order about $0.03c_{14}$.

For black holes, the condition of regularity at the spin-0 horizon selects a unique solution from the one-parameter family for a given mass.[21] When a black hole forms from collapse of matter, the spin-0 horizon develops in a nonsingular region of spacetime, where the evolution should be regular. This motivated the conjecture that collapse produces a black hole with nonsingular spin-0 horizon, which has been confirmed for some particular examples in numerical simulations of collapse of a scalar field.[22]

The black holes with nonsingular spin-0 horizons are rather close to Schwarzschild outside the horizon for a wide range of couplings; for instance, the ISCO radius differs by a factor $(1 + 0.043c_1 + 0.061c_1^2)$, in the case with $c_3 = c_4 = 0$ and c_2 fixed so that the spin-0 speed is unity.[23] (This expansion is accurate at least when $c_1 \lesssim 0.5$. No solution with regular spin-0 horizon exists in this case when $c_1 \gtrsim 0.8$.) Inside the horizon the solutions differ more, but like Schwarzschild they contain a spacelike singularity. Black hole solutions with singular spin-0 horizons have been studied in Ref. 24. These solutions can differ much more outside the horizon. Quasi-normal modes of black holes in æ-theory have been investigated in Refs. 25.

8. Special values of c_i?

The first case to be examined in detail[26,4] was $c_{13} = c_2 = c_4 = 0$, i.e., the "Maxwell action" together with the unit constraint on the vector. The PPN result for α_2 (5) is infinite in this case, and the spin-0 mode speed is zero. The perturbation series used in the PPN analysis is thus evidently not applicable. Independently of that however, other problems with this case have been identified, such as the formation of shock discontinuities[4,27] and a possibly related instability.[19]

Assuming now that $\alpha_{1,2} = 0$ and the constraints (12), (13) are satisfied, and putting aside the case $c_1 = c_3 = 0$ which is not covered by existing

PPN analyses, all but one of the cases in which one of the c_i vanishes, or in which one of c_{13}, c_{14}, or c_{123} vanishes, have the property that the spin-1 mode speed (10) diverges while the energy of that mode is nonzero. It seems very unlikely that such cases are observationally viable, although they have not been examined carefully. The exception is the special case $c_3 = c_4 = 2c_1 + 3c_2 = 0$, with $2/3 < c_1 < 1$. This large value of c_1 is probably inconsistent with the strong field constraints from orbital binaries, but as mentioned above those are not yet precisely known because the sensitivity parameters have not yet been computed for neutron stars, so this case is not yet ruled out.

9. Conclusion

Einstein-æther theory is an intriguing theoretical laboratory in which gravitational effects of possible Lorentz violation can be meaningfully studied. There is a large (order unity) two-parameter space of Einstein-æther theories for which (i) the PPN parameters are identical to those of GR, (ii) the linear perturbations are stable and carry positive energy, (iii) there is no vacuum Čerenkov radiation, (iv) the dynamics of the cosmological scale factor and perturbations differ little from GR, (v) non-rotating neutron star and black hole solutions are close to those of GR, but might be distinguishable with future observations. Radiation damping from binaries, imposes an order 0.001 constraint on one combination of the parameters. Strong self-field effects in neutron stars and black holes produce violations of the strong equivalence principle and higher order post-Newtonian effects which will constrain all the parameters c_i to be less than around 0.01, presuming that the sensitivity parameters for neutron stars (which have not yet been computed with the required precision) turn out to have the expected magnitude.

Acknowledgments

I am grateful to C.T. Eling, B.Z. Foster, B. Li, and E. Lim for helpful correspondence. This work was supported by NSF grant PHY-0601800.

References

1. C.M. Will and K. Nordvedt, Jr., Astrophys. J. **177**, 757 (1972); K. Nordvedt, Jr. and C.M. Will, Astrophys. J. **177**, 775 (1972); R.W. Hellings and K. Nordvedt, Jr., Phys. Rev. **D7**, 3593 (1973).

2. J. W. Elliott, G. D. Moore and H. Stoica, JHEP **0508**, 066 (2005) [arXiv:hep-ph/0505211].
3. See, for example, M. Gasperini, Class. Quantum Grav. **4**, 485 (1987); Gen. Rel. Grav. **30**, 1703 (1998); and references therein.
4. T. Jacobson and D. Mattingly, Phys. Rev. D **64**, 024028 (2001) [arXiv:gr-qc/0007031].
5. S. M. Carroll and E. A. Lim, Phys. Rev. D **70**, 123525 (2004) [arXiv:hep-th/0407149].
6. C. M. Will, Living Reviews in Relativity **9**, 3 (2006).
7. C. Eling and T. Jacobson, Phys. Rev. D **69**, 064005 (2004) [arXiv:gr-qc/0310044].
8. B. Z. Foster and T. Jacobson, Phys. Rev. D **73**, 064015 (2006) [arXiv:gr-qc/0509083].
9. B. Z. Foster, Phys. Rev. D **73**, 104012 (2006) [arXiv:gr-qc/0602004].
10. M. L. Graesser, A. Jenkins and M. B. Wise, Phys. Lett. B **613**, 5 (2005) [arXiv:hep-th/0501223].
11. D. Mattingly and T. Jacobson, in V.A. Kostelecký, ed., *Proceedings of the Second Meeting on CPT and Lorentz Symmetry*, World Scientific, Singapore, 2002 [arXiv:gr-qc/0112012].
12. T. Jacobson and D. Mattingly, Phys. Rev. D **70**, 024003 (2004) [arXiv:gr-qc/0402005].
13. C. Eling, Phys. Rev. D **73**, 084026 (2006) [arXiv:gr-qc/0507059].
14. E. A. Lim, Phys. Rev. D **71**, 063504 (2005) [arXiv:astro-ph/0407437].
15. B. Li, D. F. Mota and J. D. Barrow, arXiv:0709.4581 [astro-ph].
16. B. Z. Foster, arXiv:0706.0704 [gr-qc].
17. B. Z. Foster, private communication.
18. C. Eling and T. Jacobson, Class. Quant. Grav. **23**, 5625 (2006) [arXiv:gr-qc/0603058].
19. M. D. Seifert, Phys. Rev. D **76**, 064002 (2007) [arXiv:gr-qc/0703060].
20. C. Eling, T. Jacobson and M. Coleman Miller, Phys. Rev. D **76**, 042003 (2007) [arXiv:0705.1565 [gr-qc]].
21. C. Eling and T. Jacobson, Class. Quant. Grav. **23**, 5643 (2006) [arXiv:gr-qc/0604088].
22. D. Garfinkle, C. Eling and T. Jacobson, Phys. Rev. D **76**, 024003 (2007) [arXiv:gr-qc/0703093].
23. C. T. Eling, private communication.
24. T. Tamaki and U. Miyamoto, arXiv:0709.1011 [gr-qc].
25. R. A. Konoplya and A. Zhidenko, Phys. Lett. B **644**, 186 (2007) [arXiv:gr-qc/0605082]; Phys. Lett. B **648**, 236 (2007) [arXiv:hep-th/0611226].
26. V. A. Kostelecký and S. Samuel, Phys. Rev. D **40**, 1886 (1989).
27. M. A. Clayton, arXiv:gr-qc/0104103.

TESTS OF LORENTZ-INVARIANCE VIOLATION IN NEUTRINO OSCILLATIONS

K. WHISNANT

Department of Physics and Astronomy, Iowa State University, Ames, IA 50011, U.S.A.
E-mail: whisnant@iastate.edu

The Standard-Model Extension, which describes possible new interactions due to spontaneous breaking of Lorentz symmetry, allows new terms in the effective Hamiltonian for neutrino propagation that have different energy dependence than those due to neutrino mass differences. The bicycle model has no neutrino mass terms, but can reproduce the usual $1/E$ dependence of the oscillation argument at high energy due to a see-saw-like mechanism. However, we show that the bicycle model is ruled out by a combination of solar, atmospheric, long-baseline and reactor data. The requirement of a $1/E$ dependence at high energy also severely constrains more general Lorentz-violating models.

1. Introduction

The Standard-Model Extension (SME)[1] includes Lorentz-violating interactions that preserve gauge symmetries but break particle Lorentz symmetry. These new interactions can contribute to neutrino oscillations with or without neutrino mass. These new interactions can be nonisotropic, which could lead to different oscillation parameters for neutrinos propagating in different directions.

The effective hamiltonian that describes the evolution of neutrinos in vacuum in the presence of Lorentz-invariance violating interactions may be written as[1]

$$(h_{eff})_{ij} = E\ \delta_{ab} + \frac{(m^2)_{ab}}{2E} + \frac{1}{E}\left[(a_L)^{\mu} p_{\mu} - (c_L)^{\mu\nu} p_{\mu} p_{\nu}\right]_{ab} , \qquad (1)$$

where $p_{\mu} = (E, -E\hat{p})$ is the neutrino four-momentum, $\hat{p}$ the neutrino direction, and a, b are flavor indices. The coefficients a_L have dimensions of energy and the c_L are dimensionless. The Kronecker delta term on the right-hand side of Eq. (1) may be ignored since oscillations are insensitive to terms in h_{eff} proportional to the identity. For antineutrinos, $a_L \to -a_L$.

Direction dependence of the neutrino evolution enters via the space components of a_L and c_L. The coefficients may be specified in a celestial equatorial frame (T, X, Y, Z), which has Z axis along the Earth's rotation axis and X axis towards the vernal equinox. The two-parameter bicycle model[1] can be defined as follows: c_L is isotropic, with only one nonzero element $(c_L)^{TT}_{ee} \equiv 2c$, $(a_L)^{\mu}_{e\mu} = (a_L)^{\mu}_{e\tau} = (0, a\hat{Z}/\sqrt{2})$ are the only nonzero a_L, and there are no neutrino mass terms. We have generalized the model by letting $(a_L)^{\mu}_{e\mu} = (a_L)^{\mu}_{e\tau} = (0, a\hat{n}/\sqrt{2})$, where $\hat{n}$ is the preferred direction for the a_L interaction. We also consider a model which has a linear combination of direction-dependent and direction-independent a_L. We have shown[2] that all variations of the bicycle model are inconsistent with the combined data from solar, atmospheric, long-baseline and reactor neutrinos. We have also shown that there are strong constraints to more general Lorentz-violating models of massless neutrinos; further investigation is needed to determine if any can be consistent with all neutrino data.

2. Neutrino oscillations in the generalized bicycle model

2.1. *Effective Hamiltonian*

Neutrino oscillations occur due to eigenenergy differences in h_{eff} and the fact that the neutrino flavor eigenstates are not eigenstates of h_{eff}. For massless neutrinos $p_\mu = (E, -E\hat{p})$, where $\hat{p}$ is the direction of neutrino propagation. Then for the generalized bicycle model the effective Hamiltonian is

$$h_{eff} = \begin{pmatrix} -2cE & \frac{1}{\sqrt{2}} a\cos\Theta & \frac{1}{\sqrt{2}} a\cos\Theta \\ \frac{1}{\sqrt{2}} a\cos\Theta & 0 & 0 \\ \frac{1}{\sqrt{2}} a\cos\Theta & 0 & 0 \end{pmatrix}, \quad (2)$$

where $\cos\Theta = \hat{p}\cdot\hat{n}$, *i.e.*, Θ is the angle between the neutrino momentum and the preferred direction. At high energies the cE term dominates and there is a see-saw mechanism; one eigenvalue difference is $\Delta_{32} = a^2\cos^2\Theta/(2cE)$ and the only appreciable mixing is between ν_μ and ν_τ. This leads to the oscillation probability $P(\nu_\mu \to \nu_\tau) = \sin^2(\Delta_{32}L/2)$, which describes maximal mixing with $\delta m^2_{eff} = (a\cos\Theta)^2/c$. Since the measured values for δm^2_{eff} agree for atmospheric neutrinos *and* the K2K and MINOS long-baseline experiments, the effective $\cos^2\Theta$ must also have similar values in all of these experiments.

2.2. *Atmospheric and long-baseline neutrinos*

With the Earth's rotation axis chosen as the $\hat{Z}$ direction and the position of the detector given by (θ, ϕ) in a standard spherical polar coordinate system, the neutrino direction can be written as $\hat{p} = -\cos\beta\ \hat{r} + \sin\beta(-\sin\alpha\ \hat{\theta} + \cos\alpha\ \hat{\phi})$, where $\mathbf{r}$ denotes the detector position, and the unit vectors $\hat{r}$, $\hat{\theta}$ and $\hat{\phi}$ point in the upward, southerly and easterly directions, respectively. The angle β is the usual zenith angle ($\beta = 0$ for a downward event) and α denotes the compass direction of the neutrino velocity projected on the plane tangent to the Earth's surface ($\alpha = 0$ for a neutrino going in the eastward direction). We take the preferred direction to be $\hat{n} = \sin\xi\cos\chi\ \hat{X} + \sin\xi\sin\chi\ \hat{Y} + \cos\xi\ \hat{Z}$, so in our spherical polar coordinate system

$$\begin{aligned}\hat{n} = &[\sin\xi\cos(\phi+\chi)\cos\theta_L + \cos\xi\sin\theta_L]\ \hat{r} \\ &+[\sin\xi\cos(\phi+\chi)\sin\theta_L - \cos\xi\cos\theta_L]\ \hat{\theta} - \sin\xi\sin(\phi+\chi)\ \hat{\phi}\,, \end{aligned} \tag{3}$$

where the usual angle spherical polar θ has been replaced by the latitude of the detector $\theta_L = \frac{\pi}{2} - \theta$ (positive for the northern hemisphere, negative for the southern hemisphere). The azimuthal angle ϕ is chosen so that $\phi = 0$ corresponds to the preferred direction χ, so that the angle χ may be dropped. The angular dependence in the oscillation formulas is then

$$\begin{aligned}\cos\Theta = &\cos\xi(\sin\beta\sin\alpha\cos\theta_L - \cos\beta\sin\theta_L) - \sin\xi\sin\beta\cos\alpha\sin\phi \\ &- \sin\xi\cos\phi(\sin\beta\sin\alpha\sin\theta_L + \cos\beta\cos\theta_L)\,. \end{aligned} \tag{4}$$

In Eq. (4), ξ gives the orientation of the preferred axis with respect to the Earth's rotation axis, α (compass direction) and β (zenith angle) relate to the neutrino direction, and ϕ depends on the time of the sidereal day ($\phi = 0$ when the detector is facing the preferred direction).

If the preferred direction is aligned with the Earth's rotation axis, then $\xi = 0$ and

$$\cos^2\Theta = (\sin\beta\sin\alpha\cos\theta_L - \cos\beta\sin\theta_L)^2\,. \tag{5}$$

Note that in this case Θ does not depend on time of day (measured by ϕ). For accelerator experiments with relatively short baselines compared to the Earth's radius (such as K2K and MINOS), the neutrino path can be considered to be in the plane that is tangent to the Earth's surface (i.e., $\beta = \pi/2$), so that $\cos^2\Theta = \sin^2\alpha\cos^2\theta_L$. Using the values of the latitude θ_L and neutrino direction α in K2K and MINOS, a value of $a^2/c = \delta m^2/\cos^2\Theta$ can be inferred from each experiment (see Table 1). These disagree by nearly two orders of magnitude, so the case $\xi = 0$ is ruled out by a combination of K2K and MINOS data.

Table 1. Inferred value of a^2/c for K2K and MINOS.

Experiment	θ_L	α	$\cos^2\Theta$	$a^2/c = \delta m^2/\cos^2\Theta$
K2K	36°	174°	7.2×10^{-3}	4.0×10^{-1} eV^2
MINOS	48°	124°	4.5×10^{-1}	8.0×10^{-3} eV^2

If $\xi \neq 0$, then the preferred direction is not aligned with the Earth's rotation axis. For long-baseline events there will be variation in $\cos^2\Theta$ (and hence in δm^2_{eff}) with ϕ. It can be shown that the ranges of δm^2_{eff} over the sidereal day for K2K and MINOS do not overlap for $0 \leq \xi \leq 15°$ and $165° \leq \xi \leq 180°$, so the K2K and MINOS experiments cannot be simultaneously explained for these values of ξ.

For $15° \leq \xi \leq 165°$, the ranges of δm^2_{eff} for K2K and MINOS overlap, and a detailed fit is required to say more. However, twice a day the δm^2_{eff} for K2K is zero and there are no oscillations, which in principle could be detected if the data were binned versus sidereal time.

2.3. *Solar neutrinos*

In a coordinate system (X', Y', Z') where the Z' axis is perpendicular to the Earth's orbital plane (the ecliptic plane), the direction of neutrino propagation may be written as $\hat{p} = \cos\psi\, \hat{X}' + \sin\psi\, \hat{Y}'$, where ψ gives the position of the Earth in its orbit ($\psi = 0$ at the vernal equinox, $\psi = \pi/2$ at the summer solstice, etc.). The equatorial coordinates are related to the ecliptic coordinates via rotation by an angle $\eta \simeq 23°$ about the X' axis, where η is the tilt of the Earth's rotation axis from the perpendicular to the ecliptic. Then in the celestial equatorial frame the direction of propagation for solar neutrinos is $\hat{p} = \cos\psi\, \hat{X} + \sin\psi\cos\eta\, \hat{Y} - \sin\psi\sin\eta\, \hat{Z}$ and therefore

$$\cos\Theta = \hat{p}\cdot\hat{n} = \cos\psi\cos\chi\sin\xi + \sin\psi(\sin\chi\sin\xi\cos\eta - \cos\xi\sin\eta)\,; \quad (6)$$

$\cos\Theta$ for solar neutrinos is independent of detector latitude (θ_L) and time of day (ϕ).

In matter there is an additional term in the hamiltonian due to coherent forward scattering of ν_e with electrons in matter, so that $-2cE$ in the upper left element of h_{eff} is replaced by $-2cE + \sqrt{2}G_F N_e$. Assuming adiabatic propagation, the oscillation probability averaged over the year is shown in Fig. 1. There is a minimum in P at $E_{min} = G_F N_e^0/2\sqrt{2}\ c$; the minimum

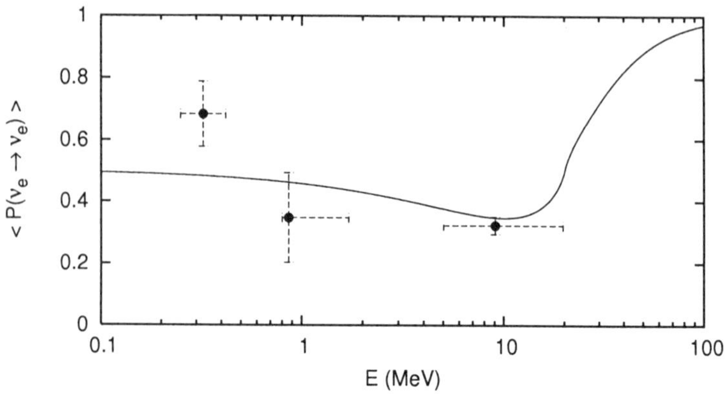

Figure 1. Solar neutrino survival probability versus energy (solid curve). Also shown are the solar neutrino data.

value of P is determined by $a \cos \Theta$. It can be shown that the propagation is adiabatic except close to the two times during the year where $\cos \Theta = 0$, which affects the binned probability by at most a few percent.

If the probability minimum lies in the middle of the ^{8}B solar neutrino region, then the approximate survival probability of the ^{8}B neutrinos will be nearly flat with energy, as required by data. This results in the constraint $c \simeq G_F N_e^0 / 2\sqrt{2}\, E_{min} = 1.7 \times 10^{-19}$, where N_e^0 is the electron number density at creation. Allowed values for a, ξ and χ are determined by the condition that P_{min} for the ^{8}B neutrinos is near the measured value of 0.34.

The measured solar neutrino survival probability does not exhibit much variation throughout the year. The SNO collaboration has tested their solar neutrino data for periodicities[3] by measuring the ratio R of the monthly event rate to the annual average event rate and found a variation during the year that is consistent with the $1/r^2$ dependence of the flux as the Earth's distance from the sun varies. The uncertainties in the rate are of order 3-5%, so there is little room for any additional annual variation. In Fig. 2 we show the best fit to the SNO periodicity data (with the $1/r^2$ dependence removed) when the ^{8}B probability is constrained to be near its measured value of 0.34; the generalized direction-dependent bicycle model is strongly ruled out by the solar neutrino data.

2.4. *Adding a direction independent term*

Since the pure direction-dependent case is ruled out, we can generalize the model further to include both direction-independent as well as direction-

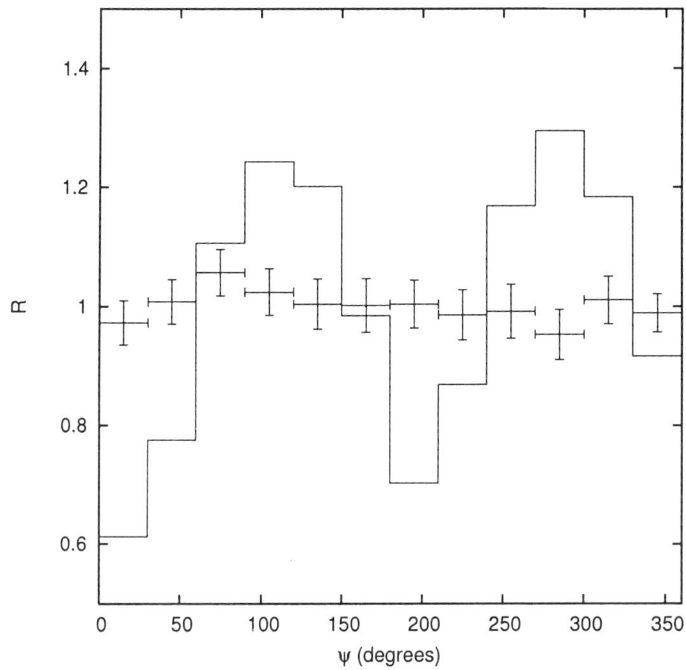

Figure 2. Best-fit prediction for R for the ^{8}B neutrinos (solid curve) and the SNO measured value for R (data points) shown versus time of year.

dependent terms in the off-diagonal elements of h_{eff}. This increases the number of parameters in the model to five. If we define $(a_L)^{\mu}_{e\mu} = (a_L)^{\mu}_{e\tau} = (a\cos\rho, a\sin\rho\;\hat{n}/\sqrt{2})$, where $\hat{n}$ is again the preferred direction, then $\cos\Theta$ is replaced by $\cos\rho + \sin\rho\cos\Theta$.

For the pure direction-independent case ($\rho = 0$ or π), $\delta m^2_{eff} = a^2/c$ for atmospheric and long-baseline neutrinos. As before, c is determined by where the minimum in P occurs for ^{8}B solar neutrinos and the value of P for ^{8}B solar neutrinos depends on a. For $P = 0.34$, we find $a \simeq 2.5\times10^{-12}$ eV, which, when combined with the value of c from before, gives $\delta m^2_{eff} = 3.6\times10^{-5}$ eV^2 for atmospheric and long-baseline neutrinos, which is clearly in contradiction with the data. Therefore, the pure direction-independent case is ruled out by the combined data.

For a mixture of direction-dependent and direction-independent terms in h_{eff}, a fit must be done to the solar data to determine an allowed region in parameter space, and then the predictions for δm^2_{eff} in long-baseline experiments can be compared to data. We use the 12 bins from the SNO periodicity data sample for the ^{8}B solar data, plus the constraints on long-baseline δm^2 from K2K to find an allowed region in the parameters a, ξ, χ, and ρ (c is still constrained by the position of the solar minimum).

For the range of parameters allowed by the SNO and K2K data, the oscillation probability for KamLAND is found to be quite flat, varying by a few percent at most over the entire spectrum. However, KamLAND data[4] shows a well-defined oscillation that is inconsistent with a uniform suppression. Therefore the bicycle model with a mixture of direction-dependent and direction-independent terms is also excluded.

3. More general Lorentz-violating models

We have also looked at the most general direction-independent Lorentz-violating model with no neutrino masses,[5] which has 16 independent parameters,[1] eight $(c_L)^{TT}_{ij}$ and eight $(a_L)^{T}_{ij}$. To be consistent with the phenomenological fact that the argument of the oscillation probability goes like $1/E$ at high energies, we expand the eigenvalues of h_{eff} in powers of E, $\lambda_i = \alpha_1 E + \alpha_0 + \alpha_{-1} E^{-1} + ...$, and require that there is a degeneracy in the eigenvalues at both leading order (E^1) and next-to-leading order (E^0), so that the leading term in an eigenvalue difference goes like E^{-1}. This puts constraints on the $(c_L)^{TT}_{ij}$ and $(a_L)^{T}_{ij}$. We have found that only three textures of the (c_L) allow such a degeneracy *and* also give maximal mixing for $\nu_\mu \to \nu_\tau$ *and* have small amplitude oscillations involving ν_e at high energies. These textures are

$$\begin{pmatrix} c_{ee} & 0 & c_{e\tau} \\ 0 & 0 & 0 \\ c_{e\tau} & 0 & c_{\tau\tau} \end{pmatrix}, \quad \begin{pmatrix} c_{ee} & c_{e\mu} & c_{e\tau} \\ c_{e\mu} & 0 & c_{\mu\tau} \\ c_{e\tau} & c_{\mu\tau} & 0 \end{pmatrix}, \quad \begin{pmatrix} c_{ee} & c_{e\mu} & c_{e\tau} \\ c_{e\mu} & 0 & c_{\mu\tau} \\ c_{e\tau} & c_{\mu\tau} & c_{\tau\tau} \end{pmatrix}. \tag{7}$$

In some cases the degeneracy requirements require fine tuning. These models must be checked for consistency with the solar and KamLAND data.[5]

References

1. V. A. Kostelecký and M. Mewes, Phys. Rev. D **70**, 031902 (2004) [arXiv:hep-ph/0308300].
2. V. Barger, D. Marfatia and K. Whisnant, arXiv:0706.1085 [hep-ph].
3. B. Aharmim *et al.* [SNO Collaboration], Phys. Rev. D **72**, 052010 (2005) [arXiv:hep-ex/0507079].
4. T. Araki *et al.* [KamLAND Collaboration], Phys. Rev. Lett. **94**, 081801 (2005) [arXiv:hep-ex/0406035].
5. V. Barger, D. Marfatia and K. Whisnant, in preparation.

SEARCH FOR CPT VIOLATION IN NEUTRAL KAONS AT KLOE: STATUS AND PERSPECTIVES

ANTONIO DI DOMENICO

Dipartimento di Fisica, Università di Roma "La Sapienza"
and I.N.F.N. Sezione di Roma
P.le A. Moro 2
I-00185 Rome, Italy
E-mail: antonio.didomenico@roma1.infn.it

AND THE KLOE COLLABORATION*

The neutral kaon system offers a unique possibility to perform fundamental tests of CPT invariance, as well as of the basic principles of quantum mechanics. The most recent limits on various types of CPT violation and decoherence mechanisms, obtained by the KLOE experiment at the DAΦNE e^+e^- collider, are reviewed, with a special focus on possible CPT violation related to Lorentz symmetry breaking in the framework of the Standard-Model Extension (SME). It is shown that all four Δa_μ CPT violation parameters of the SME could be independently measured at KLOE, complementing and completing results obtained by fixed beam experiments.

1. Introduction

DAΦNE, the Frascati ϕ-factory, is an e^+e^- collider working at a center of mass energy of $\sqrt{s} \sim 1020$ MeV, corresponding to the peak of the ϕ resonance. The ϕ production cross section is $\sim 3\mu$b, and its decay into

---

* The KLOE Collaboration: F. Ambrosino, A. Antonelli, M. Antonelli, F. Archilli, C. Bacci, P. Beltrame, G. Bencivenni, S. Bertolucci, C. Bini, C. Bloise, S. Bocchetta, F. Bossi, P. Branchini, R. Caloi, P. Campana, G. Capon, T. Capussela, F. Ceradini, S. Chi, G. Chiefari, P. Ciambrone, E. De Lucia, A. De Santis, P. De Simone, G. De Zorzi, A. Denig, A. Di Domenico, C. Di Donato, S. Di Falco, B. Di Micco, A. Doria, M. Dreucci, G. Felici, A. Ferrari, M.L. Ferrer, G. Finocchiaro, S. Fiore, C. Forti, P. Franzini, C. Gatti, P. Gauzzi, S. Giovannella, E. Gorini, E. Graziani, M. Incagli, W. Kluge, V. Kulikov, F. Lacava, G. Lanfranchi, J. Lee-Franzini, D. Leone, M. Martini, P. Massarotti, W. Mei, S. Meola, S. Miscetti, M. Moulson, S. Müller, F. Murtas, M. Napolitano, F. Nguyen, M. Palutan, E. Pasqualucci, A. Passeri, V. Patera, F. Perfetto, M. Primavera, P. Santangelo, G. Saracino, G. Saracino, B. Sciascia, A. Sciubba, F. Scuri, I. Sfiligoi, T. Spadaro, M. Testa, L. Tortora, P. Valente, G. Venanzoni, R. Versaci, G. Xu.

$K^0\bar{K^0}$ has a branching fraction of 34%. The neutral kaon pair is produced in a coherent quantum state with quantum numbers $J^{PC} = 1^{--}$:

$$|i\rangle = \frac{1}{\sqrt{2}}\{|K^0\rangle|\bar{K^0}\rangle - |\bar{K^0}\rangle|K^0\rangle\} = \frac{N}{\sqrt{2}}\{|K_S\rangle|K_L\rangle - |K_L\rangle|K_S\rangle\} \qquad (1)$$

where $N = (1+|\epsilon|^2)/(1-\epsilon^2) \simeq 1$ is a normalization factor. The detection of a kaon at large (small) times *tags* a K_S (K_L) in the opposite direction.

The KLOE detector consists mainly of a large volume drift chamber[1] surrounded by an electromagnetic calorimeter.[2] A superconducting coil provides a 0.52 T solenoidal magnetic field. At KLOE a K_S is tagged by identifying the interaction of the K_L in the calorimeter (K_L-crash), while a K_L is tagged by detecting a $K_S \to \pi^+\pi^-$ decay near the interaction point (IP).

KLOE completed the data taking in March 2006 with a total integrated luminosity of ~ 2.5 fb^{-1}, corresponding to $\sim 7.5\times10^{-9}$ ϕ-mesons produced.

2. CPT symmetry tests with $K_S \to \pi e \nu$ decays

The first measurement of the K_S semileptonic charge asymmetry has been performed by KLOE:[3]

$$\begin{aligned} A_S &= \frac{\Gamma(K_S \to \pi^- e^+ \nu) - \Gamma(K_S \to \pi^+ e^- \bar{\nu})}{\Gamma(K_S \to \pi^- e^+ \nu) + \Gamma(K_S \to \pi^+ e^- \bar{\nu})} \\ &= (1.5 \pm 9.6_{\text{stat}} \pm 2.9_{\text{syst}}) \times 10^{-3}. \end{aligned}$$

The uncertainty on A_S can be reduced at the level of $\approx 3 \times 10^{-3}$ with the analysis of the full data sample of 2.5 fb^{-1}.

From the sum and the difference of the K_S and K_L semileptonic charge asymmetries one can test CPT conservation. Using the values[a] of A_L, $\Re\delta$, and $\Re\epsilon$ from other experiments,[4] the real part of the CPT and $\Delta S = \Delta Q$ violating parameter x_-, and the real part of the CPT violating and $\Delta S = \Delta Q$ conserving parameter y, in semileptonic decay amplitudes, can be evaluated:[3]

$$\begin{aligned} \Re x_- &= \frac{A_S - A_L}{4} - \Re\delta = (-0.8 \pm 2.5) \times 10^{-3} \\ \Re y &= \Re\epsilon - \frac{A_S + A_L}{4} = (0.4 \pm 2.5) \times 10^{-3} \; . \end{aligned} \qquad (2)$$

[a] ϵ and δ are the usual complex parameters describing CP and CPT violation in the effective Hamiltonian for neutral kaons.

3. CPT test from unitarity

The unitarity relation, originally derived by Bell and Steinberger:

$$\left(\frac{\Gamma_S+\Gamma_L}{\Gamma_S-\Gamma_L}+i\tan\phi_{SW}\right)\left[\frac{\Re\epsilon}{1+|\epsilon|^2}-i\Im\delta\right]=$$
$$=\frac{1}{\Gamma_S-\Gamma_L}\sum_f A^*(K_S\to f)A(K_L\to f)\equiv\sum_f\alpha_f \tag{3}$$

can be used to bound $\Im\delta$, after having provided all the α_i parameters, Γ_S, Γ_L, and ϕ_{SW} (the superweak phase) as inputs. Using KLOE measurements, PDG[4] values, and a combined fit of KLOE and CPLEAR data, the following result is obtained:[5]

$$\Re\epsilon=(160.2\pm1.3)\times10^{-5}\ , \qquad \Im\delta=(1.2\pm1.9)\times10^{-5}\ ,$$

the main limiting factor of this result being the uncertainty on the phase ϕ_{+-} of the $\eta_{\pi^+\pi^-}\equiv A(K_L\to\pi^+\pi^-)/A(K_S\to\pi^+\pi^-)$ parameter entering in $\alpha_{\pi^+\pi^-}$.

The limits on $\Im(\delta)$ and $\Re(\delta)$[6] can be used to constrain the mass and width difference between K^0 and $\bar{K}^0$. In the limit $\Gamma_{K^0}=\Gamma_{\bar{K}^0}$, i.e., neglecting CPT-violating effects in the decay amplitudes, the best bound on the neutral kaon mass difference is obtained:

$$-5.3\times10^{-19}\ \text{GeV} < m_{K^0}-m_{\bar{K}^0} < 6.3\times10^{-19}\ \text{GeV} \quad \text{at 95 \% CL}.$$

4. Decoherence and CPT tests using Kaon interferometry

The quantum interference between the two kaon decays in the CP violating channel $\phi\to K_SK_L\to\pi^+\pi^-\pi^+\pi^-$ has been observed for the first time by KLOE.[7] The measured Δt distribution, with Δt the absolute value of the time difference of the two $\pi^+\pi^-$ decays, can be fitted with the distribution:

$$I(\pi^+\pi^-,\pi^+\pi^-;\Delta t)\propto e^{-\Gamma_L\Delta t}+e^{-\Gamma_S\Delta t}-2(1-\zeta_{SL})e^{-\frac{(\Gamma_S+\Gamma_L)}{2}\Delta t}\cos(\Delta m\Delta t)$$

where the quantum mechanical expression in the $\{K_S,K_L\}$ basis has been modified with the introduction of a decoherence parameter ζ_{SL}, and a factor $(1-\zeta_{SL})$ multiplying the interference term. Analogously, a $\zeta_{0\bar{0}}$ parameter can be defined in the $\{K^0,\bar{K}^0\}$ basis.[8] After having included resolution and detection efficiency effects, having taken into account the background due to coherent and incoherent K_S-regeneration on the beam pipe wall, the small contamination of non-resonant $e^+e^-\to\pi^+\pi^-\pi^+\pi^-$ events, and keeping fixed in the fit Δm, Γ_S and Γ_L to the PDG values, a fit is performed on the Δt distribution with the following results:

$$\zeta_{SL} = 0.018 \pm 0.040_{\text{stat}} \pm 0.007_{\text{syst}}$$
$$\zeta_{0\bar{0}} = (1.0 \pm 2.1_{\text{stat}} \pm 0.4_{\text{syst}}) \times 10^{-6}$$

compatible with the quantum mechanics prediction, i.e., $\zeta_{SL} = \zeta_{0\bar{0}} = 0$, and no decoherence effects. In particular the result on $\zeta_{0\bar{0}}$ has a high accuracy, $\mathcal{O}(10^{-6})$, due to the CP suppression present in the specific decay channel; it improves of five orders of magnitude the previous limit obtained by Bertlmann and co-workers[8] in a re-analysis of CPLEAR data. This result can also be compared to a similar one recently obtained in the B meson system,[9] where an accuracy $\mathcal{O}(10^{-2})$ can be reached.

It has been pointed out[10,11] that in the context of a hypothetical quantum gravity, CPT violation effects might occur in correlated neutral kaon states, where the resulting loss of particle-antiparticle identity could induce a breakdown of the correlation of state (1) imposed by Bose statistics. As a result the initial state (1) can be parametrized in general as:

$$|i\rangle = \frac{1}{\sqrt{2}}\left[|K^0\rangle|\bar{K}^0\rangle - |\bar{K}^0\rangle|K^0\rangle + \omega\left(|K^0\rangle|\bar{K}^0\rangle + |\bar{K}^0\rangle|K^0\rangle\right)\right] , \quad (4)$$

where ω is a complex parameter describing a completely novel CPT violation phenomenon, not included in previous analyses. Its order of magnitude could be at most

$$|\omega| \sim \left[(m_K^2/M_{\text{Planck}})/\Delta\Gamma\right]^{1/2} \sim 10^{-3}$$

with $\Delta\Gamma = \Gamma_S - \Gamma_L$. A similar analysis performed on the same data as before, including in the fit of the Δt distribution the modified initial state Eq. (4), yields the first measurement of the complex parameter ω:[7]

$$\Re(\omega) = \left(1.1^{+8.7}_{-5.3} \pm 0.9\right) \times 10^{-4} \qquad \Im(\omega) = \left(3.4^{+4.8}_{-5.0} \pm 0.6\right) \times 10^{-4} ;$$

with an accuracy that already reaches the interesting Planck's scale region. Other interesting results related to possible decoherence and CPT violation in the quantum gravity framework are discussed in Ref. 7.

5. CPT and Lorentz symmetry breaking

CPT invariance holds for any realistic Lorentz-invariant quantum field theory. However a very general theoretical possibility for CPT violation is based on spontaneous breaking of Lorentz symmetry, as developed by Kostelecký,[12,13,14] which appears to be compatible with the basic tenets of quantum field theory and retains the property of gauge invariance and renormalizability (Standard-Model Extension (SME)). In the SME for neutral kaons, CPT manifests to lowest order only in the CPT violation parameter δ, (e.g., y and x_- vanish at first order), and exhibits a dependence on the 4-momentum of the kaon:

$$\delta \approx i \sin \phi_{SW} e^{i\phi_{SW}} \gamma_K (\Delta a_0 - \vec{\beta_K} \cdot \Delta \vec{a})/\Delta m \tag{5}$$

where γ_K and $\vec{\beta_K}$ are the kaon boost factor and velocity in the observer frame, and Δa_μ are four CPT- and Lorentz-violating coefficients for the two valence quarks in the kaon.

Following Ref. 13, the time dependence arising from the rotation of the Earth can be explicitly displayed in Eq. (5) by choosing a three-dimensional basis $(\hat{X}, \hat{Y}, \hat{Z})$ in a non-rotating frame, with the $\hat{Z}$ axis along the Earth's rotation axis, and a basis $(\hat{x}, \hat{y}, \hat{z})$ for the rotating (laboratory) frame. The CPT violating parameter δ may then be expressed as:

$$\begin{aligned} \delta &= \frac{1}{2\pi} \int_0^{2\pi} \delta(\vec{p}, t) d\phi \\ &= \frac{i \sin \phi_{SW} e^{i\phi_{SW}}}{\Delta m} \gamma_K \{ \Delta a_0 + \beta_K \Delta a_Z \cos\theta \cos\chi \\ &\quad + \beta_K (\Delta a_Y \sin\chi \cos\theta \sin\Omega t + \Delta a_X \sin\chi \cos\theta \cos\Omega t) \} , \end{aligned} \tag{6}$$

where Ω is the Earth's sidereal frequency, $\cos\chi = \hat{z} \cdot \hat{Z}$, θ and ϕ are the conventional polar and azimuthal angles defined in the laboratory frame about the $\hat{z}$ axis, and an integration on the azimuthal angle ϕ has been performed, assuming a symmetric decay distribution in this variable[b]. The sensitivity to the four Δa_μ parameters can be very different for fixed target and collider experiments, showing complementary features.[13]

5.1. *Measurement of Δa_0 at KLOE*

The Δa_0 parameter can be measured through the difference $A_S - A_L$, by performing the measurement of each asymmetry with a symmetric integration over the polar angle[c] θ, thus averaging to zero any possible contribution

[b]This simplifying assumption will be maintained throughout the following; however small non-symmetric ϕ angle effects could be easily included in the formulas without significantly modifying the main conclusions in the following.

[c]In the following the $\hat{z}$ axis has been chosen along the e^+ beam at the KLOE IP, with $\chi \simeq 113°$.

from the terms proportional to $\cos\theta$ in Eq. (6):

$$A_S - A_L \simeq \left[\frac{4\Re\left(i \sin\phi_{SW} e^{i\phi_{SW}}\right)\gamma_K}{\Delta m}\right]\Delta a_0 \ . \tag{7}$$

From the measured value[3] of A_S and a preliminary evaluation of A_L by KLOE, the difference $A_S - A_L = (-2 \pm 10) \times 10^{-3}$, and a first very preliminary evaluation of the Δa_0 parameter can be obtained:

$$\Delta a_0 = (0.4 \pm 1.8) \times 10^{-17} \text{ GeV} \ . \tag{8}$$

With the analysis of the full KLOE data sample ($L = 2.5$ fb^{-1}) an accuracy $\sigma(\Delta a_0) \sim 7 \times 10^{-18}$ GeV could be reached.

An alternative method to measure Δa_0 consists in exploiting the correlation between the two kaons in double semileptonic decays $\phi \to K_S K_L \to \pi^+\ell^-\bar{\nu}, \pi^-\ell^+\nu$ with opposite lepton charges. In this case, with the analysis of 2.5 fb^{-1} an accuracy $\sigma(\Delta a_0) \sim 1 \times 10^{-17}$ GeV could be reached.

The above two methods are largely independent and could be very useful for systematics cross-checks.

5.2. *Measurement of Δa_Z at KLOE*

The Δa_Z parameter can be measured through the A_L asymmetry measured separately for K_L emitted in the forward ($\cos\theta > 0$) and backward ($\cos\theta < 0$) direction; assuming data have been uniformly taken as a function of sidereal time t, thus averaging to zero any possible contribution from the terms proportional to $\cos\Omega t$ and $\sin\Omega t$ in Eq. (6) (otherwise a proper t-dependent analysis has to be performed), one has:

$$\begin{aligned}\Delta A_L &\equiv A_L(\cos\theta > 0) - A_L(\cos\theta < 0) \\ &\simeq -\left[\frac{4\Re\left(i \sin\phi_{SW} e^{i\phi_{SW}}\right)\beta_K\gamma_K \cos\chi \langle\cos\theta\rangle}{\Delta m}\right]\Delta a_Z\end{aligned} \tag{9}$$

where $\langle\cos\theta\rangle$ is a proper average of $\cos\theta$ over the forward (backward) hemisphere. With the analysis of the full KLOE data sample an accuracy $\sigma(\Delta a_Z) \sim 3 \times 10^{-17}$ GeV could be reached.

Also for the measurement of Δa_Z an alternative and independent method exists, based on neutral kaon interferometry with $\phi \to K_S K_L \to \pi^+\pi^-, \pi^+\pi^-$ decays. In this case the intensity $I\left(\pi^+\pi^-(+), \pi^+\pi^-(-); \Delta t\right)$ can be measured, where the two identical final states are distinguished by their forward or backward emission (the symbols $+$ and $-$ represent

$\cos\theta > 0$ and $\cos\theta < 0$, respectively), and the following asymmetry evaluated:

$$A(\Delta t) = \frac{I(\pi^+\pi^-(+),\pi^+\pi^-(-);\Delta t > 0) - I(\pi^+\pi^-(+),\pi^+\pi^-(-);\Delta t < 0)}{I(\pi^+\pi^-(+),\pi^+\pi^-(-);\Delta t > 0) + I(\pi^+\pi^-(+),\pi^+\pi^-(-);\Delta t < 0)}.$$

The above asymmetry for $\Delta t \gg \tau_S$ tends to zero, because ϵ and δ are 90° out of phase (see Ref. 15 for a detailed analysis):

$$A(\Delta t \gg \tau_S) \simeq -2\Re\left(\frac{\delta}{\epsilon}\right) \sim 0 \tag{10}$$

while for $|\Delta t| \leq 5\tau_S$ it is sensitive to $\Im(\delta/\epsilon)$, and therefore to Δa_Z:

$$A(0 < \Delta t \; < 5\tau_S) \propto \Im\left(\frac{\delta}{\epsilon}\right) \simeq \left[\frac{\sin\phi_{SW}\beta_K\gamma_K \cos\chi \, \langle\cos\theta\rangle}{\Delta m|\epsilon|}\right] \Delta a_Z \; . \tag{11}$$

A preliminary analysis based on the same data used for the measurement of the decoherence parameters,[7] yields the first very preliminary result on Δa_Z:

$$\Delta a_Z = (-1 \pm 4) \times 10^{-17} \text{ GeV} \; . \tag{12}$$

With the analysis of 2.5 fb^{-1} an accuracy $\sigma(\Delta a_Z) \sim 2 \times 10^{-17}$ GeV could be reached.

Also in this case the two methods could be used for cross-checks.

5.3. *Measurement of* Δa_X, Δa_Y *at KLOE*

The Δa_X, Δa_Y and Δa_Z parameters can be all simultaneously measured by performing a proper sidereal time dependent analysis of asymmetries (9), (10). An accuracy $\sigma(\Delta a_{X,Y}) = \mathcal{O}(1 \times 10^{-17}$ GeV) could be reached with the analysis of the full KLOE data sample. However there exists a preliminary measurement performed by the KTeV collaboration[16] based on the search of sidereal time variation of the phase ϕ_{+-}, that constrains Δa_X and Δa_Y to less than 9.2×10^{-22} GeV at 90% C.L.

The possibility at KLOE to further constrain the SME parameters using the Bell-Steinberger relation is under study.

6. Conclusions and future plans

Several parameters related to possible CPT and/or QM violations have been measured at KLOE, some of them for the first time, and with a precision reaching the interesting Planck's scale region; with the analysis of the full data sample (2.5 fb^{-1}) KLOE will further improve all results.

All four Δa_μ parameters of the SME could be independently measured at KLOE, complementing and completing results obtained by fixed beam experiments (see also recent results in the B meson system[17]).

Proposals for continuing the KLOE physics program (KLOE-2) at an improved DAΦNE machine, able to deliver an integrated luminosity up to $20 - 50$ fb^{-1}, are being strongly considered. In the meanwhile R&D programs to study possible upgrades of DAΦNE and KLOE have started.

One of the main physics issues of KLOE-2 is the search for CPT violation and/or decoherence effects; limits on several parameters are expected to be improved by about one order of magnitude.

Acknowledgments

I would like to thank Alan Kostelecký for the invitation to this very interesting and successful meeting, and the pleasant stay in Bloomington.

References

1. M. Adinolfi *et al.*, Nucl. Instr. Meth. A **488**, 51 (2002).
2. M. Adinolfi *et al.*, Nucl. Instr. Meth. A **482**, 364 (2002).
3. F. Ambrosino *et al.* (KLOE Collaboration), Phys. Lett. B **636**, 173 (2006).
4. W.-M. Yao *et al.* (Particle Data Group), J. Phys. G **33**, 1 (2006).
5. F. Ambrosino *et al.* (KLOE Collaboration), JHEP **12**, 011 (2006).
6. A. Angelopoulos *et al.* (CPLEAR Collaboration), Eur. Phys. J. C **22**, 55 (2001).
7. F. Ambrosino *et al.* (KLOE Collaboration), Phys. Lett. B **642**, 315 (2006).
8. R. A. Bertlmann *et al.*, Phys. Rev. D **60**, 114032 (1999).
9. A. Go *et al.* (BELLE Collaboration), quant-ph/0702267.
10. J. Bernabeu, N. Mavromatos, J. Papavassiliou, Phys. Rev. Lett. **92**, 131601 (2004).
11. J. Bernabeu *et al.*, Nucl. Phys. B **744**, 180 (2006).
12. V. A. Kostelecký, Phys. Rev. Lett. **80**, 1818 (1998).
13. V. A. Kostelecký, Phys. Rev. D **61**, 016002 (1999).
14. V. A. Kostelecký, Phys. Rev. D **64**, 076001 (2001).
15. C. D. Buchanan *et al.*, Phys. Rev. D **45**, 4088 (1992).
16. H. Nguyen (KTeV Collaboration), in V.A. Kostelecký, ed., *CPT and Lorentz Symmetry II*, World Scientific, Singapore, 2002.
17. D. Stoker (BABAR Collaboration), these proceedings.

SEARCH FOR CPT VIOLATION IN B^0-$\overline{B}^0$ OSCILLATIONS WITH *BABAR*

D.P. STOKER

Department of Physics and Astronomy
University of California at Irvine
Irvine, CA 92697, U.S.A.
E-mail: dpstoker@uci.edu

I describe searches for CPT violation in B^0-$\overline{B}^0$ oscillations using $\Upsilon(4S) \rightarrow B\overline{B}$ decays recorded by the *BABAR* detector at the PEP-II asymmetric-energy B Factory at SLAC. Preliminary results are given for combinations of the quantities Δa_μ in the Lorentz-violating Standard-Model Extension.

1. Introduction

In the general Lorentz-violating Standard-Model Extension (SME),[1] the parameter for CPT violation in neutral meson oscillations depends on the 4-velocity of the meson.[2] We have searched[3] for this effect using $\Upsilon(4S) \rightarrow B\overline{B}$ decays recorded by the *BABAR* detector at the PEP-II asymmetric-energy e^+e^- collider. Any observed CPT asymmetry should vary with a period of one sidereal day ($\simeq 0.99727$ solar-day) as the $\Upsilon(4S)$ boost direction follows the Earth's rotation with respect to the distant stars.[4]

The "light" and "heavy" physical states of the B^0-$\overline{B}^0$ system are

$$\begin{aligned} |B_L\rangle &= p\sqrt{1-\mathsf{z}}|B^0\rangle + q\sqrt{1+\mathsf{z}}|\overline{B}^0\rangle, \\ |B_H\rangle &= p\sqrt{1+\mathsf{z}}|B^0\rangle - q\sqrt{1-\mathsf{z}}|\overline{B}^0\rangle. \end{aligned} \tag{1}$$

The complex parameter z vanishes if CPT is conserved; T invariance implies $|q/p| = 1$. In the $w\xi$ formalism of the SME, $\xi = -\mathsf{z}$ and $w = |q/p|$.

In the SME, flavor-dependent Lorentz and CPT-violating coupling coefficients for the two valence quarks in the B^0 meson are contained in quantities Δa_μ. The CPT parameter z depends on the meson 4-velocity $\beta^\mu = \gamma(1, \vec{\beta})$ in the observer frame as[2]

$$\mathsf{z} \simeq \frac{\beta^\mu \Delta a_\mu}{\Delta m - i\Delta\Gamma/2}, \tag{2}$$

where $\beta^\mu \Delta a_\mu$ is real and varies with sidereal time due to the rotation of $\vec{\beta}$ relative to the constant vector $\Delta\vec{a}$. The magnitude of the decay rate difference $\Delta\Gamma \equiv \Gamma_H - \Gamma_L$ is known to be small compared to the B^0-$\overline{B}^0$ oscillation frequency $\Delta m \equiv m_H - m_L$; hence the SME constrains

$$\Delta m \mathrm{Re}\,\mathrm{z} \simeq 2\Delta m(\Delta m/\Delta\Gamma)\mathrm{Im}\,\mathrm{z} \simeq \beta^\mu \Delta a_\mu. \qquad (3)$$

Analogous Δa_μ apply to oscillations of other neutral mesons. Limits on Δa_μ specific to $K^0\overline{K}^0$ oscillations[5] and to $D^0\overline{D}^0$ oscillations[6] have been reported by the KTeV and FOCUS collaborations, respectively. KTeV has also reported constraints on sidereal variation of the CPT parameter ϕ_{+-}.[7]

We adopt[8] the basis $(\hat{X}, \hat{Y}, \hat{Z})$ for the fixed frame containing $\Delta\vec{a}$ and the basis $(\hat{x}, \hat{y}, \hat{z})$ for the rotating laboratory frame. We take β^μ for each B meson to be the $\Upsilon(4S)$ 4-velocity and choose $\hat{z}$ to lie along $-\vec{\beta}$, so that

$$\beta^\mu \Delta a_\mu = \gamma \left[\Delta a_0 - \beta \Delta a_Z \cos\chi - \beta \sin\chi \left(\Delta a_Y \sin \Omega\hat{t} + \Delta a_X \cos \Omega\hat{t}\right)\right] \qquad (4)$$

where the sidereal time $\hat{t}$ is given by the right ascension of $\hat{z}$ as it precesses around the Earth's rotation axis $(\hat{Z})$ at the sidereal frequency Ω. The latitude (37.4° N) and longitude (122.2° W) of *BABAR* and the $\Upsilon(4S)$ Lorentz boost ($\langle\beta\gamma\rangle \simeq 0.55$ toward 37.8° east of south) yield $\hat{t} = 14.0$ sidereal-hours at the Unix epoch (00:00:00, 1 Jan. 1970) and $\cos\chi = \hat{z} \cdot \hat{Z} = 0.628$.

2. Search using Inclusive Dilepton Events

Neutral B mesons from $\Upsilon(4S)$ decay evolve in orthogonal flavor states until one decays, after which the flavor of the other continues to oscillate. We use *direct* semileptonic decays (where $b \to X\ell\nu$, with $\ell = e$ or μ) to tag the flavor of each $B^0(\overline{B}^0)$ by the charge of its daughter lepton $\ell^+(\ell^-)$. To first order in z, the decay rate for opposite-sign dilepton $(\ell^+\ell^-)$ events is

$$\begin{aligned} N^{+-} \propto e^{-|\Delta t|/\tau_{B^0}} \; & \{\cosh(\Delta\Gamma\Delta t/2) + \cos(\Delta m \Delta t) \\ & -2\,\mathrm{Re}\,\mathrm{z}\sinh(\Delta\Gamma\Delta t/2) + 2\,\mathrm{Im}\,\mathrm{z}\sin(\Delta m\Delta t)\}. \end{aligned} \qquad (5)$$

We define $\Delta t \equiv t^+ - t^-$ to be the difference of the proper times for B meson decays to ℓ^+, ℓ^-. The asymmetry A_{CPT} between the decay rates at $\Delta t > 0$ and $\Delta t < 0$ compares the probabilities $P(B^0 \to B^0)$ and $P(\overline{B}^0 \to \overline{B}^0)$:

$$A_{CPT}(\Delta t) \simeq \frac{-\mathrm{Re}\,\mathrm{z}\,\Delta\Gamma\Delta t + 2\,\mathrm{Im}\,\mathrm{z}\sin(\Delta m\Delta t)}{\cosh(\Delta\Gamma\Delta t/2) + \cos(\Delta m\Delta t)}. \qquad (6)$$

Here we make the approximation $\sinh(\Delta\Gamma\Delta t/2) \simeq \Delta\Gamma\Delta t/2$, which is valid for the range $|\Delta t| < 15\,\mathrm{ps}$ used in this analysis. We use $|\Delta\Gamma| = 6\times10^{-3}\,\mathrm{ps}^{-1}$ in the $\cosh(\Delta\Gamma\Delta t/2)$ term, consistent with the central value in Ref. 9.

The *BABAR* detector is described in detail elsewhere.[10] Any day/night variations in detector response tend to cancel over sidereal time for long data-taking periods. We use about 232 million $\Upsilon(4S) \to B\overline{B}$ decays and $16\,\mathrm{fb}^{-1}$ of off-resonance data, from 40 MeV below the $\Upsilon(4S)$ resonance, collected during 1999–2004 to search for variations in z of the form

$$\mathsf{z} = \mathsf{z}_0 + \mathsf{z}_1 \cos(\Omega t + \phi). \tag{7}$$

Each event's timestamp yields the time elapsed since the Unix epoch. We use this time, folded over one sidereal day, for t in Eq. (7).

The event selection is the same as in Ref. 11. Briefly, we suppress non-$B\overline{B}$ background by event-shape and event-topology requirements, and select events having at least two well-identified lepton candidates with momenta $0.8-2.3\,\mathrm{GeV}/c$ in the $\Upsilon(4S)$ rest frame that are not part of reconstructed $J/\psi, \psi(2S) \to e^+e^-, \mu^+\mu^-$ decays or photon conversions. Lepton candidates must have at least one z-coordinate measurement in the silicon vertex tracker (SVT) to allow Δt to be well-measured. We reject events for which a neural-network algorithm classifies either of the two highest-momentum lepton candidates (the *dilepton*) as a *cascade* lepton from a $b \to (c, \tau) \to \ell$ transition. The selected dilepton sample comprises 1.18 million opposite-sign events and 0.22 million same-sign events.

To measure Δt, we assume each lepton originates from a direct B meson decay at the point on the lepton track with the least transverse distance to the $\Upsilon(4S)$. The component Δz, along the Lorentz boost, of the distance between these two points yields $\Delta t = \Delta z/\langle\beta\gamma\rangle c$. For opposite-sign events $\Delta z = z^+ - z^-$; for same-sign events we take $\Delta z > 0$.

We model the Δt-distribution of the dilepton sample with the probability density functions (PDFs) used in Ref. 11 to represent contributions from $B^0\overline{B}^0$ and B^+B^- decays and non-$B\overline{B}$ events. The latter are estimated, using off-resonance data, to be 3.1% of the sample. The fit to data determines that 59% of the $B\overline{B}$ events are B^+B^- decays. With contributions of minor $B\overline{B}$ backgrounds fixed to values from Monte Carlo (MC) simulation, the fit to data determines the fractions of $B^0\overline{B}^0$ and B^+B^- decays that are *signal* events ($\simeq 80\%$) with two direct leptons, and the fractions that are *opposite B cascade* (*obc*) events with one direct lepton and a $b \to c \to \ell$ decay of the other B meson ($\simeq 10\%$). Same-sign dilepton events are retained to improve the determination of the *signal* and *obc* fractions.

Each PDF is a convolution of a decay rate in Δt with a resolution function that is a sum of Gaussians or, for events with a cascade lepton, its convolution with one or two double-sided exponentials accounting for the lifetimes of intermediate τ or $D_{(s)}$ meson states, respectively. For signal events, the resolution function is determined by the fit to data, with the width of the third (widest) Gaussian fixed at 8 ps. For leptons from different B mesons, we use a $B^0\overline{B}^0$ decay rate that contains z (Eq. (5)) for opposite-sign events and is $\propto e^{-|\Delta t|/\tau_{B^0}}\{\cosh(\Delta\Gamma\Delta t/2) - \cos(\Delta m \Delta t)\}$ for same-sign events; for B^+B^- decays, it is $\propto e^{-|\Delta t|/\tau_{B^\pm}}$. For leptons from the same B meson, the decay rates are exponentials with effective lifetimes determined from MC simulation. Dilution factors are included to account for wrong flavor tags in non-signal events.

We extract z from a maximum likelihood fit to the numbers of opposite-sign and same-sign dilepton events, each binned in Δt and sidereal time t.

Several sources of systematic uncertainty are considered. We vary separately τ_{B^0}, $\tau_{B^\pm}$, and Δm by their known uncertainties,[12] and $|\Delta\Gamma|$ over the range $0-0.1\,\mathrm{ps}^{-1}$. Fixed parameters in the PDF resolution functions for non-signal events are varied separately by 10%, as are fixed $B\overline{B}$ background component fractions. The effects of possible SVT internal misalignments and uncertainty in the absolute z-scale are evaluated in $B^0\overline{B}^0$ MC samples. We use $e^+e^- \to \mu^+\mu^-(\gamma)$ data events, with true $\Delta z = 0$, to check for sidereal variations in measured Δz that could mimic a Lorentz-violation signal and find a negligibly small amplitude of $(0.022 \pm 0.025)\,\mu$m.

Our preliminary results for the CPT violation parameter in Eq. (7) are:

$$\begin{aligned} \mathrm{Im}\,\mathsf{z}_0 &= (-14.1 \pm 7.3(\mathrm{stat.}) \pm 2.4(\mathrm{syst.})) \times 10^{-3}, \\ \mathrm{Re}\,\mathsf{z}_0\,\Delta\Gamma &= (-7.2 \pm 4.1(\mathrm{stat.}) \pm 2.1(\mathrm{syst.})) \times 10^{-3}\mathrm{ps}^{-1}, \\ \mathrm{Im}\,\mathsf{z}_1 &= (-24.0 \pm 10.7(\mathrm{stat.}) \pm 5.9(\mathrm{syst.})) \times 10^{-3}, \\ \mathrm{Re}\,\mathsf{z}_1\,\Delta\Gamma &= (-18.8 \pm 5.5(\mathrm{stat.}) \pm 4.0(\mathrm{syst.})) \times 10^{-3}\mathrm{ps}^{-1}. \end{aligned} \tag{8}$$

The statistical correlation between $\mathrm{Im}\,\mathsf{z}_0$ and $\mathrm{Re}\,\mathsf{z}_0\,\Delta\Gamma$ is 76%; between $\mathrm{Im}\,\mathsf{z}_1$ and $\mathrm{Re}\,\mathsf{z}_1\,\Delta\Gamma$ it is 79%. We note that $\mathsf{z} \to -\mathsf{z}$ for $\phi \to \phi+\pi$ in Eq. (7). The results are compatible with the SME constraint $\mathrm{Re}\,\mathsf{z}\,\Delta\Gamma \simeq 2\Delta m\,\mathrm{Im}\,\mathsf{z}$.

In Fig. 1 we exhibit the sidereal-time dependence of the measured asymmetry A_{CPT}^{meas} for the opposite-sign dilepton events with $|\Delta t| > 3$ ps, thereby omitting highly-populated bins where any asymmetry is predicted to be small. Figure 2 shows confidence level contours for $\mathrm{Im}\,\mathsf{z}_1$ and $\mathrm{Re}\,\mathsf{z}_1\,\Delta\Gamma$. The significance for sidereal variations in z is 2.2σ.

We use Eqs. (3), (4), (8) to extract values for the SME quantities Δa_μ:

$$\Delta a_0 - 0.30\Delta a_Z \approx -(5.2 \pm 4.0)(\Delta m/\Delta\Gamma) \times 10^{-15}\,\mathrm{GeV},$$

$$\sqrt{(\Delta a_X)^2 + (\Delta a_Y)^2} \approx (37 \pm 16)|\Delta m/\Delta\Gamma| \times 10^{-15}\,\mathrm{GeV}.$$

We now use the periodogram method[13] to compare the spectral power

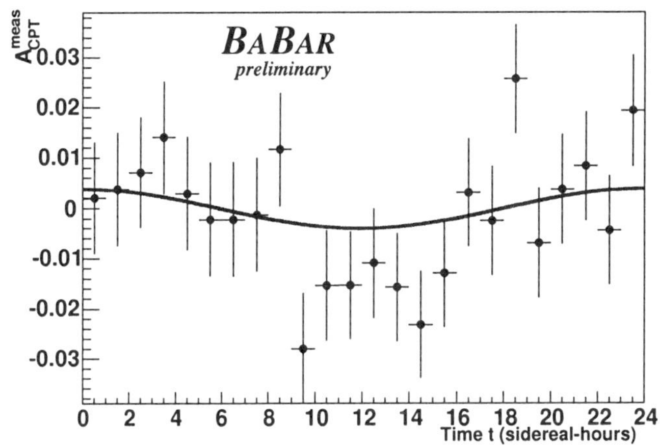

Figure 1. Asymmetry $A_{CPT}^{\rm meas}$ for opposite-sign dilepton events with $|\Delta t| > 3\,\rm ps$ versus sidereal time ($t = 0$ at Unix epoch). The curve is a projection, for $|\Delta t| > 3\,\rm ps$, using results of the two-dimensional likelihood fit for $|\Delta t| < 15\,\rm ps$.

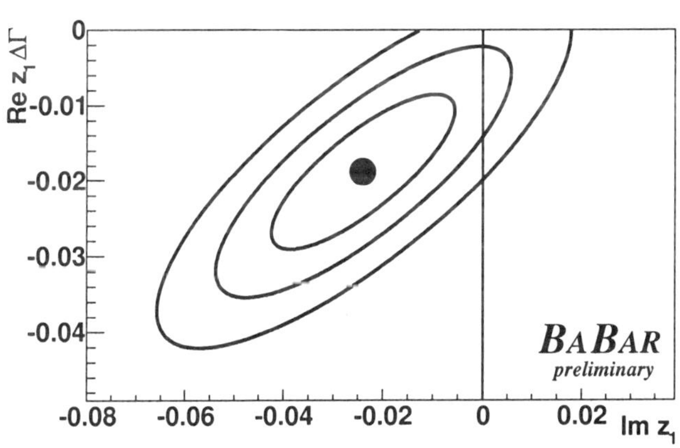

Figure 2. Contours showing 1σ, 2σ, and 3σ significance for $\mathrm{Im}\,\mathbf{z}_1$ and $\mathrm{Re}\,\mathbf{z}_1\,\Delta\Gamma$.

for variations in z at the sidereal frequency with those in a wide band of surrounding frequencies. The spectral power at a test frequency ν is

$$P(\nu) \equiv \frac{1}{N\sigma_w^2}\left|\sum_{j=1}^{N} w_j e^{2i\pi\nu T_j}\right|^2, \tag{9}$$

where N data points measured at times T_j have weights w_j with variance σ_w^2. Here, T_j is the time elapsed since the Unix epoch for opposite-sign dilepton event j. We use weights $w_j \propto \Delta m \Delta t_j - \sin(\Delta m \Delta t_j)$, obtained by applying the SME constraint $\mathrm{Re}\, \mathrm{z}\, \Delta\Gamma \simeq 2\Delta m\, \mathrm{Im}\, \mathrm{z}$ to the numerator of Eq. (6). In the absence of an oscillatory signal, the probability that $P(\nu)$ exceeds a value S at a preselected frequency is $\exp(-S)$; if M independent frequencies are tested, the largest $P(\nu)$ value exceeds S with probability

$$\Pr\left\{P_{\max}(\nu) > S; M\right\} = 1 - \left(1 - e^{-S}\right)^M . \tag{10}$$

We use 20994 test frequencies from 0.26 year^{-1} to 2.1 solar-day^{-1}, separated by 10^{-4} solar-day^{-1}. This oversamples the frequency range by a factor of about 2.2 and avoids underestimating the spectral power of a signal. The number of independent frequencies is about 9500.

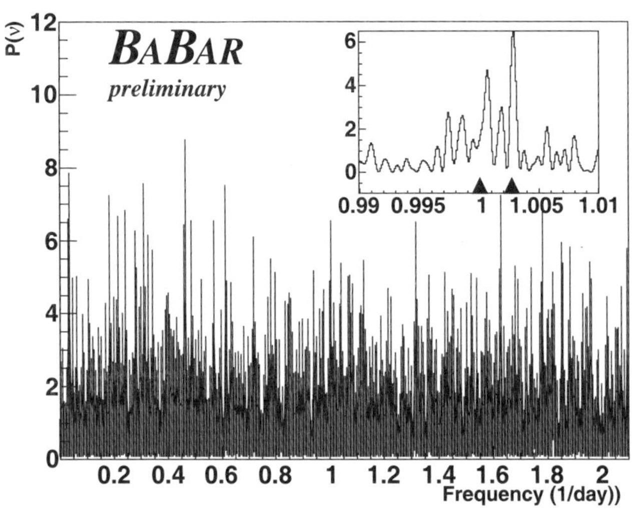

Figure 3. Periodogram for opposite-sign dilepton events. The solar-day and sidereal-day frequencies are marked by triangles in the inset.

Figure 3 shows the largest spectral power we obtain is $P_{\max}(\nu) = 8.80$. With no signal, a larger value is expected with 76% probability. At the sidereal frequency, $P(\nu) = 5.28$ — a value exceeded at 78 test frequencies. The inset of Fig. 3 shows the sidereal frequency lies 1.6 bin-widths below a peak with $P(\nu) = 6.57$. At the solar-day frequency, where any effects due to day/night variations in detector response should appear, $P(\nu) = 1.47$.

Neither the likelihood fit nor the periodogram method detect asymmetries large enough to provide strong evidence for CPT and Lorentz violation.

3. Search using Reconstructed CP and Flavor Eigenstates

A previous search[9] for CPT violation yielded a sidereal-time-integrated measurement of z using about 88 million $\Upsilon(4S) \to B\overline{B}$ decays recorded by *BABAR*. With events in which one of two neutral B mesons from an $\Upsilon(4S)$ decay is fully reconstructed as a CP or flavor eigenstate, we measure

$$\begin{aligned}(\mathrm{Re}\,\lambda_{CP}/|\lambda_{CP}|)\mathrm{Re}\,\mathsf{z} &= 0.014 \pm 0.035(\text{stat.}) \pm 0.034(\text{syst.}),\\ \mathrm{Im}\,\mathsf{z} &= 0.038 \pm 0.029(\text{stat.}) \pm 0.025(\text{syst.}),\end{aligned} \tag{11}$$

where $\lambda_{CP} = (q/p)(\overline{A}_{CP}/A_{CP})$ contains the amplitudes for $\overline{B}^0$ and B^0 decays to the reconstructed CP eigenstate, and $|\lambda_{CP}|$ is of order unity.

Acknowledgments

We are indebted to Alain Milsztajn (deceased) for his help with the periodogram analysis. This contribution was supported by the U.S. Department of Energy, and *BABAR* institutions are supported by the national funding agencies.

References

1. D. Colladay and V.A. Kostelecký, Phys. Rev. D **55**, 6760 (1997); Phys. Rev. D **58**, 116002 (1998); V.A. Kostelecký, Phys. Rev. D **69**, 105009 (2004).
2. V.A. Kostelecký, Phys. Rev. Lett. **80**, 1818 (1998).
3. B. Aubert *et al.* (*BABAR* Collaboration), hep-ex/0607103.
4. V.A. Kostelecký, Phys. Rev. D **64**, 076001 (2001).
5. H. Nguyen (KTeV Collaboration), in V.A. Kostelecký, ed., *CPT and Lorentz Symmetry II*, World Scientific, Singapore, 2002.
6. J.M. Link *et al.* (FOCUS Collaboration), Phys. Lett. B **556**, 7 (2003).
7. Y.B. Hsiung (KTeV Collaboration), Nucl. Phys. B (Proc. Suppl.) **86**, 312 (2000).
8. V.A. Kostelecký and C.D. Lane, Phys. Rev. D **60**, 116010 (1999).
9. B. Aubert *et al.* (*BABAR* Collaboration), Phys. Rev. D **70**, 012007 (2004).
10. B. Aubert *et al.* (*BABAR* Collaboration), Nucl. Instrum. Methods **A479**, 1 (2002).
11. B. Aubert *et al.* (*BABAR* Collaboration), Phys. Rev. Lett. **96**, 251802 (2006).
12. K. Anikeev *et al.* (Heavy Flavor Averaging Group), hep-ex/0505100.
13. N.R. Lomb, Astrophys. Space Sci., **39**, 447 (1976); J.D. Scargle, Astrophys. J, **263**, 835 (1982).

THEORETICAL TOPICS IN SPACETIME-SYMMETRY VIOLATIONS

R. LEHNERT
Center for Theoretical Physics
Massachusetts Institute of Technology
Cambridge, MA 02139, U.S.A.
E-mail: rlehnert@lns.mit.edu

The Lorentz- and CPT-violating Chern–Simons extension of electrodynamics is considered. In the context of $N = 4$ supergravity in four spacetime dimensions, it is argued that cosmological solutions can generate this extension. Within Chern–Simons electrodynamics, theoretical and phenomenological topics are reviewed that concern the number of the remaining spacetime symmetries and the vacuum Cherenkov effect, respectively.

1. Introduction

Spacetime-symmetry investigations owe their present popularity to the idea that Lorentz- and CPT-violation could be a signature from unknown physics possibly arising at the Planck scale. Research in this field may therefore be divided into three broad and overlapping areas: the identification of mechanisms that can generate Lorentz and CPT breakdown in underlying physics, the determination and theoretical study of the ensuing low-energy effects, and the low-energy phenomenology together with the corresponding experimental tests.

During the last two decades, a number of theoretical arguments suggesting the possibility of spacetime-symmetry breaking in underlying physics have been put forward. Examples of such arguments involve string field theory,[1] realistic noncommutative field theories,[2] spacetime-varying fields,[3] various quantum-gravity models,[4] nontrivial spacetime topology,[5] random-dynamics models,[6] multiverses,[7] and brane-world scenarios.[8] Although the underlying dynamics remains Lorentz invariant in most of the above situations, Lorentz and CPT symmetry are nevertheless violated in the ground state at low energies. These ideas provide one of key motivations for Lorentz- and CPT-violation research.

At presently attainable energies, the effects resulting from Lorentz and CPT breakdown in underlying physics are described by the Standard-Model Extension (SME)—an effective-field-theory framework containing the usual Standard Model[9] and general relativity.[10] Various theoretical investigations and consistency analyses have been performed within the context of the SME.[11,12] While some of these studies have clarified conceptual issues, non have suggested any internal inconsistencies.

The SME has also provided the basis for numerous phenomenological and experimental investigations of Lorentz and CPT violation.[13] Specific analyses include, for example, ones with photons,[14,15] neutrinos,[16] electrons,[17] protons and neutrons,[18] mesons,[19] and muons.[20] These studies have placed tight constraints on numerous SME coefficients or combinations of them. Some of the obtained bounds can be considered to probe the Planck scale.

The above remarks demonstrate that all three of the aforementioned areas of Lorentz- and CPT-violation research are active and vibrant fields of scientific inquiry spanning many physics disciplines. This talk aims at illustrating within a specific example—namely the Maxwell–Chern–Simons (MCS) model[21] contained in the SME—how the three subfields are interwoven. Section 2 shows that the Lorentz- and CPT-violating MCS model can arise in underlying Lorentz-invariant physics, more specifically in a low-energy cosmological context of $N = 4$ supergravity. In Sec. 3, some theoretical issues regarding the counting of symmetries in the MCS model are discussed at the level of the SME. A review of vacuum Cherenkov radiation, which is a phenomenological effect occuring in MCS theory, is presented in Sec. 4.

2. Emergence of the MCS model in supergravity cosmology

The discussion in this section is based upon results obtained in the first paper of Ref. 3, which considers pure $N = 4$ supergravity in four spacetime dimensions. Although unrealistic in detail, it is a limit of $N = 1$ supergravity in eleven dimensions, which is contained in M-theory. We may thus expect that the model can nevertheless illuminate generic aspects of a candidate underlying theory.

When only one of the model's graviphotons, $F^{\mu\nu}$, is excited, the bosonic part of our model is given by

$$\kappa\mathcal{L}_{\rm sg} = -\tfrac{1}{2}\sqrt{g}R + \sqrt{g}(\partial_\mu A\partial^\mu A + \partial_\mu B\partial^\mu B)/4B^2$$
$$-\tfrac{1}{4}\kappa\sqrt{g}MF_{\mu\nu}F^{\mu\nu} - \tfrac{1}{4}\kappa\sqrt{g}NF_{\mu\nu}\tilde{F}^{\mu\nu}\ , \qquad (1)$$

where M and N are known functions of the scalars A and B, $g = -\det(g_{\mu\nu})$, and $\tilde{F}^{\mu\nu} = \varepsilon^{\mu\nu\rho\sigma} F_{\rho\sigma}/2$. We can rescale $F^{\mu\nu} \to F^{\mu\nu}/\sqrt{\kappa}$ removing the explicit appearance of the gravitational coupling κ from the equations of motion. We represent the model's fermions by the energy–momentum tensor of dust $T_{\mu\nu} = \rho\, u_\mu u_\nu$ modeling, e.g., galaxies. Here, u^μ is a unit timelike vector and ρ is the fermionic energy density. At tree level, the fermionic matter is uncoupled from the scalars, so that $T_{\mu\nu}$ is conserved separately.

With the phenomenological input of an isotropic homogeneous flat Friedmann–Robertson–Walker Universe, we can take $F^{\mu\nu} = 0$ on large scales. Our cosmology then obeys the Einstein equations and the equations of motion for the scalars A and B. These equations can be solved analytically yielding a nontrivial dependence of $A = A_\mathrm{b}(t)$ and $B = B_\mathrm{b}(t)$ on the comoving time t. Consider now small localized excitations of $F_{\mu\nu}$ in the scalar background A_b and B_b. The effective Lagrangian $\mathcal{L}_\mathrm{cosm}$ for such situations in local inertial coordinates follows from Eq. (1) and is

$$\mathcal{L}_\mathrm{cosm} = -\tfrac{1}{4} M_\mathrm{b} F_{\mu\nu} F^{\mu\nu} - \tfrac{1}{4} N_\mathrm{b} F_{\mu\nu} \tilde{F}^{\mu\nu} \,, \tag{2}$$

where $A_\mathrm{b}(t)$ and $B_\mathrm{b}(t)$ imply the time dependence of M_b and N_b. Comparison with the usual Maxwell Lagrangian $\mathcal{L}_\mathrm{em} = -\frac{1}{4e^2} F_{\mu\nu} F^{\mu\nu} - \frac{\theta}{16\pi^2} F_{\mu\nu} \tilde{F}^{\mu\nu}$ establishes that $e^2 \equiv 1/M_\mathrm{b}$ and $\theta \equiv 4\pi^2 N_\mathrm{b}$. Thus, e and θ acquire time dependencies via the varying background A_b and B_b.

The time dependence of e is an interesting topic in itself, but in the present context the goal is to obtain the Lorentz- and CPT-violating Chern–Simons term $(k_{AF})^\mu A^\nu \tilde{F}_{\mu\nu}$ contained in the SME. This can be achieved at the level of the action via an integration by parts of the θ-angle term. This establishes the desired result

$$\mathcal{L}_\mathrm{cosm} \supset \tfrac{1}{2} (\partial_\mu N_\mathrm{b})\, A_\nu \tilde{F}^{\mu\nu} \,. \tag{3}$$

It is thus apparent that, starting from a Lorentz-invariant model, our supergravity cosmology has indeed generated one particular SME operator.

3. Symmetry counting in the MCS model

In addition to the usual ten Poincaré invariances (four translations, three rotations, and three boosts), conventional electrodynamics possesses five further spacetime symmetries: one dilatation and four special conformal transformations. The inclusion of our Chern–Simons term $(k_{AF})^\alpha A^\beta \tilde{F}_{\alpha\beta}$ preserves translation invariance, since $(k_{AF})^\mu$ is assumed to be constant in the minimal SME. However, $(k_{AF})^\mu$ has mass dimensions suggesting that dilatation and conformal symmetry are violated. One further expects the Lorentz group to be broken down to the appropriate three-dimensional little group associated with $(k_{AF})^\mu$. This suggests that the MCS model maintains seven spacetime symmetries—four translations and three (of the original six) Lorentz invariances. This section argues that there is, in fact, one additional spacetime symmetry if $(k_{AF})^\mu$ is lightlike: a combination of a boost and a dilatation.

We begin by streamlining our notation $(k_{AF})^\mu \to k^\mu = (k, \vec{k})$ and by recalling that a dilatation, also called a scale transformation, takes $A^\mu(x) \to e^\rho A^\mu(e^\rho x)$, where the size of the dilatation is determined by the parameter ρ. A dilatation therefore takes

$$\int d^4x\, \mathcal{L}_{\rm MCS} \to \int d^4x\, (\mathcal{L}_{\rm M} + e^{-\rho}\, \mathcal{L}_{\rm CS}) \neq \int d^4x\, \mathcal{L}_{\rm MCS}\,, \tag{4}$$

where a suitable change of integration variables is understood. It is apparent that the conventional piece $\mathcal{L}_{\rm M}$ and the Chern–Simons extension $\mathcal{L}_{\rm CS}$ transform differently. Moreover, the difference between the original and the transformed actions is not a boundary term establishing the non-invariance of $\mathcal{L}_{\rm MCS}$ under dilatations.

We next look at Lorentz transformations, which can be implemented via $\Lambda^\mu{}_\nu(\vec{\theta}, \vec{\beta})$. Here, $\vec{\theta}$ and $\vec{\beta}$ characterize rotations and boosts, respectively. Under such transformations, the MCS Lagrangian changes according to $\mathcal{L}_{\rm MCS} = \mathcal{L}_{\rm M} + \mathcal{L}_{\rm CS} \to \mathcal{L}_{\rm M} + \Lambda^\mu{}_\gamma(-\vec{\theta}, -\vec{\beta})\, k^\gamma\, A^\nu\, \tilde{F}_{\mu\nu}$. We have again suppressed the the dummy integration variables x. Next, we consider a special boost: $\vec{\beta} = \beta\hat{k}$, where $\hat{k} \equiv \vec{k}/|\vec{k}|$. Such a boost scales k^μ by a factor of e^β, so that $\Lambda^\mu{}_\gamma(\vec{0}, -\beta\hat{k})\, k^\gamma\, A^\nu\, \tilde{F}_{\mu\nu} = \exp(\beta)\, k^\mu\, A^\nu\, \tilde{F}_{\mu\nu} \neq \mathcal{L}_{\rm CS}$ yielding

$$\mathcal{L}_{\rm MCS} \to \mathcal{L}_{\rm M} + e^\beta\, \mathcal{L}_{\rm CS} \neq \mathcal{L}\,. \tag{5}$$

This shows that symmetry under boosts along $\hat{k}$ is broken, as expected.

Although each individual transformation (4) and (5) no longer determines a symmetry, the specific form of these transformations shows that a dilatation *combined* with a suitable boost along the spatial direction of a lightlike k^μ remains an invariance of $\mathcal{L}_{\rm MCS}$. We can verify this explicitly by studying the currents

$$D^\mu \equiv \theta^{\mu\nu} x_\nu \tag{6}$$

and

$$J^\mu_{\alpha\beta} \equiv \theta^\mu_\alpha x_\beta - \theta^\mu_\beta x_\alpha\,. \tag{7}$$

These are the usual dilatation and Lorentz currents, respectivly. To extract from Eq. (7) the current corresponding to a boost along $\hat{k}$, we decompose

k^μ into its purely timelike and its purely spacelike part $k^\mu = k\,(k_T^\mu + k_S^\mu)$, where $k_T^\mu = (1, \vec{0})$ and $k_S^\mu = (0, \hat{k})$. The desired current component is then given by $J^\mu_{\alpha\beta}\, k_S^\alpha\, k_T^\beta$. The divergences of these currents satisfy

$$\partial_\mu D^\mu = -\mathcal{L}_{\mathrm{CS}} \tag{8}$$

and

$$\partial_\mu J^\mu_{\alpha\beta}\, k_S^\alpha\, k_T^\beta = +\mathcal{L}_{\mathrm{CS}}\,. \tag{9}$$

It again becomes clear that D^μ and $J^\mu_{\alpha\beta}\, k_S^\alpha\, k_T^\beta$ are not conserved individually. However, their sum $Q^\mu \equiv D^\mu + J^\mu_{\alpha\beta}\, k_S^\alpha\, k_T^\beta$ is, in fact, conserved. An explicit gauge-invariant expression for Q^μ can be obtained:[12]

$$Q^\mu = \left[\tfrac{1}{4}\,\eta^\mu_\nu F^2 + F^{\mu\alpha} F_{\alpha\nu}\right]\left[x^\nu + (k_T\!\cdot\! x)\, k_S^\nu - (k_S\!\cdot\! x)\, k_T^\nu\right]. \tag{10}$$

We can thus see that in the lightlike MCS model an additional conserved current relative to the spacelike and timelike cases exists. With Killing-vector techniques, one can show that this is the only additional conservation law in the present context.[12] This extended symmetry structure is described by the Lie algebra sim(2).[12]

4. Vacuum Cherenkov radiation in the MCS model

The Lorentz- and CPT-violating SME coefficients act in many respects like a background. This analogy is particularly well suited for the electrodynamics sector of the SME because this sector exhibits many parallels to the conventional Maxwell theory in macroscopic media. It is therefore natural to ask as to whether such analogies can be exploited to identify possible phenomenological Lorentz- and CPT-breaking effects in the SME that can be employed for tests.

One conventional effect inside a macroscopic medium is that the phase speed of light c_{ph} can be slowed down relative to the vacuum $c > c_{ph}$. It then becomes possible for ordinary charges q to travel faster than light inside this medium:

$$v_q > c_{ph} \equiv \frac{\omega}{|\vec{p}|}. \tag{11}$$

Here, v_q is the charge's speed, ω the photon frequency, and $\vec{p}$ the photon wave vector. It turns out this configuration is unstable in the sense that these fast charges are decelerated rapidly through the emission of photons. This well established effect is called Cherenkov radiation. It can be employed, for example, in modern particle detectors.

Can the Cherenkov effect also occur in our MCS model? This is indeed the case, which can be seen as follows. One can verify[15] that Condition (11) continues to hold in the presence of Lorentz and CPT violation. It follows that the MCS plane-wave dispersion relation $\omega = \omega(\vec{p})$ must be investigated. This dispersion relation is determined by

$$p^4 + 4\,k^2 p^2 - 4(k \cdot p)^2 = 0\,, \tag{12}$$

where $p^\mu = (\omega, \vec{p})$. For a given $\vec{p}$, Eq. (12) determines a quartic equation in ω, so that there are four branches of solutions. Two of these branches lie inside the momentum-space lightcone where $\omega(\vec{p}) > \vec{p}$, which is inconsistent with Requirement (11). However, the other two branches are located outside the lightcone where $\omega(\vec{p}) < \vec{p}$ and the Cherenkov condition (11) is satisfied. We conclude that the Cherenkov effect can indeed occur in vacuum within the context of the MCS model.

An important criterion for experimental Lorentz tests is the rate for vacuum Cherenkov radiation. Within the MCS model, a classical calculation treating the charge as an external source yields[15]

$$\dot{P}^\mu = -\mathrm{sgn}(k_0)\,\frac{q^2}{4\pi}\frac{k_0^4}{\vec{k}^2}\,(0, \hat{k}) \tag{13}$$

in the charge's rest frame. Here, $\dot{P}^\mu$ is the rate of four-momentum radiation and q the particle's charge. It is apparent that the rate is suppressed: it is second order in the SME coefficient k^μ. In the MCS model, vacuum Cherenkov radiation is therefore phenomenologically less interesting.

References

1. See, e.g., V.A. Kostelecký and S. Samuel, Phys. Rev. D **39**, 683 (1989); V.A. Kostelecký and R. Potting, Nucl. Phys. B **359**, 545 (1991); B. Altschul and V.A. Kostelecký, Phys. Lett. B **628**, 106 (2005).
2. See, e.g., S.M. Carroll *et al.*, Phys. Rev. Lett. **87**, 141601 (2001); Z. Guralnik *et al.*, Phys. Lett. B **517**, 450 (2001); C.E. Carlson *et al.*, Phys. Lett. B **518**, 201 (2001); A. Anisimov *et al.*, Phys. Rev. D **65**, 085032 (2002).
3. V.A. Kostelecký *et al.*, Phys. Rev. D **68**, 123511 (2003); R. Jackiw and S.-Y. Pi, Phys. Rev. D **68**, 104012 (2003); O. Bertolami *et al.*, Phys. Rev. D **69**, 083513 (2004); N. Arkani-Hamed *et al.*, JHEP **0507**, 029 (2005).
4. J. Alfaro *et al.*, Phys. Rev. D **66**, 124006 (2002); D. Sudarsky *et al.*, Phys. Rev. Lett. **89**, 231301 (2002); G. Amelino-Camelia, Mod. Phys. Lett. A **17**, 899 (2002); R.C. Myers and M. Pospelov, Phys. Rev. Lett. **90**, 211601 (2003); N.E. Mavromatos, Lect. Notes Phys. **669**, 245 (2005).
5. F.R. Klinkhamer, Nucl. Phys. B **578**, 277 (2000); F.R. Klinkhamer and J. Schimmel, Nucl. Phys. B **639**, 241 (2002).

6. C.D. Froggatt and H.B. Nielsen, arXiv:hep-ph/0211106.
7. J.D. Bjorken, Phys. Rev. D **67**, 043508 (2003).
8. C.P. Burgess *et al.*, JHEP **0203**, 043 (2002); A.R. Frey, JHEP **0304**, 012 (2003); J. Cline and L. Valcárcel, JHEP **0403**, 032 (2004).
9. D. Colladay and V.A. Kostelecký, Phys. Rev. D **55**, 6760 (1997); Phys. Rev. D **58**, 116002 (1998); V.A. Kostelecký and R. Lehnert, Phys. Rev. D **63**, 065008 (2001).
10. V.A. Kostelecký, Phys. Rev. D **69**, 105009 (2004); R. Bluhm and V.A. Kostelecký, Phys. Rev. D **71**, 065008 (2005).
11. See, e.g., R. Jackiw and V.A. Kostelecký, Phys. Rev. Lett. **82**, 3572 (1999); V.A. Kostelecký *et al.*, Phys. Rev. D **65**, 056006 (2002); R. Lehnert, Phys. Rev. D **68**, 085003 (2003); J. Math. Phys. **45**, 3399 (2004); Phys. Rev. D **74**, 125001 (2006); V.A. Kostelecký and R. Potting, Gen. Rel. Grav. **37**, 1675 (2005); B. Altschul, Phys. Rev. D **73**, 045004 (2006); J. Phys. A **39**, 13757 (2006).
12. A.J. Hariton and R. Lehnert, Phys. Lett. A **367**, 11 (2007).
13. See, e.g., these proceedings; V.A. Kostelecký, ed., *CPT and Lorentz Symmetry III*, World Scientific, Singapore, 2005; D. Mattingly, Living Rev. Rel. **8**, 5 (2005).
14. See, e.g., J. Lipa *et al.*, Phys. Rev. Lett. **90**, 060403 (2003); P.L. Stanwix *et al.*, Phys. Rev. D **74**, 081101 (2006); M.E. Tobar *et al.*, Phys. Rev. D **71**, 025004 (2005); M. Hohensee *et al.*, Phys. Rev. D **75**, 049902 (2007); J.P. Cotter and B. Varcoe, physics/0603111; H. Müller *et al.*, arXiv:0706.2031.
15. See, e.g., R. Lehnert and R. Potting, Phys. Rev. Lett. **93**, 110402 (2004); Phys. Rev. D **70**, 125010 (2004); C. Kaufhold and F.R. Klinkhamer, Nucl. Phys. B **734**, 1 (2006); B. Altschul, Phys. Rev. Lett. **98**, 041603 (2007).
16. See, e.g., T. Katori and R. Tayloe, in Ref. 13; LSND Collaboration, L.B. Auerbach *et al.*, Phys. Rev. D **72**, 076004 (2005); V.A. Kostelecký and M. Mewes, Phys. Rev. D **69**, 016005 (2004); V. Barger *et al.*, arXiv:0706.1085.
17. See, e.g., H. Dehmelt *et al.*, Phys. Rev. Lett. **83**, 4694 (1999); G. Gabrielse *et al.*, Phys. Rev. Lett. **82**, 3198 (1999); L.-S. Hou, W.-T. Ni, and Y.-C.M. Li, Phys. Rev. Lett. **90**, 201101 (2003); H. Müller, Phys. Rev. D **71**, 045004 (2005).
18. See, e.g., D.F. Phillips *et al.*, Phys. Rev. D **63**, 111101(R) (2001); P. Wolf *et al.*, Phys. Rev. Lett. **96**, 060801 (2006); O. Bertolami *et al.*, Phys. Lett. B **395**, 178 (1997).
19. See, e.g., BELLE Collaboration, K. Abe *et al.*, Phys. Rev. Lett. **86**, 3228 (2001); BaBar Collaboration, B. Aubert *et al.*, Phys. Rev. Lett. **92**, 181801 (2004); arXiv:hep-ex/0607103; FOCUS Collaboration, J.M. Link *et al.*, Phys. Lett. B **556**, 7 (2003).
20. V.W. Hughes *et al.*, Phys. Rev. Lett. **87**, 111804 (2001); R. Bluhm *et al.*, Phys. Rev. Lett. **84**, 1098 (2000).
21. S.M. Carroll *et al.*, Phys. Rev. D **41**, 1231 (1990).

A SECOND-GENERATION CO-MAGNETOMETER FOR TESTING FUNDAMENTAL SYMMETRIES

S. J. SMULLIN, T. W. KORNACK, G. VASILAKIS, M. V. ROMALIS

Department of Physics, Princeton University, Princeton, NJ 08544, U.S.A.
E-mail: ssmullin@princeton.edu

We present the design and initial characterization of a second-generation co-magnetometer used to search for fields that violate CPT symmetry. This second generation experiment is designed to be quiet and robust at time scales $\sim$ 100 s. Additionally, the experiment can be rotated, allowing for a cycle time of much less than a day. These improvements are expected to substantially increase sensitivity to CPT-violating couplings.

1. Introduction

Like an atomic magnetometer, an atomic co-magnetometer is based on a measurement of the polarization of optically-pumped K. A co-magnetometer includes an additional polarizable species, ^{3}He polarized via spin-exchange collisions with K. When properly tuned, coupling between the ^{3}He spins and K spins makes the system insensitive to magnetic fields and highly sensitive to CPT-violating fields that couple to spins with a strength that is not proportional to the magnetic moment of a spin.[1]

The co-magnetometer is sensitive only to CPT-violating fields that are perpendicular to the two laser beams used to pump (polarize) and probe (measure the polarization of) K. To search for a CPT-violating field using an experiment fixed in the laboratory frame, we look for signal with a period of one sidereal day. The results from CPT-I, presented in this volume,[2] were limited by systematics at this time scale. The basic principles of the co-magnetometer and design of CPT-I are described in detail elsewhere.[3,4] Here, we describe the experimental innovations and initial characterization of the second generation experiment, CPT-II, which is built to be rotated in the laboratory frame and to have low noise on the time scale of its rotation.

The essential components of CPT-II are shown in Fig. 1. This co-magnetometer is sensitive to $\tilde{b}$ fields in the Standard-Model Extension;[5] because the K spin is primarily carried by the electron and ^{3}He has a nuclear

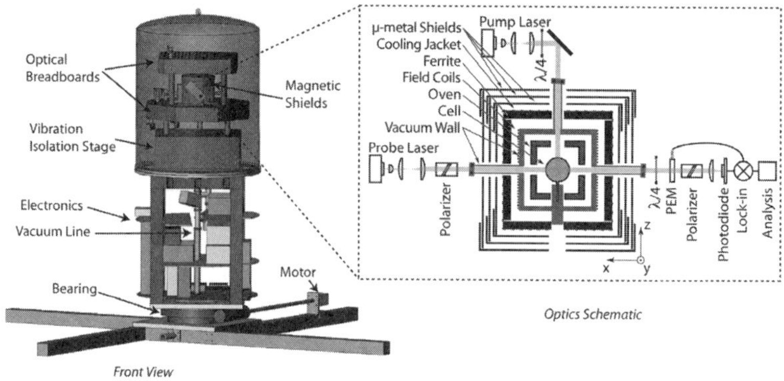

Figure 1. Drawing of the CPT-II experiment (front view) and schematic of the optics, oven, and inner vacuum space. The co-magnetometer's sensitive axis is the y direction. In CPT-II, the z direction is normal to the floor. For clarity, the bell jar is shown as transparent. The passive vibration isolation stage is made by MinusK.

spin, the co-magnetometer is sensitive to fields in the proton, neutron, and electron sectors. By rotating the experiment in the laboratory frame, as described here, we can put bounds on $\tilde{b}$ in all three celestial directions.

2. Thermal Effects and Beam Motion

Measurements from CPT-I were limited by several known sources of low-frequency noise,[3] making the minimum of the Allan variance typically less than 10 s. This limitation was presumed to be primarily due to thermal effects, including flexing of the breadboard due to differential thermal expansion and beam motion due to convection currents in the air.

Reducing these thermal effects are critical because the co-magnetometer is sensitive, to first-order, to beam motion. When all magnetic fields and lightshifts in the beam directions are properly zeroed, as detailed in Ref. 3, the leading-order terms in the co-magnetometer signal S are:

$$S = \frac{P_z^e \gamma_e}{R_{tot}} (\tilde{b}_y^n - \tilde{b}_y^e + L_y + \frac{\Omega_y}{\gamma_n} + \frac{s_m R_m + \alpha R_p}{\gamma_e P_z^e}). \tag{1}$$

Here, the constant of proportionality is a function of the K polarization P_z^e, gyromagnetic ratio γ_e, and relaxation rate R_{tot}. The subscript y refers to the direction perpendicular to the pump and probe beams; $\tilde{b}_y^n$ and $\tilde{b}_y^e$ are CPT-violating fields in the y direction, experienced by the ^{3}He nucleus and by the K electron, respectively, and L_y is the lightshift. The co-magnetometer is sensitive to rotation at angular velocity Ω_y about the

y direction, and γ_n is the gyromagnetic ratio of the ^{3}He nuclear spin.

The final two terms in Eq. 1 are related to beam position. The co-magnetometer is sensitive to the lack of perpendicularity α between the two laser beams. This sensitivity depends on the pumping rate R_p of the pump beam. Beam motion can also cause a varying signal if the probe beam has a circular polarization s_m that varies with beam position. Birefringence of the glass windows and glass cell in the beam path can add some circular polarization to the nominally linearly-polarized probe beam. This birefringence can be cancelled by putting in the optical path a glass plate stressed by a piezoelectric stack. However, because birefringence varies across the glass surfaces, beam motion can cause noise proportional to the probe beam pumping rate R_m.

Beam motion can also create polarimetry noise, which then is read as noise on the co-magnetometer. Dichroism in the glass vapor cell causes a signal that looks the same as rotation of the K spins; because dichroism varies across the cell, beam motion can cause noise on the signal. In CPT-I, this sensitivity required laborious steering of the probe beam to the "sweet spot" of the vapor cell.[3]

3. CPT-II Design and Characterization

CPT-II is designed to have low noise at frequencies < 1 Hz and to increase the cycle frequency of the experiment to take advantage of the stability at shorter time scales. CPT-II is smaller than CPT-I; shorter optical path lengths reduce beam motion due to flexing of the optical breadboards. A millitorr vacuum space surrounds the oven and the beam path inside the magnetic shields, eliminating convection noise in this high temperature measurement region. To further reduce convection noise, a bell jar surrounding the optical breadboards allows for evacuation of air in the entire optical path. The experiment is mounted on a rotation stage to allow for cycling at a period of ~ 100 s ($\ll 1$ sidereal day); this cycling can improve statistics, reduce noise, and allow for better determination and removal of systematics. Improved magnetic shielding, featuring an inner shield made of ferrite, increases the shielding factor and reduces magnetic noise, making the system an order of magnitude more sensitive when run as a normal magnetometer.[6] We aim for the co-magnetometer to be limited by noise from the probe beam alone (polarimetry noise) and this in turn is ideally limited by photon shot noise. Reducing beam position noise is an important step.

3.1. *Optics*

The basic optics of CPT-II, shown in Fig. 1, are the same as in CPT-I, with important differences chosen for stability, sensitivity, and size. We use distributed feedback lasers, cooled to reach the desired wavelength. Lasers are mounted in homemade mounts designed for thermal stability. A photoelectric modulator (PEM) is used to make a highly sensitive polarimetry measurement at a frequency of $\sim$ 50 kHz; this was found to provide a quieter polarimetry measurement than a balanced polarimeter or a Faraday modulator. Additionally, the PEM is compact and generates no heat. Currently, we have not implemented optics for probe beam steering and position and power feedback on the lasers.

3.2. *Oven*

Enclosing the oven in a vacuum space not only eliminates noise due to air convection; it also eliminates vibration that can be caused by forced air heating. For heating, twisted pairs of high resistance wire are embedded in panels of the boron nitride oven. The heaters are driven at frequencies $>$ 20 kHz, much larger than the Larmor frequencies of the atomic species in the cell. Achievement of high and stable ^{3}He polarization indicates that the heater wire is not dramatically perturbing the magnetic environment. To thermally insulate the oven, the inside of the G10 coil form that forms the vacuum space is painted with low emissivity paint from Solec.

3.3. *Bell Jar*

A stainless steel bell jar, of volume $\sim$ 700 L, fits over the optical breadboards and the vibration isolation stage, sealing with an o-ring on an aluminum baseplate that contains feedthroughs for electrical signals and water cooling lines. Simply enclosing the experiment in the bell jar reduces air convection. Evacuating air from the bell jar eliminates convection noise. To improve thermal control of the system, the bell jar is backfilled with He gas. Compared to vacuum, the index of refraction of He $n_{He} - 1$ is an order of magnitude less than that of air; convection noise is not a concern. Additionally, helium's high thermal conductivity ($\sim$ 6 times that of air) keeps the system at thermal equilibrium. The effectiveness of the bell jar and He is shown in Fig. 2. The probe beam position noise, measured with a four quadrant photodiode, is reduced orders of magnitude. This translates directly into a reduction in polarimetry and co-magnetometer noise.

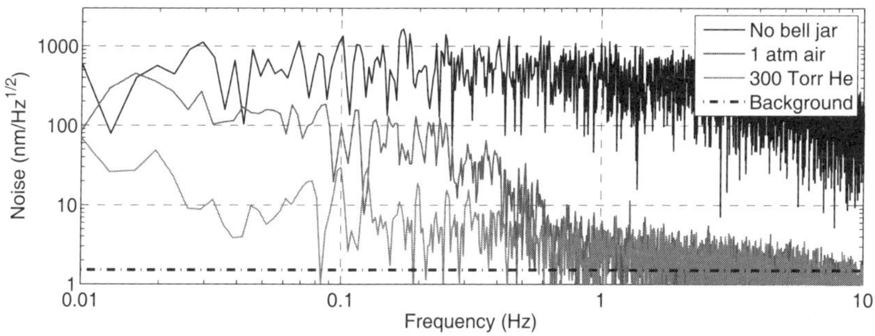

Figure 2. Position noise of the probe beam, measured ~ 0.76 m from the laser, before the beam enters the magnetic shields. Noise that exists when there is no bell jar (top spectrum) is reduced when the bell jar is in place and filled with air (middle spectrum). Noise is reduced even more when air is removed from the bell jar and replaced with 300 Torr He. Detector calibration has uncertainty on the order of 20%. The background line shows the typical measured electrical noise limit of the detector.

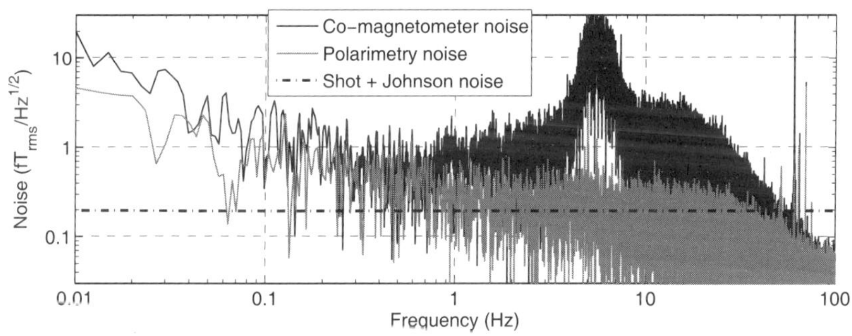

Figure 3. Co-magnetomer signal spectrum, with the bell jar filled with 300 Torr of He and with magnetic fields zeroed. Signal is measured from a lockin at the PEM frequency with a time constant of 3 ms. A line shows the calculated shot and Johnson noise of the photodiode measurement, which corresponds to 18 nrad/Hz$^{1/2}$ on a polarimetry measurement or 0.2 fT/Hz$^{1/2}$ of co-magnetometer noise. Polarimetry noise is recorded with the pump beam blocked and applied field far above the compensation point. The broad peak near 6 Hz on the co-magnetometer is due to the ^{3}He resonance. Additional low-frequency co-magnetometer noise is most likely due to drift of ^{3}He polarization.

The noise spectra of the polarimetry measurement and co-magnetometer signal with He in the bell jar are shown in Fig. 3. Note that the noise is given in magnetic field units; our goal is to measure 1 fT/Hz$^{1/2}$ at 0.01 Hz. Conversion from magnetic field to energy units is detailed elsewhere.[2,3]

3.4. *Rotation*

To exploit the stability of the system at time scales of 1–100 s, we want to rotate the experiment roughly every 100 s. We will record data, rotate the experiment and stop, wait for the experiment to settle, and record data again. In initial tests, the ringdown time of the system (a combination of the time constant of the vibration isolation stage and the ringdown of the magnetometer's gyroscopic signal due shaking during rotation) was on the order of 5 s. During rotation, there was no indication of probe beam motion, which shows that while the entire experiment may shake on the vibration isolation stage, the optical assembly is mechanically rigid and robust.

Since the sensitive y axis of the system is parallel to the floor, the co-magnetometer signal varies with the experiment's angular position, due to the component of the earth's rotation in the y direction. Converting the rotation signal into an equivalent magnetic field signal, as the system rotates there will be a sinusoidally varying signal of magnitude $\sim$ 270 fT, much larger that the 0.1 fT sensitivity we hope to have in our search for anomalous fields. Additionally upon rotation of the experiment, the direction of earth's magnetic field with respect to the experiment changes; even with the 10^8 shielding factor of the magnetic shields, this would change the zeroing of the magnetic fields after rotation, making it hard to maintain stability. External Helmholtz coils fixed in the laboratory frame will be used to cancel the local magnetic field at the experiment.

4. Upcoming Challenges

The laser mounts, the magnetic shields, the vibration isolation system, and the bell jar work as designed, achieving low noise at sub-Hz frequencies. Making the rotation of the system robust, repeatable, and calibrated is the large remaining challenge for this experiment. Additional damping of breadboard motion during rotation will reduce the settling time of the system after rotating and increase the repeatability of the rotation. This could be accomplished using eddy-current damping driven via feedback from position sensors.

To separate a CPT-violating signal of 0.1 fT from earth's rotation signal of 270 fT, we need great precision and accuracy in our measurement. We define the angular position of the experiment by the direction of the sensitive axis. If we compare measurements with the experiment's sensitive axis pointing directly east and west, earth's rotation will create no signal; in this case, we need to know very well the position of the experiment. With viewports added to the bell jar, this position measurement can be accomplished with laser pointers mounted on the breadboard and photodiode arrays outside the bell jar. If instead we rotate between north and south positions, the signal due to earth's rotation will vary as cosine of the angular position and thus the position need not be well known; in this case, the calibration of the experiment must be well known so that the large signal from earth's rotation can be subtracted. It is expected that the features that make this experiment low noise will also keep the calibration stable. The final signal will be analyzed by looking for a signal that varies with the rotation of the experiment and with the sidereal day. These two different rotations will allow us to set bounds on anomalous fields in all three celestial dimensions.

In sum, smaller, more robust mechanical design and evacuation of the air in the optical path reduces noise at time scales ~ 100 s in CPT-II. Rotation of the experiment will suppress long time scale systematics, potentially rendering CPT-II many orders of magnitude more sensitive to sidereal $\tilde{b}$ variation than CPT-I.

Acknowledgments

This works is supported by NSF and DARPA.

References

1. T. W. Kornack and M. V. Romalis, Phys. Rev. Lett. **89** 253002 (2002).
2. T. W. Kornack, G. Vasilakis, and M. V. Romalis, these proceedings.
3. T. W. Kornack, Ph.D. Thesis, Princeton University, 2005.
4. T. W. Kornack and M. V. Romalis, in V.A. Kostelecký, ed., *CPT and Lorentz Symmetry III*, World Scientific, Singapore, 2005.
5. D. Colladay and V.A. Kostelecký, Phys. Rev. D **58**, 116002 (1998); V.A. Kostelecký and C. D. Lane, Phys. Rev. D **560**, 1160010 (1999).
6. T. W. Kornack, S. J. Smullin, S.-K. Lee, and M. V. Romalis, Appl. Phys. Lett. **90**, 223501 (2007).

NAMBU-GOLDSTONE AND MASSIVE MODES IN GRAVITATIONAL THEORIES WITH SPONTANEOUS LORENTZ BREAKING

R. BLUHM

Physics Department
Colby College
Waterville, ME 04901, U.S.A.
E-mail: rtbluhm@colby.edu

Spontaneous breaking of local Lorentz symmetry is of interest as a possible mechanism originating from physics at the Planck scale. If such breaking occurs, however, it raises the questions of what the fate of the Nambu-Goldstone modes is, whether a Higgs mechanism can occur, and whether additional massive modes (analogous to the Higgs particle) can appear. A summary of some recent work looking at these questions is presented here.

1. Introduction

The idea that Lorentz symmetry might be spontaneously broken began to catch on when it was shown that mechanisms in string theory might lead to this form of symmetry breaking.[1] Since then, spontaneous Lorentz breaking has been examined in its own right in a number of contexts, including investigating its phenomenological effects and its effects on gravity. However, as soon as a theory allows spontaneous breaking of a symmetry, well-known consequences from particle physics must be considered and addressed. The first is the Goldstone theorem, which states that when a continuous symmetry is spontaneously broken, massless Nambu-Goldstone (NG) modes appear. The second is the possibility of a Higgs mechanism, resulting in massive gauge fields, for the case when the symmetry is local. The third is the possibility that additional massive modes might appear (analogous to the Higgs boson in the case of the electroweak model). Clearly, all three of these can have physical implications and must be accounted for in any theory with spontaneous symmetry breaking.

In this work, these processes are examined for the case where it is Lorentz symmetry that is spontaneously broken.[1,2,3,4] First, the fate of the

NG modes is examined. Then, since Lorentz symmetry is a local symmetry in the context of gravity, the possibility of a Higgs mechanism is considered. Lastly, the possibility of additional massive modes (analogous to the Higgs particle) is considered as well. An explicit illustration of these processes is given for the case of a bumblebee model, in which a vector field acquires a nonzero vacuum value.

2. Spontaneous Lorentz Breaking

In a gravitational theory, Lorentz symmetry acts in local frames, transforming tensor components with respect to a local basis, e.g., T_{abc} (where Latin indices denote components with respect to a local frame). Similarly, diffeomorphisms act in the spacetime manifold, transforming components with respect to the spacetime coordinate system, e.g., $T_{\lambda\mu\nu}$ (denoted using Greek indices). These local and spacetime tensor components are linked by a vierbein. For example, the spacetime metric and local Minkowski metric are related by

$$g_{\mu\nu} = e_\mu{}^a e_\nu{}^b \eta_{ab}. \tag{1}$$

With a vierbein formalism, spinors can naturally be incorporated into a theory. A vierbein formalism also parallels gauge theory, with Lorentz symmetry acting as a local symmetry group. The spin connection $\omega_\mu{}^{ab}$ enters in covariant derivatives that act on local tensor components and plays the role of the gauge field for the Lorentz symmetry. In contrast, the metric excitations, e.g., $h_{\mu\nu} = g_{\mu\nu} - \eta_{\mu\nu}$, act as the gauge fields for the diffeomorphism symmetry. In the context of a vierbein formalism, there are primarily two geometries that can be distinguished. In a Riemannian geometry (with no torsion), the spin connection is nondynamical and does not propagate. However, in a Riemann-Cartan geometry (with nonzero torsion), the spin connection must be treated as independent degrees of freedom that in principle can propagate.

Local Lorentz symmetry is spontaneously broken when a local tensor field acquires a nonzero vacuum expectation value (vev), e.g., for the case of a three-component tensor,

$$< T_{abc} > = t_{abc}. \tag{2}$$

Spontaneous Lorentz breaking can be introduced into a theory dynamically by adding a potential term V to the Lagrangian. For example, a potential of the form

$$V \sim (T_{\lambda\mu\nu}\, g^{\lambda\alpha} g^{\mu\beta} g^{\nu\gamma}\, T_{\alpha\beta\gamma} \pm t^2)^2, \tag{3}$$

consisting of a quadratic function of products of the tensor components $T_{\lambda\mu\nu}$, has a minimum when

$$T_{\lambda\mu\nu}\, g^{\lambda\alpha} g^{\mu\beta} g^{\nu\gamma}\, T_{\alpha\beta\gamma} = \mp t^2. \tag{4}$$

Solutions of Eq. (4) span a degenerate space of possible vacuum solutions. Spontaneous Lorentz breaking occurs when a particular vacuum value t_{abc} in the local frame is chosen, obeying $\mp t^2 = t_{abc}\, \eta^{pa} \eta^{qb} \eta^{rc}\, t_{pqr}$, where the sign depends on the timelike or spacelike nature of the tensor.

3. Nambu-Goldstone Modes

Consider a theory with a tensor vev in a local Lorentz frame, $< T_{abc} > = t_{abc}$, which spontaneously breaks Lorentz symmetry. Since the vacuum value for the vierbein is also a constant or fixed function, e.g., $< {e_\mu}^a > = {\delta_\mu}^a$, the spacetime tensor therefore has a vev as well,

$$< T_{\lambda\mu\nu} > = t_{\lambda\mu\nu}. \tag{5}$$

This means that when Lorentz symmetry is spontaneously broken, diffeomorphisms are spontaneously broken as well. This implies that NG modes should appear (in the absence of a Higgs mechanism) for both of these broken symmetries. In general, the NG modes consist of field excitations that stay within the minimum of the potential V. They therefore obey the condition (4). A solution of this condition is given in terms of the vierbein and the local vev,

$$T_{\lambda\mu\nu} = e_{\lambda a} e_{\mu b} e_{\nu c}\, t^{abc}. \tag{6}$$

As a general rule, there can be up to as many NG modes as there are broken symmetries. Since the maximal case corresponds to six broken Lorentz generators and four broken diffeomorphisms, there can therefore be up to ten NG modes. Where do the NG modes reside? In general, the answer depends on the choices of gauge. However, one natural choice is to put all of the NG modes into the vierbein. A counting argument shows this is possible. The vierbein ${e_\mu}^a$ has 16 components. With no spontaneous Lorentz violation, typically the six Lorentz and four diffeomorphism degrees are used to gauge away ten components, leaving up to six independent degrees of freedom. (Note that a general gravitational theory can have up to six propagating metric modes, but general relativity is special in that there are only two). In contrast, in a theory with spontaneous Lorentz breaking, up to all ten NG modes can potentially propagate as additional degrees of freedom in the vierbein.

4. Gravitational Higgs Mechanisms

With two sets of broken symmetries, local Lorentz transformations and diffeomorphisms, there are potentially two types of Higgs mechanisms. Furthermore, there is the possibility that additional massive modes can exist as excitations that do not stay in the minimum of the potential V.

For the case of the broken diffeomorphisms. it was shown that the conventional Higgs mechanism involving the metric does not occur.[1] This is because the mass term that is generated by covariant derivatives involves the connection, which consists of derivatives of the metric and not the metric itself. As a result, no mass term for the metric is generated according to the usual Higgs mechanism. However, it was also shown that because of the form of the potential, e.g., as in Eq. (3), quadratic terms involving the metric can arise. This results in an alternative form of the Higgs mechanism[1] that has no direct analogue in nonabelian gauge theory. (In nonabelian gauge theory, the potential V involves only the scalar Higgs fields and not the gauge fields. In contrast here, both the metric and tensor fields enter in the massive-field excitations). The additional mass terms that arise in this alternative Higgs mechanism can potentially modify gravity in a way that avoids the van Dam, Veltmann, and Zakharov discontinuity.[5] They are therefore potentially interesting in studies of modified gravity theory.

In contrast, for the case of the broken Lorentz symmetry, it is found that a conventional Higgs mechanism can occur.[3] In this case, the relevant gauge field is the spin connection. This field appears directly in covariant derivatives acting on local tensor components, and for the case where the local tensors acquire a vev, quadratic mass terms for the spin connection can be generated. However, a viable Higgs mechanism involving the spin connection can occur only if the spin connection is a dynamical (i.e., propagating) field. This then requires that there is nonzero torsion and that the geometry is Riemann-Cartan. As a result, a conventional Higgs mechanism for the spin connection is possible, but only in a Riemann-Cartan geometry. However, even if torsion is permitted, constructing a viable model with a massive propagating spin connection that is ghost- and tachyon-free remains a challenging and open problem.[3] Therefore, for simplicity in the remainder of this work, a Riemann spacetime (with no torsion) is assumed. In this restricted context, the only possible process giving rise to massive modes is the alternative Higgs mechanism, in which massive modes are due to excitations that do not stay in the minimum of the potential V.

5. Bumblebee Models

To investigate further the effects of NG and massive modes in theories with spontaneous Lorentz violation, it is useful to work in the context of a definite model. The simplest example involves a vector field that acquires a nonzero vev. Models of this type are known as bumblebee models.[1,2] Examples have been studied in various forms by a number of authors.[1–4,6–17]

Bumblebee models are defined as field theories with a vector field B_μ that acquires a nonzero vev, $< B_\mu > = b_\mu$. The vev is induced by a potential V in the Lagrangian that has a minimum when the vacuum solution holds. Bumblebee models can be defined with generalized kinetic terms for the vector and gravitational fields. However, for brevity, an example with a Maxwell kinetic term is considered here. The Lagrangian then has the form $\mathcal{L} = \mathcal{L}_{\rm G} + \mathcal{L}_{\rm B} + \mathcal{L}_{\rm M}$, where $\mathcal{L}_{\rm G}$ describes the pure-gravity sector, $\mathcal{L}_{\rm M}$ describes the matter sector (including possible interactions with B_μ), and

$$\mathcal{L}_{\rm B} = -\tfrac{1}{4} B_{\mu\nu} B^{\mu\nu} - V(B_\mu B^\mu \pm b^2), \tag{7}$$

describes the bumblebee field. (For simplicity, additional possible interactions between the curvature tensor and B_μ are neglected here as well). The bumblebee field strength in Riemann spacetime is $B_{\mu\nu} = \partial_\mu B_\nu - \partial_\nu B_\mu$.

A noteworthy feature of all bumblebee models is that they do not have local $U(1)$ gauge symmetry. This symmetry is broken explicitly by the presence of the potential V. However, it is common to include couplings to matter that involve the notion of charge in the matter sector. For example, terms involving current couplings with charged matter can be included by defining, $\mathcal{L}_{\rm M} = B_\mu J^\mu$ with $D_\mu J^\mu = 0$. In this case, the theory has a global $U(1)$ symmetry that gives rise to charge conservation in the matter sector. This assumption also implies that initial values can be chosen that maintain stability of the Hamiltonian.[4]

Different forms of the potential V can be considered. One example is a smooth quadratic potential,

$$V = \tfrac{1}{2}\kappa (B_\mu B^\mu \pm b^2)^2, \tag{8}$$

where κ is a constant (of mass dimension zero). This type of potential allows both NG excitations (that stay within the potential minimum) as well as massive excitations (that do not). An alternative would be to consider a linear Lagrange-multiplier potential

$$V = \lambda (B_\mu B^\mu \pm b^2), \tag{9}$$

where the Lagrange-multiplier field λ imposes a constraint that only allows NG excitations in B_μ and excludes massive-mode excitations. However, for definiteness here, the smooth potential (8) is chosen, which allows a massive-mode excitation.

For such a bumblebee model, three Lorentz symmetries and one diffeomorphism are broken. Therefore, up to four NG modes can appear. However, the diffeomorphism NG mode is found not to propagate.[3] It drops out of the kinetic terms and is purely an auxiliary field. In contrast, the Lorentz NG modes do appear in the form of two massless transverse modes and one auxiliary mode. These have properties similar to the photon in electrodynamics, which raises the interesting possibility that photons might be described as NG modes in theories with spontaneous Lorentz violation.[3] Previous links between QED gauge fields, fermion composites, and NG modes have been uncovered in flat spacetime (with global Lorentz symmetry).[18] However, bumblebee models are different. They consist of theories with a noncomposite vector field, have no local U(1) gauge symmetry, and give rise to photons as NG modes in the presence of gravity. Note that bumblebee models also include possible couplings between the vacuum value b_μ and a matter current J^μ. Such an interaction can provide an unmistakable signature of physical Lorentz violation that would distinguish it from any gauge-fixed form of QED. Note as well that any such signal would be contained in the Standard-Model Extension (SME).[19] Thus, on-going investigations of Lorentz breaking using the SME have sensitivity to all signals of spontaneous Lorentz breaking involving couplings between matter and the background vevs.

To determine more thoroughly whether conventional Einstein-Maxwell solutions can emerge from bumblebee models, the role of the massive mode must be investigated.[4] This mode constitutes an additional degree of freedom beyond those of the NG modes. It also alters the form of the initial-value problem. For simplicity, only the case of a purely timelike vacuum vector $b_\mu = (b, 0, 0, 0)$ is considered here. In this case, in the weak-field limit, it is found that the massive mode does not propagate as a free field. Instead, it remains purely an auxiliary field that has no time dependence. As a result, its value is fixed by the initial conditions at $t = 0$. Although it does not propagate, the massive mode can nevertheless alter the form of the static potentials. An example of this can be seen by solving for the modified static potentials in the presence of a point particle with mass m and charge q. It is found that both the electromagnetic and gravitational potentials are modified by the presence of the massive mode, where the specific forms of the modified potentials depend on the assumed initial value of the static massive mode. There are therefore numerous cases that could be explored, including examples that might be relevant in considering alternative explanations of dark matter. However, in the large-mass limit (e.g., approaching the Planck scale), excitation of the massive mode is highly suppressed, and the static potentials approach the conventional Coulomb and Newtonian forms. In the limit of a vanishing massive mode, these become exact expressions. As a result, it is found that the usual Einstein-Maxwell solutions (describing both propagating photons and the usual static potentials) can emerge from a bumblebee model (without local U(1) symmetry), in which local Lorentz symmetry is spontaneously broken.

Acknowledgments

This work was supported by NSF grant PHY-0554663.

References

1. V.A. Kostelecký and S. Samuel, Phys. Rev. D **40**, 1886 (1989); Phys. Rev. D **39**, 683 (1989); Phys. Rev. Lett. **63**, 224 (1989).
2. V.A. Kostelecký, Phys. Rev. D **69**, 105009 (2004).
3. R. Bluhm and V.A. Kostelecký, Phys. Rev. D **71**, 0065008 (2005).
4. R. Bluhm, S.-H. Fung, and V.A. Kostelecký, to appear.
5. H. van Dam and M. Veltman, Nucl. Phys. B **22**, 397 (1970); V.I. Zakharov, JEPT Lett. **12**, 312 (1970).
6. V.A. Kostelecký and R. Lehnert, Phys. Rev. D **63**, 065008 (2001).
7. B. Altschul and V.A. Kostelecký, Phys. Lett. B **628**, 106 (2005).
8. V.A. Kostelecký and Q.G. Bailey, Phys. Rev. D **74**, 045001 (2006).
9. C. Eling, T. Jacobson, and D. Mattingly, gr-qc/0410001.
10. P. Kraus and E.T. Tomboulis, Phys. Rev. D **66**, 045015 (2002).
11. J.W. Moffat, Intl. J. Mod. Phys. D **12**, 1279 (2003).
12. B.M. Gripaios, JHEP **0410**, 069 (2004).
13. S.M. Carroll and E.A. Lim, Phys. Rev. D **70**, 123525 (2004).
14. O. Bertolami and J. Paramos, Phys. Rev. D **72**, 044001 (2005).
15. J.W. Elliott, G.D. Moore, and H. Stoica, JHEP **0508**, 066 (2005).
16. J.L. Chkareuli, C.D. Froggatt, and H.B. Nielsen, hep-th/0610186.
17. H.-C. Cheng, M. Luty, S. Mukohyama, and J. Thaler, JHEP **0605**, 076 (2006).
18. J.D. Bjorken, Ann. Phys. **24**, 174 (1963); Y. Nambu, Prog. Theor. Phys. Suppl. Extra 190 (1968).
19. For a review of the Standard-Model Extension, see R. Bluhm, in *Special Relativity: Will It Survive the Next 101 Years?* eds., J. Ehlers and C. Lämmerzahl, Springer, Berlin, 2006 [hep-ph/0506054].

THE ALPHA ANTIHYDROGEN EXPERIMENT

N. MADSEN

Physics Department, Swansea University, Singleton Park, Swansea, U.K.
E-mail: N.Madsen@swan.ac.uk

G. ANDRESEN, W. BERTSCHE, A. BOSTON, P. D. BOWE, E. BUTLER, C. L. CESAR, S. CHAPMAN, M. CHARLTON, M. CHARTIER, J. FAJANS, R. FUNAKOSHI, D. R. GILL, J. S. HANGST, W. N. HARDY, R. S. HAYANO, M. HAYDEN, R. HYDOMAKO, M. J. JENKINS, L. V. JØRGENSEN, L. KURCHANINOV, P. NOLAN, K. OLCHANSKI, A. OLIN, R. D. PAGE, A. POVILUS, F. ROBICHEAUX, E. SARID, S. SEIF EL NASR, D. M. SILVEIRA, J. W. STOREY, R. I. THOMPSON, D. P. VAN DER WERF, J. S. WURTELE, Y. YAMAZAKI

(ALPHA COLLABORATION*)

ALPHA is a new experiment at the CERN Antiproton Decelerator (AD). The short term goal of ALPHA is trapping of cold antihydrogen, with the long term goal of conducting precise spectroscopic comparisons of hydrogen and antihydrogen. Here we present the current status of ALPHA and the physics considerations and results leading to its design as well as recent progress towards trapping.

1. Introduction

Antihydrogen is the simplest atomic antimatter system, and it offers great opportunities for studies of symmetries between matter and antimatter. In 2002 the ATHENA experiment was the first to produce cold antihydrogen.[1] This result was followed by similar observations by the ATRAP collaboration.[2] Since then efforts have been focused on understanding the formation mechanism and determining the state of the antiatoms

*ALPHA consists of people from: Aarhus University, Denmark; Auburn University, U.S.A.; University of British Columbia, Canada; University of California Berkeley, U.S.A.; University of Calgary, Canada; University of Liverpool, U.K.; NRCN, Israel; RIKEN, Japan; Federal University of Rio de Janeiro, Brazil; Simon Fraser University, Canada; Swansea University, U.K.; University of Tokyo, Japan; TRIUMF, Canada.

formed as well as their temperature. An important finding is that the antihydrogen formed using the technique of merged plasmas of positrons and antiprotons (which, to date, is the most efficient method of antihydrogen formation) is very warm compared to the depths of state-of-the-art atom traps.[3,4] ALPHA believes that cold and trapped antihydrogen offers the best route to conduct precision measurements on the antiatoms. ALPHA has therefore designed, built and commissioned an apparatus to do exactly that. Here we give an overview of the ALPHA experiment, and discuss some of the physics behind various design decisions, as well as some of the most recent results from ALPHA's first AD runs in 2006.

2. The ALPHA Apparatus

An overview of the ALPHA apparatus is shown in Fig. 1. The ALPHA apparatus has been designed with the intention to trap antihydrogen atoms. Further to what is shown on the figure, ALPHA has inherited the ATHENA positron accumulator.[5]

The ALPHA apparatus uses a versatile Penning-Malmberg trap for trapping and manipulation of the charged particles used for antihydrogen production. The left-hand part of the trap (see Fig. 1) is enclosed by high voltage electrodes to catch the incoming antiprotons, whereas the right-hand section is used for transfer and manipulations of positrons transferred from the positron accumulator. The center of the trap is the so-called mixing trap where the antihydrogen is formed.

An antihydrogen atom can be trapped in a three dimensional magnetic minimum due to its magnetic dipole moment. If antihydrogen can be created cold enough in the magnetic trap, or be cooled once inside it, it can be trapped. Such a magnetic trap, where the transverse minimum is formed by using a multipole magnet, and the axial minimum by a pair of pinch or mirror coils (see Fig. 1) will, when superimposed on the solenoid field of the Penning-Malmberg trap, have a transverse well depth U (in K) according to:

$$U = \frac{\mu}{k_B}\left[\sqrt{B_W^2 + B_S^2} - B_S\right]. \tag{1}$$

Here μ is the antihydrogen magnetic moment, k_B is Boltzmann's constant, B_W is the multipole field at the trap wall (inner radius of the Penning-Malmberg trap electrodes) and B_S the solenoid field. We note that a lower solenoid field will allow a deeper trap for the same multipole field at the wall. In the axial direction the fields add linearly, such that there are no

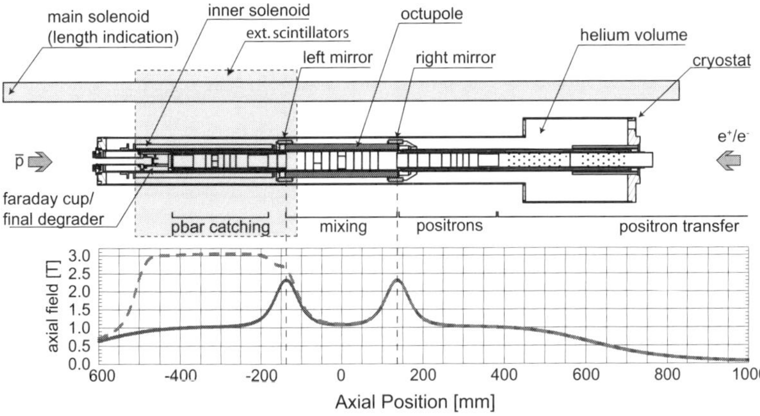

Figure 1. Overview of the ALPHA apparatus. The plot of the axial B-field shows the field with the main solenoid and the mirrors (full) and the added inner solenoid (dashed).

limitations to the trap depth by implementing a high solenoid field.

Trapping of antiprotons is significantly more efficient using a high solenoid field. The ALPHA apparatus therefore consists of a main solenoid which delivers a homogenous axial magnetic field throughout and an internal short solenoid around the antiproton catching region of the trap. This provides a low axial field in the (anti)atom trap to increase the trap depth while keeping the catching efficiency of antiprotons high.

In typical atom traps the transverse magnetic minimum is supplied by a quadrupole. However, recent measurements have demonstrated that strong transverse field components can significantly deteriorate the trapping efficiency and particle lifetime of a Penning trap.[6] Higher order multipoles will, for the same trap depth, result in much smaller transverse field components near the center of the trap. The rapid fall in field away from the wall for a higher order multipole means that extra effort has to be made to make the magnet inner coil as close to the trap vacuum as possible. ALPHA settled for an octupole design as a compromise between trap depth and particle survival.[7] The state-of-the-art ALPHA atom trap produces an effective depth of 1.13 T at B_S= 1 T corresponding to 0.74 K for ground state antihydrogen (Eq. (1)).

3. Antihydrogen Formation Alternatives

Antihydrogen is normally formed by merging two cold plasmas of antiprotons and positrons. This method has, as mentioned earlier, been shown to produce large amounts of antihydrogen before the antiprotons are in thermal equilibrium with the positrons. The antiatoms are thus at a higher temperature than ambient, as the antiproton carries most of the momentum of the antihydrogen.[3,4] Thus, even with, as expected for ALPHA, a 4 K ambient temperature, further measures are needed to create antihydrogen that is cold enough to trap.

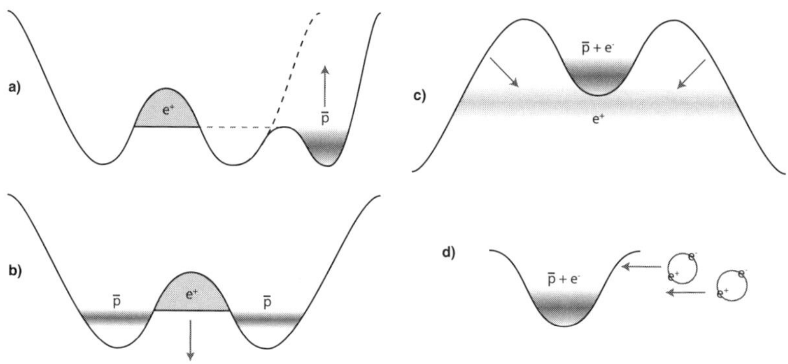

Figure 2. Alternative antihydrogen formation methods for creating antihydrogen cold enough to be trapped. See text for discussion.

The merging-scheme has the advantage of being very easy to implement, and of forming antihydrogen at impressive rates of up to[8] 400 s^{-1}. ALPHA has thus chosen to focus first on variations of this scheme as the basis for methods for forming trappable antihydrogen. Figure 2 shows four approaches that are being discussed. To avoid positrons interacting with hot antiprotons, the antiprotons can be dribbled into the positron plasma, (a). However, (a) will be difficult to implement in practice, as the positron space charge typically varies from load to load with more than the 0.1 mV precision needed such that the antiprotons have low enough energy in the positron plasma. In (b) the antiprotons are injected below the positron space-charge level and the positrons are then slowly moved into contact with the antiprotons. A simple rule of thumb here is that the positron well must move slow enough that the energy change should be less than 0.1 mV ($\sim$1 K) in the time it takes an antiproton to make a bounce in the side-well in which it is trapped. For typical wells this time is $\sim$1 μs, which gives an upper limit on the rate of 100 Vs^{-1}. This limit will of course also depend on the formation probability, i.e., the antiproton may have to make more passes through the positron plasma in order to form an antihydrogen atom.

If so, this will significantly lower the aforementioned upper limit. ALPHA plans to study this technique in 2007. Alternatively, it might be possible to hold the antiprotons in the middle and either perform a so-called inverted mixing (c) or make positronium formed nearby collide with them (d).[9,10] The positronium method has so far only produced low numbers of antihydrogen in a proof-of-principle experiment carried out by ATRAP[11] and the inverted-mixing technique will require recycling the positrons which will cool down into the side wells. Both techniques have the ambient temperature as the lower limit for the antihydrogen temperature, a limit method (b) does not necessarily have, as antiprotons can be injected at almost arbitrarily low energy into the positron plasma. Furthermore, in order to keep the antiprotons at the ambient temperature some electrons must be kept with them as ejecting them all is likely to lead to heating.

4. Recent Results

ALPHA took its first antiprotons from the CERN AD from September to November 2006 when the apparatus was fully commissioned, apart from the silicon vertex detector. The latter is to facilitate ATHENA-style antiproton annihilation imaging.[3,12] Furthermore, the first important steps towards cold, trapped antihydrogen were taken.

4.1. *Particle survival in octupole fields*

As mentioned earlier, ALPHA decided to use an octupole magnet to create the transverse magnetic minimum in order to avoid the limitations on charged particle trapping that a quadrupole could induce.

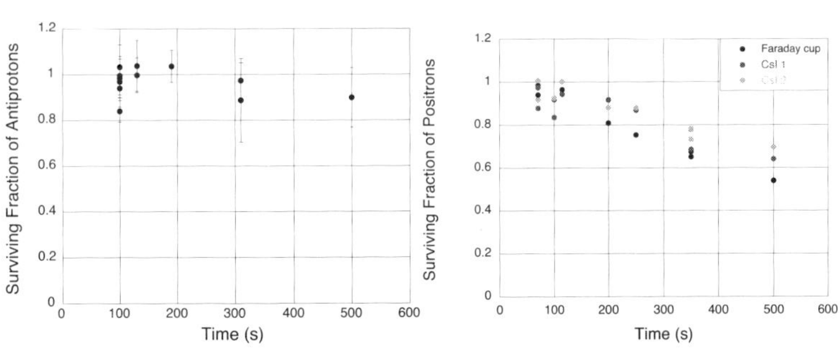

Figure 3. The ratio of the number of antiprotons (positrons) stored in the octupole field to the number stored without the field is plotted versus holding time.

Figure 3 shows the numbers of particles (antiprotons and positrons respectively) as a function of time held trapped in an octupole field of 1.2 T at the inner wall (radius 22.3 mm).[13] In these conditions, there is no evidence for fast, so-called ballistic loss, which occurs when the particles follow the diverging field-lines induced by the multipole. This effect was observed in the earlier quadrupole measurements.[6] Furthermore, confinement lifetimes in excess of ~100 sec were observed, which is more than sufficient for antihydrogen formation. The lifetime, and the ballistic loss, depend strongly on the radii and the lengths of the plasmas. The radii of the plasmas used here were unknown, but we note that the antiprotons are captured in an axial field of 3 T, and subsequently transferred to 1 T, causing the cloud to expand by a factor $\sqrt{3}$. The results show that our magnetic trap is compatible with antihydrogen formation using the standard schemes.

4.2. *Mixing in low magnetic fields*

A second cause for concern was how the antiproton positron interaction would change when taking place in a field of only 1 T. The reduced magnetic field will cause the synchrotron cooling of the positrons to slow, in addition to the expansion of the antiproton cloud mentioned above. An outstanding issue from ATHENA was concern over the spatial overlap between the positrons and the antiprotons.

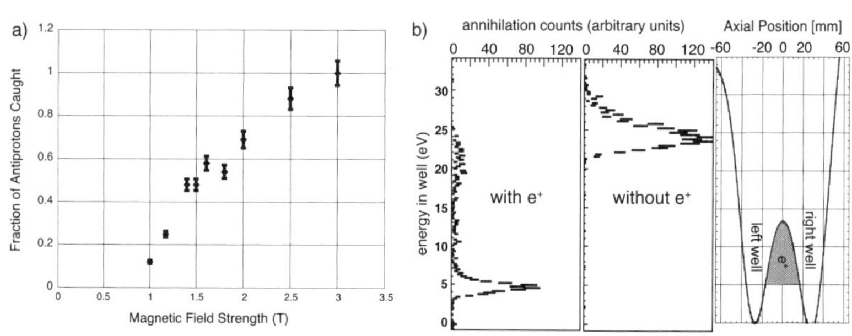

Figure 4. a) Antiproton catching efficiency vs. axial magnetic field. b) Antiproton cooling by positrons in a 1 T axial magnetic field.

Figure 4a shows, first, that it is indeed an advantage to use a 3 T axial field for antiproton catching as the efficiency is about an order of magnitude higher than at 1 T. Secondly, Fig. 4b shows antiprotons being cooled by positrons in 1 T, making it likely that antihydrogen will form.

5. Summary and Outlook

We have presented the new ALPHA antihydrogen experiment at the CERN Antiproton Decelerator. The ALPHA experiment was successfully commissioned in 2006, where we demonstrated that the neutral (anti)atom trap is compatible with confinement of the antiprotons and positrons used for forming antihydrogen. We also demonstrated positron cooling of antiprotons at the low 1 T axial field necessary for the neutral trap.

In 2007 we plan to make the first attempt at using variations of the standard mixing techniques to produce antihydrogen cold enough to trap. In order to increase our chances of making enough antihydrogen, we are also developing the so-called rotating wall technique[14] in an attempt to sympathetically compress the antiproton plasma to significantly enhance the radial overlap of the positrons and the antiprotons when the two species are merged. Reducing the radial size of both species should also lessen the influence of the octupole magnetic field on the plasma confinement. An increased radial overlap of the two species should significantly increase the likelihood of first trapping of the antihydrogen by 2008.

Acknowledgments

We would like to thank the AD team at CERN for always being helpful and providing a high quality antiproton beam. This work was supported by CNPq, FAPERJ, CCMN/UFRJ (Brazil), NSERC, NRC/TRIUMF (Canada), FNU (Denmark), ISF (Israel), MEXT, RIKEN (Japan), EPSRC, The Leverhulme Trust (U.K.), DOE, NSF (U.S.A.).

References

1. M. Amoretti *et al.* (ATHENA), Nature (London) **419**, 456 (2002).
2. G. Gabrielse *et al.* (ATRAP), Phys. Rev. Lett. **89**, 213401 (2002).
3. N. Madsen *et al.* (ATHENA), Phys. Rev. Lett. **94**, 033403 (2005).
4. G. Gabrielse *et al.* (ATRAP), Phys. Rev. Lett. **93**, 073401 (2004).
5. L.V. Jørgensen *et al.* (ATHENA), Phys. Rev. Lett. **95**, 025002 (2005).
6. J. Fajans *et al.*, Phys. Rev. Lett. **95**, 155001 (2005).
7. W. Bertsche *et al.* (ALPHA), Nucl. Instr. Meth. A **566**, 746 (2006).
8. M. Amoretti *et al.* (ATHENA), Nucl. Instr. & Meth. B **247**, 133 (2006).
9. J.W. Humberston *et al.*, J. Phys. B. **20**, L25-29 (1987).
10. M. Charlton, Phys. Lett. A **143**, 143 (1990).
11. C.H. Storry *et al.* (ATRAP), Phys. Rev. Lett. **93**, 263401 (2004).
12. M.C. Fujiwara *et al.* (ATHENA), Phys. Rev. Lett. **92**, 065005 (2004).
13. G. Andresen *et al.* (ALPHA), Phys. Rev. Lett. **98**, 023402 (2007).
14. X.-P. Huang *et al.*, Phys. Rev. Lett. **78**, 875 (1997).

ATOM INTERFEROMETRY TESTS THE ISOTROPY OF POST-NEWTONIAN GRAVITY

H. MÜLLER, S.-W. CHIOW, S. HERRMANN, AND S. CHU
Physics Dept., Varian 226, 382 Via Pueblo Mall, Stanford, CA94305, U.S.A.
E-mail: holgerm@stanford.edu

K.-Y. CHUNG
Physics Dept., National University of Singapore
2 Science Drive 3, Singapore 117542

We present a test of the local Lorentz invariance of post-Newtonian gravity by monitoring Earth's gravity with a Mach-Zehnder atom interferometer that features a resolution of about $8 \times 10^{-9}\,\mathrm{g}/\sqrt{\mathrm{Hz}}$, the highest reported thus far. Limits on four of the $\bar{s}^{JK}$ coefficients of the Standard-Model Extension at a level of parts in 10^9 and on three $\bar{s}^{0J}$ coefficients at parts in 10^5 are obtained.

1. Introduction

The description of gravitation by a dynamic geometry of space-time, Einstein's General Relativity, is based on the Einstein Equivalence Principle. This encompasses local Lorentz invariance (LLI), universality of free fall (UFF), and local position invariance (LPI). Attempts to unify gravity with the non-gravitational Standard Model have failed so far. This suggests that one of its foundations might be violated at some level of precision.[1] Several tests of the UFF[2] and LPI[3] have been performed.[1] Violations of LLI in all sectors of the Standard Model and gravity are comprehensively described by the Standard-Model Extension (SME).[4] In the non-gravitational Standard Model (for example, for photons, electrons, protons, and neutrons), they have been studied extensively.[1,5] Far less attention, however, has been paid on the LLI of the gravitational interaction, in spite of the pioneering work of Nordtvedt and Will in the 1970s.[6,7,8]

The relative weakness of the gravitational force means that only a few experiments of exceptional sensitivity can hope to detect tiny deviations from Newtonian gravity, including lunar laser ranging and torsion balances. A relatively recent addition to this list is precision atom interfer-

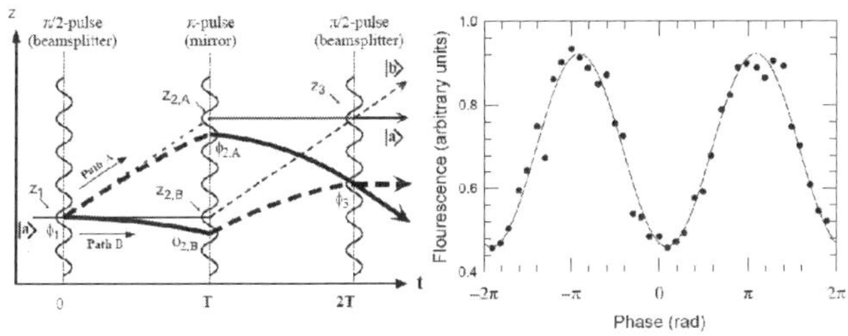

Figure 1. Left: schematic of a Mach Zehnder atom interferometer. In our experiment, a π pulse takes about $100\,\mu s$. Right: typical fringe obtained in our experiment.

ometry. Since the first realization of an atom interferometer in an atomic fountain,[9,10] it has been serving in precision measurements, e.g., of the fine structure constant,[11] the local gravitational acceleration g,[2] and Newton's constant[12] G. One reason for its outstanding precision is that the motion of neutral atoms realizes a freely falling frame to a high accuracy and that this motion can be interrogated by laser radiation in a tremendously precise way. As a result, tests of post-Newtonian gravity with atom interferometry have been proposed that will rival or exceed the precision of classical ones.[13]

Here, we report on a first step in this direction: we describe a test of the LLI of post-Newtonian gravity by testing its isotropy using an atom interferometry-based gravimeter. We use the highest resolution atomic gravimeter reported thus far,[14] which surpasses the resolution of its predecessor[2] by a factor of about three.

2. Experiment

In precision atom interferometers, the beam splitters and mirrors are not material gratings but laser radiation. In our Mach-Zehnder atom interferometer (Fig. 1, left), cesium atoms in the $F = 3, m_F = 0$ hyperfine state enter on a freely falling trajectory, moving vertically upwards. A first laser pulse is applied which transfers them into a superposition of the $F = 3$ and $F = 4$ hyperfine ground states by a two-photon Raman transition. Two counterpropagating laser beams, overlapped with the trajectory of the atoms, are used. The $F = 3$ and $F = 4$ states move vertically relative to each other because of the momentum of two photons that is transferred by the laser radiation. A total of three light pulses split, reflect, and then recombine the paths to form an interferometer.

The matter waves in both paths acquire a relative phase difference $\Delta\phi$:

the phase of the free evolution is given by the classical action $S_{Cl}/\hbar$. If we restrict attention to the $\Delta\phi$ due to a constant gravitational acceleration g, its contribution to the phase difference vanishes. However gravity shifts downwards the location at which the paths interact with the light by $\Delta z = -gt^2/2$, where t denotes time and z the vertical coordinate. This gives rise to a phase difference[2]

$$\Delta\phi = k_{\text{eff}} g T^2 - [\phi_1 - 2\phi_2 + \phi_3]. \tag{1}$$

T is the time between light pulses and ϕ_{1-3} denote the phases of the laser fields at an arbitrary reference point $z = 0$. To high accuracy, the laser radiation can be modelled as a plane wave, which results in an effective wavevector $k_{\text{eff}} = 2k$. For our experiment, $T = 0.4\,\text{s}$, $k \simeq 2\pi/(852\,\text{nm})$ and $g \simeq 9.8\,\text{m/s}^2$, so $\Delta\phi \simeq 2.3 \times 10^7\,\text{rad}$. If this can be measured to e.g., 1 mrad, we obtain a resolution of $\Delta g/g \sim 10^{-10}$.

About 10^9 Cs atoms are assembled within 650 ms from a background vapor pressure of $\sim 10^{-9}$ mbar in a 3-dimensional magneto-optical trap. A moving optical molasses accelerates them upwards to 4-5 m/s within $\sim$ 2 ms. They are then adiabatically released at a temperature of 1.2-2 μK. Raman sideband cooling in an optical lattice co-moving with the atoms further cools them to 150 nK.[15] The atoms then enter a $\sim$ 90 cm long, 3.8 cm inner diameter tube magnetically shielded by three layers of hypernom. A solenoid generates a small magnetic bias field to set the quantization axis.

The off-resonant Raman pulses at a wavelength of 852 nm are generated by two extended cavity diode lasers based on 100-mW laser diodes SDL-5411. The first is frequency stabilized ('locked') so that it arrives at the experiment with a detuning of -1030MHz from the $6S_{1/2}, F = 3 \rightarrow 6P_{3/2}, F' = 4$ transition in Cs. The second one is phase locked to the first one with a frequency difference close to the hyperfine splitting of $\simeq 9192$ MHz. Both laser beams are combined on a beam splitter and coupled into a common single-mode, polarization maintaining optical fiber.

After the fiber, each beam has a power of about 20 mW. A sample of the light is detected for deriving the error signal for the phase-lock. All non-optical frequencies are referenced to a LORAN-C frequency standard. An acousto-optical modulator (Isomet 1205) after the fiber switches the overlapped beams and controls their intensity. The beams are expanded to a diameter of about 2.5 cm. They pass the vacuum chamber with linear polarization and are retro-reflected at the top of the chamber with two passes through a quarter-wave retardation plate to form a lin$\perp$lin polarized counterpropagation geometry. Since any vibration of that mirror shows up as noise in our gravity measurement, it is held by a sophisticated active vibration isolator.[16]

Because of the free fall of the atoms, the resonance condition for the difference frequency of the Raman beams in the laboratory frame changes at a rate of $\sim 23\,\text{MHz/s}$. This is accounted for by switching the frequencies between the pulses in a way that the phase due to the laser light cancels the phase shift due to gravity.

Atoms are detected by their flourescence with a Hamamatsu R943-02 photomultiplier tube. Detection of both the $F = 3$ and $F = 4$ outputs of the interferometer allows normalization of the signals to take out fluctuations in the number of launched atoms.

Figure 1 (right) shows a typical gravity fringe, taken with 40 launches that take 75 s total. A sinewave-fit has a phase uncertainty of 0.031 rad, and determines g to an uncertainty of $\sim 1.3 \times 10^{-9}$ g. This corresponds to $11 \times 10^{-9}\, g/\sqrt{\text{Hz}}$. An improved short-term resolution of $8 \times 10^{-9}\, g/\sqrt{\text{Hz}}$ can be reached by taking data at the 50% points of the fringes only. However, as this method is more sensitive to systematic effects such as drift of the PMT sensitivity, this approach was not followed. Our resolution is more than 3 times better than the best previous one, which was reported by Peters *et al.*[2] This is mainly a consequence of our increased interaction time $T = 400\,\text{ms}$. It also surpasses the best classical absolute gravimeter, the FG-5 falling corner cube gravimeter, by a factor of about 20.

3. Analysis

In this work, we use the SME as a description of violations of LLI in gravity. It describes the gravitational two-body interaction in the post-Newtonian limit by a Lagrangian[17]

$$\mathcal{L} = \frac{1}{2}\sum_a m_a v_a^2 + \frac{1}{2}G\sum_{ab}\frac{m_a m_b}{r_{ab}}\left(1 + \frac{3}{2}\bar{s}^{00} + \frac{1}{2}\bar{s}^{jk}\hat{r}^j_{ab}\hat{r}^k_{ab}\right) - \frac{1}{2}G\sum_{ab}\frac{m_a m_b}{r_{ab}}\left(3\bar{s}^{0j}v^j_a + \bar{s}^{0j}\hat{r}^j_{ab}v^k_a\hat{r}^k_{ab}\right). \tag{2}$$

The indices a and b enumerate the particles and j, k the spatial coordinates. $\bar{s}^{\mu\nu}$ is assumed to be decomposed in an inertial reference frame that is usually chosen as a Sun-centered celestial equatorial reference frame. As a consequence of this model, periodic variations of the local gravitational acceleration

$$\frac{\delta g}{g_0} = \sum_m C_m \cos(\omega_m t + \phi_m) + D_m \sin(\omega_m t + \phi_m) \tag{3}$$

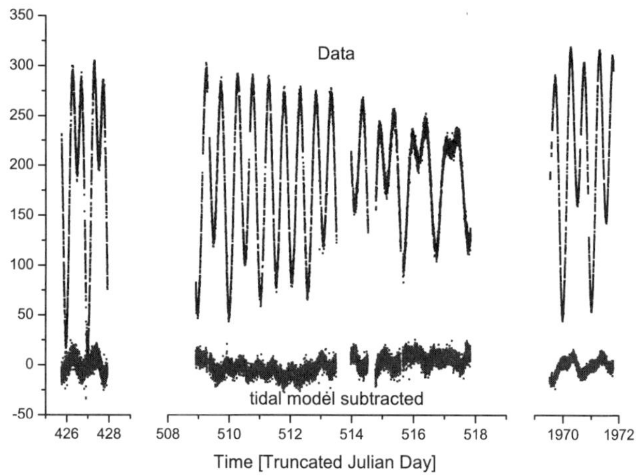

Figure 2. Data in $10^{-9}g$. Each point represents a 60-s scan of one fringe (75-s after TJD1900).

will arise, which are related to the motion of the system relative to the inertial frame, as given by Earth's orbital velocity $v/c \simeq 10^{-4}$ and the frequencies of Earth's orbit $\Omega = 2\pi/(1\,\mathrm{y})$ and rotation $\omega_\oplus \simeq 2\pi/(23.93\,\mathrm{h})$. The Fourier coefficients C_m, D_m for the six frequencies $\omega_\oplus, 2\omega_\oplus, \omega_\oplus \pm \Omega, 2\omega_\oplus \pm \Omega$ are functions of the components of $\bar{s}^{\mu\nu}$. The model goes beyond the usual parametrized post-Newtonian theory, as a preferred frame, in which gravity is isotropic, typically fails to exist.

For bounding post-Newtonian gravity, we use $\sim 60\,\mathrm{h}$ of data taken with this setup, as well as a $\sim 60\,\mathrm{h}$ and a $\sim 10\,\mathrm{d}$ run reported previously,[2] see Fig. 2. Periodic changes having an amplitude of around $100\,\mu\mathrm{gal} \simeq 10^{-7}g$ are mainly caused by the lunar and solar tides. Subtraction of a Newtonian model of these influences[18] based on the relative positions of the Sun, the Moon, and the planets, yields the graph shown at the bottom of Fig. 2. Use of an additional model of the local tides, as caused by the San Francisco Bay, would reduce the residuals, which would translate into lower bounds on the SME-parameters. However, such models are typically obtained by fitting observations and are thus not suitable for our purpose of comparing our data to a Newtonian model.

The combined data spans about 1500 d, but fragmented into three relatively short segments. A Fourier analysis yields the components given in the second column of Table 1. However, the fragmentation of the data means that the Fourier analysis yields, in fact, linear combinations $\mathcal{C}_m^{\rm exp} = \sum_n M_{mn}\mathcal{C}_n$ of the coefficients, where we denoted $\mathcal{C}_n = (C_n, D_n)$ and the results from the fragmented data carry a superscript exp. M_{mn} is a 12×12 covariance matrix given by the normalized overlap integrals of the signal components. We can solve this relation for C_n, D_n by using the inverted matrix M^{-1} and obtain the disentangled results listed in the third column of Table 1. The errors of the estimates are determined as the geometric sum of the errors entering the linear combinations. They increase substantially in this process, which is the penalty for such fragmented data.

Comparing the modulations of g arising from anisotropic gravity model[17] to our measurement, we obtain the estimates listed in the fourth column of Table 1. Some components of $\bar{s}^{\mu\nu}$ are multiply determined. These could be combined to a weighted average, but since in all cases one limit strongly outweighs the others we retain only that and discard the other limits. Finally, $\bar{s}^{TZ}$ is measured as a linear combination with $\bar{s}^{TY}$, into which we insert $\bar{s}^{TY}$ as previously determined. Again, one estimate outweighs the other and is solely retained. Our final results for the SME coefficients are (parts in 10^9)

$$\begin{aligned} \bar{s}^{XX} - \bar{s}^{YY} = -5.6(2.1), \quad \bar{s}^{XY} = -0.09(79), \\ \bar{s}^{XZ} = -13(37), \quad \bar{s}^{YZ} = -61 \pm 38 \end{aligned} \tag{4}$$

and (parts in 10^5)

$$\bar{s}^{TY} = -2.0(4.4), \quad \bar{s}^{TX} = 5.4(4.5), \quad \bar{s}^{TZ} = 1.1(26). \tag{5}$$

This is the first determination of bounds on the gravitational coefficients of the SME. However, Nordtvedt derived an order-of-magnitude limit

Table 1. Results.

Component	Measured 10^{-9}	Disentangled 10^{-9}	Result 10^{-9} for $\bar{s}^{JK}$ / 10^{-5} for $\bar{s}^{TJ}$
$C_{2\omega}$	0.342(88)	-0.44(17)	$\bar{s}^{XX} - \bar{s}^{YY} = -5.6(2.1)$
$D_{2\omega}$	-0.942(89)	-0.02(19)	$\bar{s}^{XY} = -0.09(79)$
C_{ω}	3.668(88)	3.1(8.8)	$\bar{s}^{XZ} = -13(37)$
D_{ω}	-0.267(85)	14.8(9.0)	$\bar{s}^{YZ} = -61(38)$
$C_{2\omega+\Omega}$	-1.378(87)	-1.11(65)	$\bar{s}^{TY} = 172(100)$
$D_{2\omega+\Omega}$	-1.051(89)	-1.08(68)	$\bar{s}^{TX} = -167(104)$
$C_{2\omega-\Omega}$	1.438(89)	-0.30(66)	$\bar{s}^{TY} = -2.0(4.4)$
$D_{2\omega-\Omega}$	0.536(88)	0.82(67)	$\bar{s}^{TX} = 5.4(4.5)$
$C_{\omega+\Omega}$	0.647(94)	-12.4(6.2)	$\bar{s}^{TX} = 258(129)$
$D_{\omega+\Omega}$	-2.020(82)	-3.63(6.2)	$\bar{s}^{TY} - 0.21\bar{s}^{TZ} = 76(130)$
$C_{\omega-\Omega}$	1.610(82)	9.56(6.3)	$\bar{s}^{TX} = -200(130)$
$D_{\omega-\Omega}$	2.840(92)	0.11(6.1)	$\bar{s}^{TZ} + 0.21\bar{s}^{TY} = -00(26)$

$|\Omega| \leq 10^{-9}$ on a matrix entering the anisotropic universe model.[8] If we *assume* $\bar{s}^{00} = 0$, we may translate[17] it into a bound $|\bar{s}^{JK}| \leq 4 \times 10^{-9}$. Our limits compare well with this, surpassing it by a factor of five for $\bar{s}^{XY}$. No previous limits on $\bar{s}^{TJ}$, neither direct nor translated from other formalisms, are known to us. Future work may lead to alternative bounds from lunar laser ranging and torsion balances, or the use of global g-data.

Acknowledgments

This material is based upon work supported by the National Science Foundation under Grant No. 0400866, the Air Force Office of Scientific Research, and the Multi-University Research Initiative.

References

1. G. Amelino-Camelia *et al.*, in A.Macias, C. Lämmerzahl, and D. Nunez, eds., *Gravitation and Cosmology,* AIP Conference Proceedings **758**, Melville, N.Y., 2005, p. 30.
2. A. Peters *et al.*, Nature **400,** 849 (1999); Metrologia **38,** 25 (2001).
3. C. Braxmaier *et al.*, Phys. Rev. Lett. **88,** 010401 (2002).
4. V.A. Kostelecký, Phys. Rev. D**69,** 105009 (2004).
5. See, for example, V.A. Kostelecký, ed., *CPT and Lorentz Symmetry III,* World Scientific, Singapore, 2005.
6. C.M. Will, Astrophys. J. **169,** 141 (1971).
7. K. Nordtvedt and C.M. Will, Astrophys. J. **177,** 775 (1972).
8. K. Nordvedt, Phys. Rev. D**14,** 1511 (1976).
9. M. Kasevich and S. Chu, Phys. Rev. Lett. **67,** 181 (1991).
10. S. Chu, Nature (London) **416,** 206 (2002).
11. A. Wicht *et al.*, Physica Scripta **T102,** 82 (2002); P. Clade *et al.*, Phys. Rev. Lett. **96,** 033001 (2006); H. Müller *et al.*, Appl. Phys. B **84,** 633-642 (2006).
12. J. B. Fixler *et al.*, Science **315,** 74 (2007).
13. S. Dimopoulos *et al.*, Phys. Rev. Lett. **98,** 111102 (2007).
14. K. Chung, *Applications of Atom Interferometry unsing an improved laser cooling method,* Ph.D. thesis, Stanford University, 2001.
15. P. Treutlein, K.Y. Chung, and S. Chu, Phys. Rev. A **63** 051401(r) (2001).
16. J.M. Hensley, A. Peters, and S. Chu, Rev. Sci. Instrum. **70**, 2735 (1999).
17. Q.G. Bailey and V.A. Kostelecký, Phys. Rev. D**74,** 045001 (2006).
18. Calculated from the Tamura potentials, Y. Tamura, Bulletin d'information marrées terrestres, P. Melchior (ed.), Brussels, **99,** 6813 (1987), using the ETGTAB v3.0 solid earth tide data generation software by H.G. Wenzel, Univ. Karlsruhe. See, for example, http://www.geo.uni-jena.de/geophysik/etc/etcdat/etgtab/.

PROBING LORENTZ SYMMETRY WITH GRAVITATIONALLY COUPLED MATTER

JAY D. TASSON
Physics Department, Indiana University
Bloomington, IN 47405, U.S.A.

Methods for obtaining additional sensitivities to Lorentz violation in the fermion sector of the Standard-Model Extension using gravitational couplings are discussed.

1. Introduction

The Standard Model of particle physics together with Einstein's General Relativity provide a remarkably successful description of known phenomena. General Relativity describes gravitation at the classical level, while all other interactions are described down to the quantum level by the Standard Model. However, a single quantum-consistent theory at the Planck scale remains elusive.

Ideally, experimental information would guide the development of the underlying theory; however, directly probing the Planck scale is impractical at present. A feasible alternative is to search for suppressed effects arising from Planck-scale physics in sensitive experiments that can be performed at presently accessible energies. Relativity violations arising from Lorentz-symmetry violation in the underlying theory provide a candidate suppressed effect.[1,2] The Standard-Model Extension (SME) is an effective field theory that describes Lorentz violation at our present energies.[3,4]

A large number of experimental tests of Lorentz symmetry have been performed in the context of the minimal SME. Those test include, in the Minkowski-spacetime limit, experiments with electrons,[5] protons and neutrons,[6] photons,[7] mesons,[8] muons,[9] neutrinos,[10] and the Higgs.[11] The pure-gravity sector is also being investigated in the post-Newtonian limit.[12,13] Although no compelling experimental evidence for Lorentz violation has been found to date, much remains unexplored. For example, only about half of the coefficients for Lorentz violation involving light and ordinary matter (protons, neutrons, and electrons) have been investigated experimentally, and other sectors remain nearly unexplored.

In the remainder of this proceedings, a theoretical basis for extending SME studies with ordinary matter into the post-Newtonian regime, developed with Alan Kostelecký, will be discussed.[14] The goal of this work is to obtain new sensitivities to Lorentz violation in the fermion sector using couplings to gravity. These couplings introduce new operator structures that provide sensitivities to coefficients for Lorentz violation that are unobservable in Minkowski spacetime.

2. Relativistic Theory

Gravitational effects are incorporated into the SME action in Ref. 4. The general geometric framework assumed is Riemann-Cartan spacetime, which allows for a nonzero torsion tensor $T^{\lambda}{}_{\mu\nu}$ as well as the Riemann curvature tensor $R^{\kappa}{}_{\lambda\mu\nu}$. Due to the need to incorporate spinor fields, the vierbein formalism is adopted. The vierbein also allows one to easily distinguish general coordinate and local Lorentz transformations, a feature convenient in studying Lorentz violation.[4] The spin connection $\omega_{\mu}{}^{ab}$ along with the vierbein $e^{\mu}{}_{a}$ are taken as the fundamental gravitational objects, while the basic non-gravitational fields are the photon A_{μ} and the Dirac fermion ψ.

The minimal-SME action can be expanded in the following way:

$$S = S_G + S_\psi + S'. \tag{1}$$

The first term, S_G, is the action of the pure-gravity sector, which contains the dynamics of the gravitational field and can also contain coefficients for Lorentz violation in that sector.[4,12] The Einstein-Hilbert action of General Relativity is recovered in the limit of zero torsion and Lorentz invariance. The action for the fermion sector is provided by the term S_ψ in Eq. (1).

$$S_\psi = \int d^4x (\tfrac{1}{2} i e e^{\mu}{}_{a} \overline{\psi} \Gamma^a \overleftrightarrow{D}_\mu \psi - e \overline{\psi} M \psi). \tag{2}$$

Here, Γ^a and M take the form shown in the following definitions.

$$\begin{aligned} \Gamma^a \equiv {} & \gamma^a - c_{\mu\nu} e^{\nu a} e^{\mu}{}_{b} \gamma^b - d_{\mu\nu} e^{\nu a} e^{\mu}{}_{b} \gamma_5 \gamma^b \\ & - e_\mu e^{\mu a} - i f_\mu e^{\mu a} \gamma_5 - \tfrac{1}{2} g_{\lambda\mu\nu} e^{\nu a} e^{\lambda}{}_{b} e^{\mu}{}_{c} \sigma^{bc}. \end{aligned} \tag{3}$$

$$M \equiv m + a_\mu e^{\mu}{}_{a} \gamma^a + b_\mu e^{\mu}{}_{a} \gamma_5 \gamma^a + \tfrac{1}{2} H_{\mu\nu} e^{\mu}{}_{a} e^{\nu}{}_{b} \sigma^{ab}. \tag{4}$$

The symbols a_μ, b_μ, $c_{\mu\nu}$, $d_{\mu\nu}$, e_μ, f_μ, $g_{\mu\nu\lambda}$, $H_{\mu\nu}$ are the coefficient fields for Lorentz violation of the minimal fermion sector. In general, they vary with position and differ for each species of particle. For additional discussion of the fermion-sector action, see Ref. 4.

The final portion, S', of the action (1) contains the dynamics associated with the coefficient fields for Lorentz violation and is responsible for spontaneous breaking of Lorentz symmetry. Through symmetry breaking, the coefficient fields for Lorentz violation are expected to acquire vacuum values. Thus, it is possible to write

$$t_{\lambda\mu\nu\ldots} = \bar{t}_{\lambda\mu\nu\ldots} + \widetilde{t}_{\lambda\mu\nu\ldots}, \tag{5}$$

where $t_{\lambda\mu\nu\ldots}$ represents an arbitrary coefficient field for Lorentz violation, $\bar{t}_{\lambda\mu\nu\ldots}$ is the corresponding vacuum value, and $\widetilde{t}_{\lambda\mu\nu\ldots}$ represents the fluctuations about that vacuum value. Note that a subset of these fluctuations are the massless Nambu-Goldstone modes associated with Lorentz-symmetry breaking.[15] It is possible to develop the necessary tools to analyze fermion experiments in the presence of gravity and Lorentz violation without specifying S'.[14] Those results are summarized here.

The relativistic quantum-mechanical hamiltonian obtained from action (2) provides a first step toward the goal of obtaining experimental access to Lorentz violation. The hamiltonian may be obtained perturbatively since gravitational and Lorentz-violating effects are small in the regimes of interest. Gravity may be considered perturbatively small in the laboratory and solar-system tests that will be of interest, and Lorentz violation is assumed small since it has not been observed in nature. In what follows, orders in these small quantities will be denoted $O(m, n)$, where m and n are the order of the given term in coefficients for Lorentz violation (vacuum values) and the metric fluctuation respectively.

After expanding to order $m + n = 2$ and making a field redefinition required to define the hamiltonian,[14,16] a hamiltonian of the form,

$$H = H^{(0,0)} + H^{(0,1)} + H^{(1,0)} + H^{(1,1)} + H^{(0,2)}, \tag{6}$$

is found. Here, the Lorentz invariant contributions consist of the conventional Minkowski-spacetime hamiltonian, $H^{(0,0)}$, and the first and second order gravitational corrections denoted $H^{(0,1)}$ and $H^{(0,2)}$ respectively. The first order correction to the hamiltonian due to Lorentz violation, $H^{(1,0)}$, is the same as that found in the flat-spacetime SME.[16] The $O(1,1)$ correction to the hamiltonian is the term of interest since it contains coefficients for Lorentz violation coupled to gravity. It can be written as follows:

$$H^{(1,1)} = H_h^{(1,1)} + H_a^{(1,1)} + H_b^{(1,1)} + \cdots + H_g^{(1,1)} + H_H^{(1,1)}, \tag{7}$$

where $H_h^{(1,1)}$ is the perturbation to the hamiltonian from Lorentz-violating corrections to the metric. The relevant corrections to the metric can be obtained though investigations of the classical limit in Sec. 4. The terms denoted $H_t^{(1,1)}$ are perturbations to the hamiltonian due to Lorentz-violating effects on the test particle at $O(1,1)$. As a sample of $O(1,1)$ effects, $H_a^{(1,1)}$ takes the following form:

$$H_a^{(1,1)} = \widetilde{a}_0 - \overline{a}^j h_{j0} + \Big[\widetilde{a}_j - \tfrac{1}{2}\overline{a}_j h_{00} - \tfrac{1}{2}\overline{a}^k h_{jk}\Big]\gamma^0\gamma^j. \tag{8}$$

The explicit form of the relativistic hamiltonian including all of the coefficients of the minimal fermion sector of the SME, along with additional details of its derivation, can be found in Ref. 14.

3. Non-relativistic Theory

The relativistic hamiltonian above is most useful as a tool to derive the non-relativistic hamiltonian, rather than to analyze experiments directly, because experiments most sensitive to the position operator are non-relativistic. A Foldy-Wouthuysen transformation can be applied to obtain the non-relativistic hamiltonian taking care to maintain the desired order in the small quantities.

After performing the transformation, the non-relativistic hamiltonian, h, can be written,

$$h = h^{(0,0)} + h^{(0,1)} + h^{(1,0)} + h^{(1,1)} + h^{(0,2)}. \tag{9}$$

Here, as in the relativistic case, $h^{(0,0)}$ is the conventional Minkowski-spacetime hamiltonian, $h^{(0,1)}$ and $h^{(0,2)}$ contain the leading and sub-leading gravitational corrections, and $h^{(1,0)}$ are the leading corrections due to Lorentz violation, which match Ref. 16. Again, the leading couplings of Lorentz violation to gravitational effects, $h^{(1,1)}$, are the contributions of interest. In a manner analogous to the relativistic case, this term contains contributions from the modified metric fluctuation, written $h_h^{(1,1)}$, as well as perturbations from each of the coefficients for the test particle, $h_t^{(1,1)}$.

A result of the Foldy-Wouthuysen transformation is that at leading order in the coefficients for Lorentz violation at the non-relativistic level, coefficients a_μ and e_μ always appear in the combination $(a_{\rm eff})_\mu \equiv a_\mu - m e_\mu$. This result is consistent with those of Ref. 4 indicating that e_μ can be redefined into a_μ at leading order in the coefficients for Lorentz violation via an appropriate field redefinition. As a sample of $O(1,1)$ contributions to h, the correction to the hamiltonian due to a_μ and e_μ can be written

$$h_{a_{\rm eff}}^{(1,1)} = (\widetilde{a}_{\rm eff})_0 + (\overline{a}_{\rm eff})_k h^{0k} + \tfrac{1}{m}\Big[(\widetilde{a}_{\rm eff})_j - \tfrac{1}{2}(\overline{a}_{\rm eff})_j h_{00}\Big]p^j, \tag{10}$$

to second order in momentum. Additional details of the derivation of h, along with an explicit form for the remaining spin-independent contributions to h can be found in Ref. 14.

4. Classical Theory

As in the relativistic case, the classical action can be written as the sum of partial actions.

$$S = S_G + S_u + S'. \tag{11}$$

Here, S_G and S' contain the dynamics of the gravitational field and the dynamics associated with the coefficients for Lorentz violation respectively just as in Eq. (1). The partial action S_u is the point-particle limit of S_ψ appearing in Eq. (1). The classical theory is useful for several applications including obtaining the equations of motion for a classical particle, analyzing non-relativistic quantum mechanical experiments via the path integral approach, and obtaining the modified metric.

Upon inspection of the non-relativistic hamiltonian discussed in Sec. 3, a point-particle action which corresponds to this hamiltonian can be found. The spin-independent contributions to this action can be written

$$S_u = \int d\tau \left(-m\sqrt{-(g_{\mu\nu} + 2c_{\mu\nu})u^\mu u^\nu} - (a_{\rm eff})_\mu u^\mu \right), \tag{12}$$

where u^μ is the four velocity as usual. The validity of this action has been established here only to within the assumptions made up to the presentation of the non-relativistic hamiltonian, h, and in performing calculations with this action one should not exceed the order in small quantities or the order in momentum discussed in the previous section. In addition to the match between the classical action above and the non-relativistic hamiltonian, the fact that the dispersion relation generated by the classical action matches the relativistic theory confirms the validity of this action. Note also that the $c_{\mu\nu}$ contributions to this action have been discussed previously in the context of work done in the photon sector.[17] After some consideration,[14] this action can be extended to address the case in which particles are bound within macroscopic matter as well.

5. Experimental Tests

With the theory developed above, experiments in any regime, from relativistic quantum mechanics to classical mechanics, can be analyzed, provided the contributions from S' are known. These contributions can be established directly within a model of spontaneous Lorentz violation or determined for a large class of models by examining the general form of the contributions along with the constraints available from conservation laws.

Upon a general analysis, effects are found in a number of experiments. These effects include annual and sidereal variations in the newtonian gravitational acceleration as well as variations in the gravitational force based on the proton, neutron, electron content of the bodies involved. These effects lead to signals in gravimeter tests as well as in some experiments designed to test the weak equivalence principle. These tests are described in detail in Ref. 14 and will provide the first direct sensitivities to the a_μ coefficients for the proton, neutron, and electron.

References

1. See, for example, V.A. Kostelecký, ed., *CPT and Lorentz Symmetry I-IV*, World Scientific, Singapore, 1999-2008; R. Bluhm, hep-ph/0506054; D.M. Mattingly, Living Rev. Rel. **8**, 5 (2005); G. Amelino-Camelia *et al.*, AIP Conf. Proc. **758**, 30 (2005).
2. V.A. Kostelecký and S. Samuel, Phys. Rev. D **39**, 683 (1989); Phys. Rev. D **40**, 3923 (1989); Phys. Rev. Lett. **63**, 224 (1989); V.A. Kostelecký and R. Potting, Nucl. Phys. B **359**, 545 (1991).
3. V.A. Kostelecký and R. Potting, Phys. Rev. D **51**, 3923 (1995); D. Colladay and V.A. Kostelecký, Phys. Rev. D **55**, 6760 (1997); Phys. Rev. D **58**, 116002 (1998); V.A. Kostelecký and R. Lehnert, Phys. Rev. D **63**, 116002 (2001).
4. V.A. Kostelecký, Phys. Rev. D **69**, 105009 (2004).
5. H. Dehmelt *et al.*, Phys. Rev. Lett. **83**, 4694 (1999); R. Mittleman *et al.*, Phys. Rev. Lett. **83**, 2116 (1999); G. Gabrielse *et al.*, Phys. Rev. Lett. **82**, 3198 (1999); R. Bluhm *et al.*, Phys. Rev. Lett. **82**, 2254 (1999); Phys. Rev. Lett. **79**, 1432 (1997); Phys. Rev. D **57**, 3932 (1998); D. Colladay and V.A. Kostelecký, Phys. Lett. B **511**, 209 (2001); B. Altschul, Phys. Rev. D **74**, 083003 (2006); G.M. Shore, Nucl. Phys. B **717**, 86 (2005); B. Heckel *et al.*, Phys. Rev. Lett. **97**, 021603 (2006); L.-S. Hou *et al.*, Phys. Rev. Lett. **90**, 201101 (2003); R. Bluhm and V.A. Kostelecký, Phys. Rev. Lett. **84**, 1381 (2000); H. Müller *et al.*, Phys. Rev. D **70**, 076004 (2004); H. Müller, Phys. Rev. D **71**, 045004 (2005).
6. D. Bear *et al.*, Phys. Rev. Lett. **85**, 5038 (2000); D.F. Phillips *et al.*, Phys. Rev. D **63**, 111101 (2001); M.A. Humphrey *et al.*, Phys. Rev. A **68**, 063807 (2003); Phys. Rev. A **62**, 063405 (2000); F. Canè *et al.*, Phys. Rev. Lett. **93**, 230801 (2004); P. Wolf *et al.*, Phys. Rev. Lett. **96**, 060801 (2006); M. Romalis, these proceedings; V.A. Kostelecký and C.D. Lane, Phys. Rev. D **60**, 116010 (1999); J. Math. Phys. **40**, 6245 (1999); C.D. Lane, Phys. Rev. D **72**, 016005 (2005); D. Colladay and P. McDonald, Phys. Rev. D **73**, 105006 (2006); R. Bluhm *et al.*, Phys. Rev. Lett. **88**, 090801 (2002); Phys. Rev. D **68**, 125008 (2003); O. Bertolami *et al.*, Phys. Lett. B **395**, 178 (1997).

7. J. Lipa *et al.*, Phys. Rev. Lett. **90**, 060403 (2003); H. Müller *et al.*, Phys. Rev. Lett. **99**, 050401 (2007); P. Wolf *et al.*, Phys. Rev. D **70**, 051902 (2004); S. Herrmann *et al.*, Phys. Rev. Lett. **95**, 150401 (2005); P.L. Stanwix *et al.*, Phys. Rev. D **74**, 081101(R) (2006); M. Hohensee *et al.*, Phys. Rev. D **71**, 025004 (2007); S. Reinhardt *et al.*, Nature Physics, doi:10.1038/nphys778 (2007); S.M. Carroll *et al.*, Phys. Rev. D **41**, 1231 (1990); V.A. Kostelecký and M. Mewes, Phys. Rev. D **66**, 056005 (2002); Phys. Rev. Lett. **87**, 251304 (2001); Phys. Rev. D **66**, 056005 (2002); Phys. Rev. Lett. **97**, 140401 (2006); Phys. Rev. Lett. **99**, 011601 (2007); Q.G. Bailey and V.A. Kostelecký, Phys. Rev. D **70**, 076006 (2004); R. Jackiw and V.A. Kostelecký, Phys. Rev. Lett. **82**, 3572 (1999); T. Mariz *et al.*, JHEP **0510**, 019 (2005); V.A. Kostelecký *et al.*, Phys. Rev. D **65**, 056006 (2002); V.A. Kostelecký and A.G.M. Pickering, Phys. Rev. Lett. **91**, 031801 (2003); R. Lehnert and R. Potting, Phys. Rev. Lett. **93**, 110402 (2004); Phys. Rev. D **70**, 125010 (2004); T. Jacobson *et al.*, Phys. Rev. D **67**, 124011 (2003); B. Altschul, Phys. Rev. Lett. **96**, 201101 (2006); Phys. Rev. Lett. **98**, 041603 (2007); C.D. Carone *et al.*; Phys. Rev. D **74**, 077901 (2006).
8. KTeV Collaboration, in *CPT and Lorentz Symmetry II*, Ref. 1; BELLE Collaboration, Phys. Rev. Lett. **86**, 3228 (2001); KLOE Collaboration, these proceedings; BaBar Collaboration, Phys. Rev. Lett. **92**, 142002 (2004); hep-ex/0607103; arXiv:0711.2713; FOCUS Collaboration, Phys. Lett. B **556**, 7 (2003); V.A. Kostelecký, Phys. Rev. Lett. **80**, 1818 (1998); Phys. Rev. D **61**, 016002 (2000); Phys. Rev. D **64**, 076001 (2001); N. Isgur *et al.*, Phys. Lett. B **515**, 333 (2001).
9. *g*–2 Collaboration, arXiv:0709.4670; V.W. Hughes *et al.*, Phys. Rev. Lett. **87**, 111804 (2001); R. Bluhm *et al.*, Phys. Rev. Lett. **84**, 1098 (2000).
10. LSND Collaboration, Phys. Rev. D **72**, 076004 (2005); Super-K Collaboration, in *CPT and Lorentz Symmetry III*, Ref. 1; MINOS Collaboration, these proceedings; V.A. Kostelecký and M. Mewes, Phys. Rev. D **69**, 016005 (2004); Phys. Rev. D **70**, 031902(R) (2004); Phys. Rev. D **70**, 076002 (2004); T. Katori *et al.*, Phys. Rev. D **74**, 105009 (2006); V. Barger *et al.*, Phys. Lett. B **653**, 267 (2007).
11. D.L. Anderson *et al.*, Phys. Rev. D **70**, 016001 (2004); E.O. Iltan, Mod. Phys. Lett. A **19**, 327 (2004).
12. Q.G. Bailey and V.A. Kostelecký, Phys. Rev. D **74**, 045001 (2006).
13. J. Battat *et al.*, arXiv:0710.0702; H. Müller, arXiv:0710.3768; W.M. Jensen *et al.*, these proceedings; J.M. Overduin, these proceedings; V.A. Kostelecký and R. Potting, Gen. Rel. Grav. **37**, 1675 (2005).
14. V.A. Kostelecký and J.D. Tasson, in preparation.
15. R. Bluhm and V.A. Kostelecký, Phys. Rev. D **71**, 065008 (2005); R. Bluhm *et al.*, in preparation; B. Altschul and V.A. Kostelecký, Phys. Lett. B **628**, 106 (2005).
16. V.A. Kostelecký and C.D. Lane, J. Math. Phys. **40**, 6245 (1999).
17. Q. Bailey and V.A. Kostelecký, Phys. Rev. D **70**, 076006 (2004).

TORSION BALANCE TEST OF PREFERRED-FRAME AND WEAK COUPLING TO POLARIZED ELECTRONS

B.R. HECKEL, C.E. CRAMER, E.G. ADELBERGER, AND T.S. COOK

Physics Box 351560
University of Washington
Seattle, WA 98195, U.S.A.
E-mail: heckel@phys.washington.edu

We used a torsion balance containing polarized electrons to search for new weak forces that couple to electron spins and for a violation of Lorentz Symmetry as may arise from the presence of a preferred-frame. We used the gyro-compass torque of the Earth's rotation acting on the angular momentum of the pendulum to calibrate the spin content of the pendulum. Data taken over several years allows us to test rotational invariance at the level of 10^{-31} GeV which should be compared to a benchmark value of $m_e^2/M_{Planck} = 2 \times 10^{-26}$ GeV. Preliminary results will be presented.

1. Introduction

We have developed a torsion balance containing $\approx 1 \times 10^{23}$ polarized electrons to search for new spin coupled interactions. This paper will discuss the status of our results for two types of interactions: a CP-violating "monopole-dipole" interaction between a polarized electron and an unpolarized atom and our results for the $\tilde{b}^e$ parameters of the Standard-Model Extension (SME). An analysis of boost invariance violation and other velocity dependent potentials is in progress but will not be discussed in this report.

Moody and Wilczek[1] considered the forces produced by the exchange of low-mass, spin-0 particles and showed that the exchange of a boson that contained a CP-violating admixture of 0^+ and 0^- couplings would produce a macroscopic, CP-violating "monopole-dipole" interaction between a polarized electron and an unpolarized atom with mass and charge numbers A and Z,

$$V_{eA}(r) = g_P^e g_S^A \frac{\hbar}{8\pi m_e c} \boldsymbol{\sigma}_e \cdot \left[\hat{\boldsymbol{r}} \left(\frac{1}{r\lambda} + \frac{1}{r^2} \right) e^{-r/\lambda} \right] , \qquad (1)$$

where $m_\phi = \hbar/(\lambda c)$ is the mass of the spin-0 particle, $g_{\rm P}$ and $g_{\rm S}$ are its pseudoscalar and scalar couplings, and $g_{\rm S}^A = Z(g_{\rm S}^e + g_{\rm S}^p) + (A - Z)g_{\rm S}^n$. For simplicity, we will assume that $g_{\rm S}^p = g_{\rm S}^n = g_s^N$ and $g_{\rm S}^e = 0$ so that $g_{\rm S}^A = Ag_{\rm S}^N$. We analyze our spin pendulum results for the coupling of electron spins to unpolarized masses in and near to our laboratory, to the Earth, and to the Sun.

The SME, developed by Kostelecký and coworkers,[2] gives rise to a preferred-frame that violates Lorentz and CPT invariance. In the electron sector of the theory, the Lorentz-violating terms in the lagrangian produce an effective field, $\tilde{b}_j^e$, fixed in space that couples to electron spin. In the non-relativistic limit, the potential can be written as:[3]

$$V = \tilde{b}_j^e \, \sigma_e^j . \tag{2}$$

A recent Letter[4] reported our first results for $g_{\rm P}^e g_{\rm S}^A/(\hbar c)$ and $\tilde{b}^e$. Since that time, we have taken additional data and have calibrated the spin content of our pendulum. In the process, we identified and eliminated a systematic error that limited the precision of our earlier results.

2. Rotating Torsion Balance Apparatus

The spin pendulum is mounted within a rotating torsion balance apparatus, described in detail elsewhere.[4,5] The pendulum is suspended from a 75 cm long, 30 μm diameter tungsten fiber, and centered within four layers of high permeability magnetic shields. The innermost shield is gold-coated to minimize electrostatic coupling. The pendulum and shields are located within a vacuum vessel that is held at approximately 10^{-6} torr by an ion pump. The vacuum vessel is mounted on a turntable that rotates at a constant rate of approximately five rev/hr.

Other important components of the apparatus include tilt sensors that allow the rotation axis to be monitored and aligned vertically, constant temperature water-cooled Cu shields that provided thermal isolation, and nearby machined Pb compensator masses used to cancel gradients in local gravitational field. A set of three-axis Helmholtz coils surround the apparatus and are adjusted to cancel the local magnetic field to 2%. The most significant improvement to the apparatus since Ref. 5 has been the installation of a feedback system that holds the rotation axis to within 10 nrad of vertical (the laboratory floor tilt drifts by μrads daily).

Diode laser light is doubly reflected from one of the four mirrors mounted on the pendulum and the reflected beam is focused onto a linear position sensitive photodiode to monitor the angular position of the pendulum, Θ. A rotary stage at the top of the torsion fiber allows the light to be centered onto any one of the four symmetrically placed mirrors on the spin pendulum.

3. Spin pendulum

The spin pendulum, designed to contain a large number of polarized electrons with minimal magnetic and gravitational moments, is constructed from 4 octagonal "pucks" as shown in Fig. 1. One side of each puck is Alnico 5, a Fe/Co based ferromagnet for which the magnetic field is created almost entirely by electron spins, and the other side is grade 22 $\mathrm{Sm\,Co_5}$, in which the orbital magnetic moment of the electrons in the Sm^{3+} ion nearly cancels their spin moment.

After each puck was assembled, the Alnico was magnetized to the same degree as the $\mathrm{Sm\,Co_5}$ by applying appropriate external fields to the Alnico segments resulting in a 9.6 kG field that ran toroidally within the pucks. By stacking 4 such pucks as shown in Fig. 1, we placed the effective center of the spin dipole in the middle of the pendulum and averaged out the small density differences between Alnico and $\mathrm{Sm\,Co_5}$.

The magnet ring stack was mounted within a small mu-metal magnetic shield to reduce the leakage of magnetic fields from imperfections in the magnets. The shield supports four symmetrically placed mirrors. We estimate the number of net polarized spins in our pendulum to be:[6] $N_{\rm p} = (0.96 \pm 0.17) \times 10^{23}$.

Because the magnetization of the pendulum vanishes, it follows that the pendulum should have a net angular momentum $\boldsymbol{J} = -N_{\rm p}\hbar/2\,\hat{z}$, where $\hat{z}$ is the direction of the electron spin polarization. $\boldsymbol{J}$ will cause the pendulum to act as a gyro-compass, forcing the spins to experience a South-pointing torque that corresponds to an energy per spin of $\beta^{\rm gyro} = 1.62 \times 10^{-20}$ eV.

4. Data and Analysis

The angular twist of the pendulum, $\Theta(\phi)$, ($\phi = \omega t$ being the angle of the turntable) and other sensor values were recorded every 10 s over 1-2 day data runs. The data runs were divided into "cuts", each lasting two complete turntable revolutions. Data points one half of a torsional cycle apart in time were averaged to filter out the free torsion amplitude, and the time sequence of filtered $\Theta(\phi)$ values in each cut was fit to a Fourier series out to the eighth harmonic of the turntable frequency in addition to

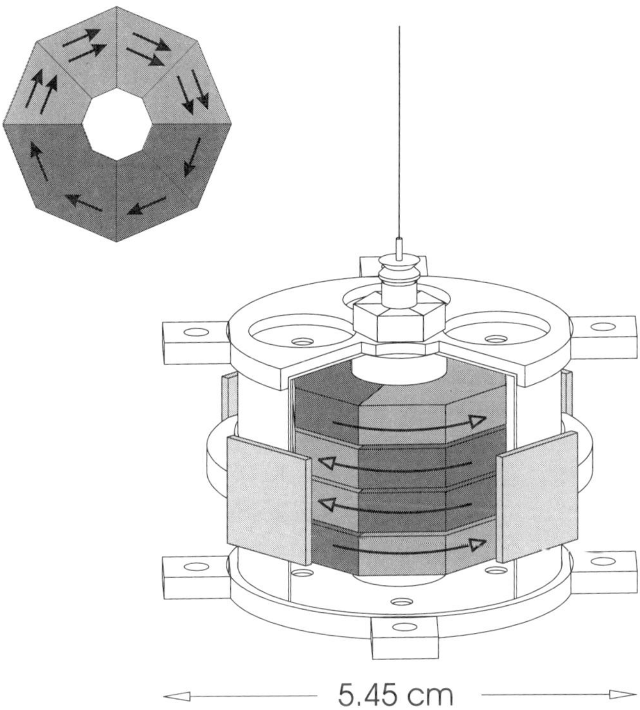

Figure 1. Scale drawing of the spin pendulum. The light and darker volumes are Alnico and $SmCo_5$, respectively. Upper left: top view of a single "puck"; the spin moment points to the right. Lower right: the assembled pendulum with the magnetic shield shown cut away to reveal the 4 pucks inside. Two of the 4 mirrors used to monitor the pendulum twist are prominent. Arrows with filled heads show the relative densities and directions of the electron spins, open-headed arrows show the directions of $\boldsymbol{B}$.

an offset and linear drift term.[5]

To an excellent approximation, an external potential, V, that couples to spin will appear in the first harmonic of the Fourier series. We write the energy of the pendulum as:

$$V = -N_{\rm p}\,\boldsymbol{\sigma}_{\rm p}\cdot\boldsymbol{\beta}\ , \tag{3}$$

where $\boldsymbol{\sigma}_{\rm p}$ represents the orientation of the pendulum's spin, and $\boldsymbol{\beta}$, the energy per spin, has a value in the laboratory frame that may depend on time. It follows that the first harmonic of the Fourier series is given by:

$$\begin{aligned}\Theta(\phi) &= (\beta_{\perp} N_{\rm p}/\kappa)\sin(\phi_0-\phi)\\ &\equiv -(N_{\rm p}/\kappa)(\beta_{\rm N}\sin\phi+\beta_{\rm E}\cos\phi)\ ,\end{aligned} \tag{4}$$

where $\beta_\perp$ is the projection of $\boldsymbol{\beta}$ on the horizontal plane, ϕ_0 is its azimuthal angle, and $\beta_{\rm N}$ and $\beta_{\rm E}$ are the North and East components of $\boldsymbol{\beta}_\perp$. The only correction applied to the $\Theta(\phi)$ Fourier amplitudes was an attenuation correction that accounts for the effects of the pendulum inertia, electronic time constants, and filtering.

4.1. *Laboratory-fixed signals*

The expected gyro-compass signal, a spin dependent interaction that couples to local masses or the Earth, and $\tilde{b}^e_Z$ (where $\hat{Z}$ points along the Earth's spin axis) all lead to $\boldsymbol{\beta}$ that are constant in time. We searched for such signals by taking data at four different orientations, ϕ_d, of the pendulum within the rotating vessel. Our first results, reported in Ref. 4, indicated an unknown systematic error that limited the precision of our measurements. We identified the error as coming from a torque that acted on the upper suspension system of the fiber. By using a ball-in-cone attachment of the fiber to the pendulum, we were able to rotate the upper suspension system with respect to the pendulum by an angle ϕ_s to separate a true torque on the pendulum from one acting on the suspension system.

We took data at all four ϕ_d settings for seven different ϕ_s values in two data sets. The results are shown in Table 1.

Table 1. Laboratory-fixed signals extracted from ball-in-cone data. Units of β are 10^{-20} eV. Dates refer to the mean time of the data sets.

signal	15/10/06 value	15/6/07 value	combined
$\beta_{\rm N}$	-1.62 ± 0.30	-1.67 ± 0.30	-1.65 ± 0.29
$\beta^{\rm gyro}$	-1.62	-1.62	-1.62
$\beta_{\rm N} - \beta^{\rm gyro}$	$+0.00 \pm 0.30$	-0.05 ± 0.30	-0.03 ± 0.29
$\beta_{\rm E}$	$+0.03 \pm 0.07$	-0.04 ± 0.05	-0.01 ± 0.04

In Table 1, $\beta_{\rm N,E}$ are the measured signals acting on the pendulum whose errors include the uncertainty in our estimated value for $N_{\rm p}$. $\beta_{\rm N} - \beta^{\rm gyro}$ and $\beta_{\rm E}$ are then the signals we can use to set limits on new laboratory-fixed spin coupled interactions.

4.2. *Astronomically modulated signals*

A spin interaction that is coupled to the Sun produces a signal with solar day modulation while components of $\tilde{b}^e$ perpendicular to the Earth's spin axis produce signals with a sidereal day modulation. We follow the convention of Bluhm and Kostelecký[3] to define the nonrotating coordinate axes of

$\tilde{b}^e_j$. The $\hat{Z}$ axis is taken to lie along the rotational north pole of the Earth. The $\hat{X}$ axis points from the Earth towards the Sun at the vernal equinox. For each cut within a data set, we compute the angular coordinates of our spin pendulum relative to the celestial axes. We then fit all of the first harmonic signals from the cuts to a function that describes the amplitude and phase of a coupling to any of the celestial axes.[4] We took 75 data sets sets, each of 4-5 day duration, over a span of 3 years. For the astronomically modulated signals, we take $N_{\rm p}^{\rm exp} = (9.80 \pm 0.27) \times 10^{22}$, the number of electrons spins we infer by taking the laboratory-fixed $\beta_{\rm N} = \beta^{\rm gyro}$.

Preliminary results for our search for $\tilde{b}^e$ are given in Table 2. Preliminary limits for a "monopole-dipole" interaction that couples to the Sun or to laboratory-fixed sources are shown in Fig. 2. In Fig. 2, our results and previous work by Youdin *et al.*,[8] Ni *et al.*,[9] and Wineland *et al.*[10] are indicated by solid, dash-dotted, dotted and dashed lines, respectively.

Table 2. 1σ constraints from our work and from Hou *et al.* on the Lorentz-violating rotation parameters $\tilde{b}^e$. Units are 10^{-22} eV. The larger $\tilde{b}^e_{\rm Z}$ errors, compared to those for $\tilde{b}^e_{\rm X}$ $\tilde{b}^e_{\rm Y}$, reflect the larger systematic uncertainty in laboratory-fixed signals.

parameter	this work	Hou *et al.*[7]
$\tilde{b}^e_{\rm X}$	-1.27 ± 1.38	-108 ± 112
$\tilde{b}^e_{\rm Y}$	$+0.39 \pm 1.38$	-5 ± 156
$\tilde{b}^e_{\rm Z}$	-3.7 ± 29.2	107 ± 2610

Acknowledgments

Michael Harris, Stefan Baeßler, Ulrich Schmidt, and Erik Swanson helped develop earlier versions of this apparatus. Jens Gundlach, C.D. Hoyle, and Stefan Schlamminger contributed helpful discussions about this experiment. We thank David Kaplan and Alan Kostelecký for inspiring conversations, and Tom Murphy, Jr. for advice on the astronomical calculations. This work was supported by NSF Grant PHY0355012 and by the DOE Office of Science. CEC is grateful for an NSF Fellowship.

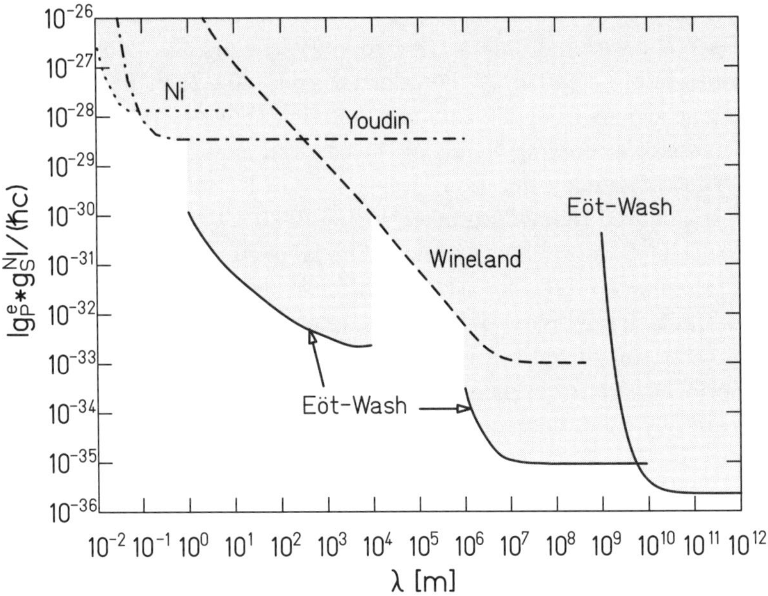

Figure 2. Upper limits on $|g_P^e g_S^N|/(\hbar c)$ as a function of interaction range λ; the shaded region is excluded at 95% confidence. Our work does not provide constraints for 10 km $< \lambda < 10^3$ km because integration over the terrestrial surrounding is not reliable in this regime.

References

1. J.E. Moody and F. Wilczek, Phys. Rev. D **30**, 130 (1984).
2. D. Colladay and V.A. Kostelecký, Phys. Rev. D **55**, 6760 (1997); Phys. Rev. D **58**, 116002 (1998).
3. R. Bluhm and V.A. Kostelecký, Phys. Rev. Lett. **84**, 1381 (2000).
4. B.R. Heckel *et al.*, Phys. Rev. Lett. **97**, 021603 (2006).
5. Y. Su, E.G. Adelberger, B.R. Heckel, J.H. Gundlach, G. Smith, M. Harris, and E. Swanson, Phys. Rev. D **50**, 3614 (1994).
6. C.E. Cramer, Ph.D. thesis, Univ. of Washington, unpublished (2007).
7. L.-S. Hou, W.-T. Ni, and Y.-C. Li, Phys. Rev. Lett. **90**, 201101 (2003).
8. A. N. Youdin *et al.*, Phys. Rev. Lett. **77**, 2170 (1996).
9. W. T. Ni *et al.*, Phys. Rev. Lett. **82**, 2439 (1999).
10. D. J. Wineland *et al.*, Phys. Rev. Lett. **67**, 1735 (1991).

SEEKING A SOLUTION OF THE PIONEER ANOMALY

MICHAEL MARTIN NIETO
Theoretical Division (MS-B285)
Los Alamos National Laboratory
Los Alamos, New Mexico 87545, U.S.A.
E-mail: mmn@lanl.gov

JOHN D. ANDERSON
121 S. Wilson Ave.
Pasadena, CA 9110, U.S.A.
E-mail: jdandy@earthlink.net

The 1972 and 1973 launched Pioneer 10 and 11 were the first missions to explore the outer solar system. They achieved stunning breakthroughs in deep-space exploration. But around 1980 an unmodeled force of $\sim 8 \times 10^{-8}$ cm/s^2, directed approximately towards the Sun, appeared in the tracking data. It later was unambiguously verified as not being an artifact. The origin remains unknown (although radiant heat remains a likely cause). Increasing effort has gone into understanding this anomaly. We review the situation and describe programs to resolve the issue.

1. The Pioneer Missions and their Navigation

The first missions to fly to deep space were the Pioneers. By using flybys they were able to obtain heliocentric velocities that were unfeasible at the time with only chemical fuels. Pioneer 10 was launched on 2 March 1972 local time, aboard an Atlas/Centaur/TE364-4 launch vehicle. It was the first craft launched into deep space, the first to reach an outer giant planet, Jupiter,[1] and the first spacecraft to leave the "solar system."

While in its Earth-Jupiter cruise, Pioneer 10 was still bound to the solar system. With the Jupiter flyby, Pioneer 10 reached escape velocity from the solar system. Pioneer 10 has an asymptotic escape velocity from the Sun of 11.322 km/s (2.388 AU/yr). (An Astronomical Unit (AU) is the mean Sun-Earth distance, about 150,000,000 km.)

Pioneer 11 followed soon after Pioneer10, with a launch on 6 April 1973. It, too, cruised to Jupiter on an approximate heliocentric ellipse. On 2 Dec.

1974, when Pioneer 11 reached Jupiter, it underwent a Jupiter gravity assist that sent it back inside the solar system to catch up with Saturn on the far side. It was then still on an ellipse, but a more energetic one. Pioneer 11 reached Saturn on 1 Sept. 1979. The trajectory took the craft under the ring plane on approach. After passing through the plane – again without catastrophic consequences – Pioneer 11 embarked on an escape hyperbolic trajectory with an asymptotic escape velocity from the Sun of 10.450 km/s (2.204 AU/yr), in roughly the opposite direction as Pioneer 10.

The Pioneer navigation was carried out at the Jet Propulsion Laboratory. It was ground-breaking in its advances – no craft had delved so far out into the solar system. NASA's Deep Space Network (DSN) was used to transmit and obtain the raw radiometric data. An S-band signal ($\sim$2.11 Ghz) was sent up via a DSN antenna located either in California, in Spain or in Australia. The signal was transponded back with a (240/221) frequency ratio ($\sim$2.29 Ghz), and received back (at another station if, during the radio round trip, the original station had rotated out of view). There the signal was compared with 240/221 times the recorded transmitted frequency and any Doppler frequency shift was measured directly by cycle count compared to an atomic clock. This produced a data record of Doppler frequency shift as a function of time, and from this a trajectory was calculated.

However, to obtain the spacecraft velocity as a function of time from this Doppler shift is not easy. The codes must include gravitational and time effects of general relativity, the effects of the Sun, planets and their large moons. The positions of the receiving stations and the effects of the environment, atmosphere, solar plasma are included.

Given the above tools, precise navigation was possible because the Pioneers were spin-stabilized. With spin-stabilization the craft are rotated at a rate of $\sim$(4-7) rpm about the principal moment-of-inertia axis. Thus, the craft is a gyroscope and attitude maneuvers are needed only when the motions of the Earth and the craft move the Earth from the antenna's line-of-sight. Thus, especially in the later years, only a few orientation maneuvers were needed every year to keep the antenna pointed towards the Earth, and these could be easily modeled.

Even so, there remained one relatively large effect on this scale that had to be modeled: the solar radiation pressure of the Sun. It produced an acceleration of $\sim 20 \times 10^{-8}$ cm/s^2 on the Pioneer craft at the distance of Saturn (9.38 AU from the Sun at encounter). Therefore, any "unmodeled force" on the craft could not be seen very well below this level at Jupiter. However, beyond Jupiter it became possible.

2. The Anomaly is Observed

In 1969 one of us (JDA) had become PI of the radioscience celestial mechanics experiment for the Pioneers. Eventually the Pioneer Doppler data going back to 1976 for Pioneer 11 and 1981 for Pioneer 10 (but also including the Jupiter flyby) was archived at the National Space Science Data Center (NSSDC), something that later was to prove extremely helpful. Part of the celestial mechanics effort, working together with the navigation team, was to model the trajectory of the spacecraft very precisely and to determine if there were any unmodeled effects.

After 1976 small time-samples (approximately 6-month to 1-year averages) of the data were periodically analyzed. These data points were obtained individually by a number of very-qualified investigators. At first nothing significant was found. But when the analysis was done around Pioneer 11 's Saturn flyby, things dramatically changed. (See the first three data points in Fig. 1.) So people kept following Pioneer 11. They also started looking more closely at the incoming Pioneer 10 data.

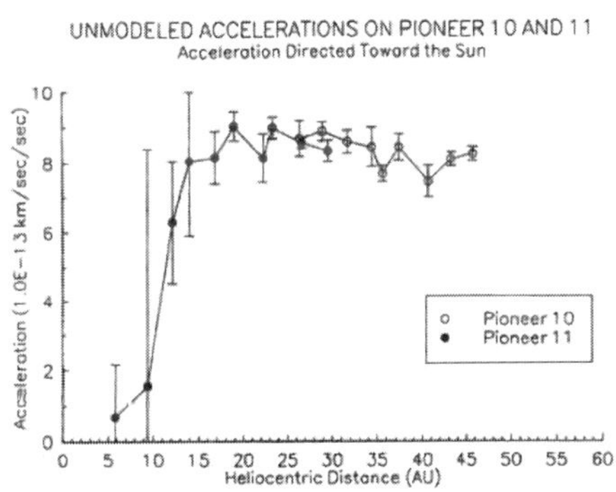

Figure 1. A JPL Orbital Data Program (ODP) plot of the early unmodeled accelerations of Pioneer 10 and Pioneer 11, from about 1981 to 1989 and 1977 to 1989, respectively.

By 1987 it was clear an anomalous acceleration $\sim 8 \times 10^{-8}$ cm/s^2 appeared to be acting, approximately towards the Sun. Although of concern, the effect was small and did not affect the necessary precision of the navigation. By 1992 it was clear a more detailed look would be useful.

In 1994 MMN contacted JDA to ask how well we understood gravity far out in the solar system. JDA responded that, "The biggest systematic in the (Pioneer) acceleration residuals is a bias of 8×10^{-13} km/s^2 directed toward the Sun." The result was an announcement in a 1994 proceedings. (More information and many references can be found in Ref. 2.)

The strong reaction was that the anomaly might be due to JPL's ODP program, so an independent code had to test it. JDA put together a team that included two former Pioneer co-workers Phil Laing and Tony Liu, of The Aerospace Corporation. Laing and Liu used the independent CHASMP navigation code to look at the Pioneer data. To within small uncertainties, their result was the same as that obtained by JPL's ODP.

This resulted in the Pioneer anomaly Collaboration's discovery paper of 1998.[3] Thereafter a detailed analysis began using the Pioneer 10 data spanning 3 January 1987 to 22 July 1998 (when the craft was 40 AU to 70.5 AU from the Sun) and Pioneer 11 data spanning 5 January 1987 to 1 October 1990 (when Pioneer 11 was 22.4 to 31.7 AU from the Sun). It was during this period that we presented our report to the CPT'01 conference.[4]

The most significant biases and errors (all in units of 10^{-8} cm/s^2) are radio beam reaction force (1.10 ± 0.11); RTG heat reflected off the craft (-0.55 ± 0.55); differential emissivity of the RTGs (± 0.85); non-isotropic radiative cooling of the craft (± 0.48); gas leakage (± 0.56). Note that except for the radio beam reaction force, which mainly is a bias, and the gas leak uncertainty, the other major systematics are due to heat.

When added to experimental residuals, this yields the final result, that there is an unmodeled acceleration, approximately towards the Sun, of

$$a_P = (8.74 \pm 1.33) \times 10^{-8} \text{cm/s}^2 \tag{1}$$

which we reported in 2002.[5] The conclusion, then, is that this "Pioneer anomaly" is in the data. The question is, "What is its origin?"[a]

3. Proposed Origins of the Anomaly

On board systematics: It is tempting to assume that radiant heat must be the cause of the acceleration, since only 63 W of directed power could cause the effect (and much more heat than that is available).

[a]One can note a small apparent annual oscillation on top of the constancy in Fig. 1. Careful analysis of clean, late-time data showed[5] that this signal is indeed significant, *as well as* a diurnal signal. These are believed to be independent of the main anomaly.

There is another anomaly that is associated with Earth flybys. A spacecraft's total *geocentric* orbital energy per unit mass *should* be the same before and after the flyby. But the data indicate this is not always true.[7]

The heat on the craft ultimately comes from the Radioisotope Thermoelectric Generators (RTGs), which yield heat from the radioactive decay of ^{238}Pu. Before launch, the four RTGs had a total thermal fuel inventory of 2580 W ($\approx$ 2070 W in 2002). Of this heat 165 W was converted at launch into electrical power emanating from around the main bus ($\approx$ 65 W in 2002), significantly from the louvers at the bottom of the bus.

Therefore, it has been argued whether the anomalous acceleration is due to anisotropic heat reflection off of the back of the spacecraft high-gain antennas, to the heat emanating from the louvers (open or closed), or if a combination of both these sources must be considered.

In fact, the craft was designed so that the heat was radiated out in a very fore/aft symmetric manner. Further, the heat from electric power went down by almost a factor of 3 during the mission. With all these points in mind, no one as yet has been able to firmly tie this down, despite heated controversy.[6] Therefore, heat as a mechanism remains to be clearly resolved, but studies are underway (see Section 4).

Other physics: Drag from normal matter dust as well as gravity from the Kuiper belt have been ruled out. Also, if this is a modification of gravity, it is not universal; i.e., it is not a scale independent force that affects planetary bodies in bound orbits.[5] The anomaly could, in principle be i) some modification of gravity, ii) drag from dark matter or a modification of inertia, or iii) a light acceleration. (Remember, the signal is a Doppler shift which is only interpreted as an acceleration.) Future understanding of the anomaly will determine which, if any, of these proposals are viable.

In the above circumstances the true direction of the anomaly should be i) towards the Sun, ii) along the craft velocity vector, or iii) towards the Earth. If the origin is heat, or any other spacecraft-generated force, the acceleration would be iv) along the spin axis. (Any internal systematic forces normal to the spin axis are canceled out by the rotation.)

4. Possibilities for Progress

Studying the entire data set: The major analysis[5] used data from 1987.0 to 1998.5. However, a long data span would help discern whether the anomaly is truly constant or if there is a time-dependence. Further, one might be able to do 3-dimensional tracking precisely enough to determine the exact direction of the anomaly. Perhaps most intriguingly, by closely studying the data around the planetary flybys, it might be determined if there was an onset of the anomaly near these encounters.

If all the Doppler data, from launch to last contact, could be precisely analyzed together *and* all systematics external to the craft could be separated, the above tasks could be accommodated. The data itself was stored using obsolete formats on obsolete platforms. Further, it was not to be found in one place, although the NSSDC did have a sizable segment. The untangling and interpretation of the archival Doppler data is an ongoing and important project. Further, there is the telemetry which contains the engineering data, including such things as temperatures, voltages, spin rates, etc. This data has also now been transferred to modern format.

In the long run the study of the telemetry might prove to be of most "use." The Collaboration has long observed that, even if the anomaly turns out to be due to systematics, a thorough anomaly inquiry would still result in a win. One would obtain a better understanding of how spacecraft behave in deep space and therefore how to build, model, and track craft there.

Ultimately things will be decided by the analysis of the Doppler data, if it can show both the size of the anomaly with time and also, with luck, its direction. This latter will be most difficult to discern because of the large systematics close in to the Sun. But the results could be very rewarding.

The New Horizons mission to Pluto: On 19 Jan 2006 the New Horizons mission to Pluto and the Kuiper Belt was launched from Cape Canaveral. Although of relatively low mass ($\sim$478 kg), this craft was not designed for precision tracking but it might yield useful information.

The main reason is that for much of its life New Horizons will be in spin-stabilization mode, for example for much of the period after June 2007 (Jupiter encounter was on 28 Feb. 2007) until soon before the Pluto encounter on 14 July 2015. This is designed to save fuel so that as much fuel as possible will be available after Pluto encounter to aim at a Kuiper Belt Object. The Doppler and range data from these periods could supply a test, at some level, of the Pioneer anomaly. With luck something could be learned from the New Horizons data by 2010 or soon thereafter.

In summary, we await to see if this mission can be of use to the study of the anomaly. The analysis of data from the Pluto mission will be a challenge, but though it might not aid our goals, we encourage it.

A Dedicated Mission? If the above efforts are not able to yield a *negative* resolution of the anomaly, then a new experimental test might be needed, either as an attached experiment or probe, or even as a dedicated mission. In Europe a collaboration arose. The aim was to present a proposal for ESA's Cosmic Vision program, with launches during the period 2015-2025. Its timing would be perfect if the two investigations described above indicate a dedicated test of the anomaly is called for.

The collaboration decided to submit a two-stage proposal. The first stage would use modern 3-axis, drag-free accelerometers developed by ONERA in France. This mission would be a slower mission using flybys for gravity assists out to ~ 10 AU. This mission would get its main power from solar cells. As such this mission could be flown cheaper and quicker. But it would have power limitations and entail large solar radiation pressure effects from the solar-cell assemblies that are needed at larger heliocentric distances. A later mission would be equipped with modern quantum-technology instruments to measure precise accelerations at longer distances. If called for, a dedicated mission could be very exciting and definitive.

5. Conclusion

Although it is unlikely that the Pioneer anomaly is caused by new physics, that is not ruled out. It is possible that the anomaly could be something importantly new. For that reason, and also because we want to be able to account for every source of systematic error in navigation, we continue to pursue this study. This anomaly will be resolved eventually.

Acknowledgments

The work described in this manuscript was supported by the U.S. Department of Energy (MMN) and by the National Aeronautics and Space Administration (JDA).

References

1. M. Wolverton, *The Depths of Space*, Joseph Henry, Washington, 2004.
2. M. M. Nieto and J. D. Anderson, Contemp. Phys. **48**, No. 1. 41 (2007).
3. J. D. Anderson, P. A. Laing, E. L. Lau, A. S. Liu, M. M. Nieto, and S. G. Turyshev, Phys. Rev. Lett. **81**, 2858 (1998).
4. M. M. Nieto, J. D. Anderson, P. A. Laing, E. L. Lau, and S. G. Turyshev, in V.A. Kostelecký, ed., *CPT and Lorentz Violation II*, World Scientific, Singapore, 2002.
5. J. D. Anderson, P. A. Laing, E. L. Lau, A. S. Liu, M. M. Nieto, and S. G. Turyshev, Phys. Rev. D **65**, 082004 (2002).
6. J. D. Anderson, E. L. Lau, S. G. Turyshev, P. A. Laing, and M. M. Nieto, Mod. Phys. Lett. A **17**, 875 (2002).
7. J. D. Anderson, J. K. Campbell, and M. M. Nieto, New Astron. **12** 383 (2007).

TESTING LORENTZ SYMMETRY WITH GRAVITY

QUENTIN G. BAILEY
Physics Department
Embry-Riddle Aeronautical University
3700 Willow Creek Road
Prescott, AZ 86301, U.S.A.
E-mail: baileyq@erau.edu

In this talk, results from the gravitational sector of the Standard-Model Extension (SME) are discussed. The weak-field phenomenology of the resulting modified gravitational field equations is explored. The application of the results to a variety of modern gravity experiments, including lunar laser ranging, Gravity Probe B, binary pulsars, and Earth-laboratory tests, shows promising sensitivity to gravitational coefficients for Lorentz violation in the SME.

1. Introduction

At the present time, a comprehensive and successful description of nature is provided by general relativity and the Standard Model of particle physics. It is expected, however, that a single underlying unified theory would merge them at the Planck scale. To date, a completely satisfactory theory remains elusive. Experimental clues about this underlying theory are lacking since direct measurements at the Planck scale are infeasible at present.

An alternative approach is to look for suppressed new physics effects coming from the underlying theory that are potentially detectable in modern sensitive experiments. One promising class of signals satisfying this criteria are minuscule violations of Lorentz symmetry.[1] For describing the observable signals of Lorentz violation, the effective field theory known as the Standard-Model Extension (SME) provides a useful tool.[2,3]

Much of the theoretical and experimental work on the SME has involved the the Minkowski-spacetime limit. Experimental studies have included ones with photons,[4] electrons,[5] protons and neutrons,[6] mesons,[7] muons,[8] neutrinos,[9] and the Higgs.[10] Though no compelling evidence for Lorentz violation has been found, only about half of the possible signals involving light and ordinary matter have been experimentally investigated, while some other sectors remain largely unexplored. The subject of the talk will be a recent SME-based study of gravitational experiments searching for violations of local Lorentz invariance. For a more detailed discussion, the reader is referred to Ref. 11.

2. Theory

The gravitational couplings in the SME action are presented in Ref. 3. The geometric framework assumed is a Riemann-Cartan spacetime, allowing for torsion. For simplicity, attention is restricted to the Riemann-spacetime limit. In this limit, the effective action of the pure-gravity minimal SME is written

$$S = \frac{1}{16\pi G}\int d^4x\sqrt{-g}[(1-u)R + s^{\mu\nu}R^T_{\mu\nu} + t^{\kappa\lambda\mu\nu}C_{\kappa\lambda\mu\nu}] + S'. \quad (1)$$

Here R is the Ricci scalar, $R^T_{\mu\nu}$ is the trace-free Ricci tensor, and $C_{\kappa\lambda\mu\nu}$ is the Weyl conformal tensor. The leading Lorentz-violating gravitational couplings are controlled by the coefficients for Lorentz violation u, $s^{\mu\nu}$, and $t^{\kappa\lambda\mu\nu}$. Equation (1) contains 20 independent coefficients, of which 1 is in u, 9 are in the traceless $s^{\mu\nu}$, and 10 are in the totally traceless $t^{\kappa\lambda\mu\nu}$.

It is known that explicit Lorentz violation, whereupon the coefficients for Lorentz violation in Eq. (1) are nondynamical functions of spacetime, is generally incompatible with Riemann spacetime.[3] Spontaneous Lorentz violation, however, evades this problem[12] and is the approach adopted to analyze Eq. (1). In this scenario the coefficients u, $s^{\mu\nu}$, and $t^{\kappa\lambda\mu\nu}$ are dynamical fields that acquire vacuum expectation values denoted $\overline{u}$, $\overline{s}^{\mu\nu}$, and $\overline{t}^{\kappa\lambda\mu\nu}$. The general matter action S' in Eq. (1) therefore includes the dynamics for ordinary matter as well as the coefficients for Lorentz violation.

To construct the field equations associated with the action (1), while taking into account the unknown dynamics of the coefficient fields u, $s^{\mu\nu}$, and $t^{\kappa\lambda\mu\nu}$, represents a challenging theoretical task. In the case of weak-field gravity, however, a set of modified field equations can be obtained under mild assumptions,[11] which then determine the leading corrections to general relativity arising from Lorentz violation. In particular, the dominant terms in the post-newtonian metric can be determined. From the post-newtonian metric an effective classical lagrangian for N point-like bodies can be derived. This lagrangian provides the basis for studies of orbital experiments probing the coefficients $\overline{s}^{\mu\nu}$, while the post-newtonian metric is used to describe experiments probing spacetime geometry.

It is standard to compare a given post-newtonian metric with the

Parametrized Post-Newtonian (PPN) metric.[13,14] It turns out that the match can only be achieved when the SME coefficients $\overline{s}^{\mu\nu}$ are assumed isotropic in a special coordinate frame, resulting in only one rotational scalar coefficient (taken as $\overline{s}^{00} = \overline{s}^{jj}$) remaining. This isotropic assumption is not generally adopted in SME studies and so the relationship between the SME and the PPN is one of partial overlap.

3. Lunar laser ranging

The primary observable in lunar laser ranging experiments are oscillations in the Earth-Moon distance. High sensitivity is achieved by timing laser pulses reflecting off of one or more of the five reflectors on the lunar surface.[15] Appropriate application of the effective classical lagrangian yields the Lorentz-violating corrections to the Earth-Moon coordinate acceleration. Ideally, a computer code would be used that includes the standard dynamics of the Earth-Moon system and effects from the pure-gravity sector of the minimal SME.

It is useful, however, to perform a perturbative analysis that extracts the dominant oscillation frequencies and corresponding amplitudes for Earth-Moon separation oscillations driven by Lorentz violation. The radial corrections δr arising from the Lorentz-violating terms in the acceleration take the generic form

$$\delta r = \sum_n [A_n \cos(\omega_n T + \phi_n) + B_n \sin(\omega_n T + \phi_n)]. \tag{2}$$

The dominant amplitudes are denoted A_n and B_n and the corresponding phases are ϕ_n. For example, one oscillation occurs at twice the mean orbital frequency ω with amplitudes given by $A_{2\omega} = -\frac{1}{12}(\overline{s}^{11} - \overline{s}^{22})r_0$ and $B_{2\omega} = -\frac{1}{6}\overline{s}^{12}r_0$ where r_0 is the mean Earth-Moon distance. The coefficients $\overline{s}^{11} - \overline{s}^{22}$ and $\overline{s}^{12}$ are combinations of the standard Sun-centered frame coefficients $\overline{s}^{JK}$, and depend on them through angles describing the orbit. This angular dependence indicates that it may be useful to consider artificial satellite orbits of varying orientation, in order to attain sensitivity to coefficients that may elude the lunar orbit.

For lunar laser ranging, at least 5 independent combinations of coefficients for Lorentz violation can be measured. Using standard lunar values and assuming ranging precision at the centimeter level,[15] the estimated experimental sensitivities are parts in 10^{10} on combinations of coefficients in $\overline{s}^{JK}$ and parts in 10^7 on the coefficients $\overline{s}^{TJ}$. An analysis studying the dominant Earth-Moon oscillations using 30+ years of data has recently been performed and has achieved roughly this level of sensitivity.[16] The new Apache Point Observatory Lunar Laser-ranging Operation (APOLLO), may substantially improve these sensitivities.[17]

4. Gyroscope experiment

In general relativity there are two well-known types of precession of the spin of a freely falling test body in the presence of a massive spinning body like the Earth.[18] These two types of spin precession are the geodetic precession about an axis perpendicular to the body's orbit and the gravitomagnetic precession about the spin axis of the Earth. In the context of the pure-gravity sector of the minimal SME there is an additional precession effect that occurs due to Lorentz violation.

Ultimately the dominant measurable effects controlled by the SME coefficients reveal themselves in the secular evolution of the gyroscope spin $\vec{S}$, described by $d\vec{S}/dt = gv_0\vec{\Omega}\times\vec{S}$ where $g = GM_\oplus/r_0^2$ is the mean value of the gravitational acceleration at the orbital radius r_0 and v_0 is the mean orbital velocity. The precession vector $\vec{\Omega}$ is split into two pieces via $\Omega^J = \Omega^J_E + \Omega^J_{\overline{s}}$, with the first term containing precession due to conventional effects in general relativity, and the second term containing contributions from the coefficients for Lorentz violation. The latter is given by

$$\Omega^J_{\overline{s}} = \tfrac{9}{8}(\overline{s}^{TT} - \overline{s}^{KL}\hat{\sigma}^K\hat{\sigma}^L)\hat{\sigma}^J + \tfrac{5}{4}\overline{s}^{JK}\hat{\sigma}^K, \tag{3}$$

where the result is written in the Sun-centered frame, $\hat{\sigma}$ is a unit vector normal to the orbital plane, and contributions from the Earth's inertia have been suppressed.

The result (3) gives contributions to the precession about the orbital angular momentum axis $\hat{\sigma}$ and the Earth's spin axis $\hat{J}$. In addition, however, there is a qualitatively new precession about the axis defined by $\hat{n} = \hat{\sigma}\times\hat{J}$, that is due entirely to Lorentz violation controlled by the $\overline{s}^{JK}$ coefficients. Data from the Gravity Probe B (GPB) experiment could potentially measure the combinations of coefficients occurring in Eq. (3).[19] If the spin precession vector in three orthogonal directions can be extracted, including the $\hat{n}$ direction, then attainable sensitivity to $\overline{s}^{JK}$ coefficients is expected to be at the 10^{-4} level, given GPB projected sensitivities.

5. Binary pulsars

A particularly useful testing ground for general relativity is the binary-pulsar system.[20,13] In particular, such systems contain compact objects and high orbital velocities which make them appropriate for studies of strong-field gravity. Pulsar timing data from binary pulsar systems also offers the possibility of probing SME coefficients for Lorentz violation.

The Einstein-Infeld-Hoffman (EIH) lagrangian describes the post-newtonian dynamics of such systems and represents a standard approach.[13,21] To obtain the key features arising from Lorentz violation, however, a point-mass approximation suffices and appropriate use can be made of the effective classical lagrangian. The basic orbit can be modeled as a perturbed elliptic two-body problem, where six standard orbital elements are used to describe the orbit: a, e, l_0, i, Ω, and ω. Ultimately a pulsar timing formula is used to model the number of pulses received as a function of arrival time. The timing formula receives modifications due to Lorentz violation from two sources. First, the orbital elements, with the exception of a, acquire secular Lorentz-violating corrections. Second, the timing formula itself involves an explicit dependence on combinations of coefficients for Lorentz violation.

Some simple estimates of sensitivities reveal that, for example, data from the binary pulsar system PSR 1913 + 16 could yield sensitivities[22] to Lorentz violation at the level of $\overline{s}_e \lesssim 10^{-9}$ and $\overline{s}_\omega \lesssim 10^{-11}$ where $\overline{s}_e$ and $\overline{s}_\omega$ are the combinations of coefficients relevant for the orbital elements e and ω. These combinations of coefficients will also change with the orientation of the binary pulsar system.

6. Other tests

Other types of gravitational experiments have been explored for their merits in probing the SME coefficients $\overline{s}^{\mu\nu}$.[a] In particular, Earth-laboratory tests studying gravitational interactions between either two controlled masses or between a test body and the Earth could be used.

One prediction is a newtonian potential between two point masses that is modulated by an anisotropic term $\hat{x}^{\hat{j}}\hat{x}^{\hat{k}}\overline{s}^{\hat{j}\hat{k}}$, where the unit vector $\hat{x}$ points between the two masses. The SME coefficients $\overline{s}^{\hat{j}\hat{k}}$ are taken in the Earth-laboratory frame of reference. Although the inverse-distance behavior of the usual newtonian potential is maintained, the associated force is generally misaligned relative to the unit vector $\hat{x}$. It is conceivable that experiments studying short-range tests of gravity might be used to probe these coefficients.[23] Currently, an analysis of this type is underway.[24]

[a]Analysis of the classic tests is in Ref. 11.

When considering the effects of the Earth's gravity on test bodies near the surface of the Earth, a modified local gravitational acceleration arises. In the local laboratory frame of reference this acceleration has a vertical ($\hat{z}$) component which is time dependent on sidereal day and year time scales. Experiments with gravimeters are ideally suited for probing such a time variation.[13] An analysis using gravimeter data to extract measurements on combinations of coefficients occurring in this modified acceleration is currently underway.[25] In addition, the local acceleration in the horizontal directions $\hat{x}$ (south) and $\hat{y}$ (east) receives modifications from Lorentz violation.

References

1. See, for example, V.A. Kostelecký, ed., *CPT and Lorentz Symmetry III*, World Scientific, Singapore, 2005; H. Müller *et al.*, AIP Conf. Proc. **758**, 30 (2005).
2. V.A. Kostelecký and R. Potting, Phys. Rev. D **51**, 3923 (1995); D. Colladay and V.A. Kostelecký, Phys. Rev. D **55**, 6760 (1997); **58**, 116002 (1998).
3. V.A. Kostelecký, Phys. Rev. D **69**, 105009 (2004).
4. J. Lipa *et al.*, Phys. Rev. Lett. **90**, 060403 (2003); H. Müller *et al.*, Phys. Rev. Lett. **91**, 020401 (2003); Phys. Rev. Lett. **99**, 050401 (2007); P. Wolf *et al.*, Phys. Rev. D **70**, 051902 (2004); M. Tobar *et al.*, Phys. Rev. D **71**, 025004 (2005); S. Herrmann *et al.*, Phys. Rev. Lett. **95**, 150401 (2005); S.M. Carroll *et al.*, Phys. Rev. D **41**, 1231 (1990); V.A. Kostelecký and M. Mewes, Phys. Rev. D **66**, 056005 (2002); Phys. Rev. Lett. **87**, 251304 (2001); Phys. Rev. Lett. **97**, 140401 (2006); Phys. Rev. Lett. **99**, 011601 (2007); M. Hohensee *et al.*, Phys. Rev. D **71**, 025004 (2007); P.L. Stanwix *et al.*, Phys. Rev. D **74**, 081101 (2006); R. Jackiw and V.A. Kostelecký, Phys. Rev. Lett. **82**, 3572 (1999); V.A. Kostelecký *et al.*, Phys. Rev. D **65**, 056006 (2002); T. Jacobson *et al.*, Phys. Rev. D **67**, 124011 (2003); B. Altschul, Phys. Rev. D **72**, 085003 (2005); Phys. Rev. Lett. **96**, 201101 (2006).
5. H. Dehmelt *et al.*, Phys. Rev. Lett. **83**, 4694 (1999); R. Mittleman *et al.*, Phys. Rev. Lett. **83**, 2116 (1999); G. Gabrielse *et al.*, Phys. Rev. Lett. **82**, 3198 (1999); R. Bluhm *et al.*, Phys. Rev. Lett. **82**, 2254 (1999); **79**, 1432 (1997); Phys. Rev. D **57**, 3932 (1998); D. Colladay and V.A. Kostelecký, Phys. Lett. B **511**, 209 (2001); G.M. Shore, Nucl. Phys. B **717**, 86 (2005); B. Heckel *et al.*, Phys. Rev. Lett. **97**, 021603 (2006); L.-S. Hou *et al.*, Phys. Rev. Lett. **90**, 201101 (2003); R. Bluhm and V.A. Kostelecký, Phys. Rev. Lett. **84**, 1381 (2000); H. Müller *et al.*, Phys. Rev. D **68**, 116006 (2003); Phys. Rev. D **70**, 076004 (2004); H. Müller, Phys. Rev. D **71**, 045004 (2005).
6. L.R. Hunter *et al.*, in V.A. Kostelecký, ed., *CPT and Lorentz Symmetry*, World Scientific, Singapore, 1999; D. Bear *et al.*, Phys. Rev. Lett. **85**, 5038 (2000); D.F. Phillips *et al.*, Phys. Rev. D **63**, 111101 (2001); M.A. Humphrey *et al.*, Phys. Rev. A **68**, 063807 (2003); F. Canè *et al.*, Phys. Rev. Lett. **93**,

230801 (2004); P. Wolf *et al.*, Phys. Rev. Lett. **96**, 060801 (2006); V.A. Kostelecký and C.D. Lane, Phys. Rev. D **60**, 116010 (1999); C.D. Lane, Phys. Rev. D **72**, 016005 (2005); D. Colladay and P. McDonald, Phys. Rev. D **73**, 105006 (2006); R. Bluhm *et al.*, Phys. Rev. Lett. **88**, 090801 (2002); Phys. Rev. D **68**, 125008 (2003); O. Bertolami *et al.*, Phys. Lett. B **395**, 178 (1997); G. Lambiase, Phys. Rev. D **72**, 087702 (2005).

7. KTeV Collaboration, in V.A. Kostelecký, ed., *CPT and Lorentz Symmetry II*, World Scientific, Singapore, 2002; OPAL Collaboration, Z. Phys. C **76**, 401 (1997); DELPHI Collaboration, preprint DELPHI 97-98 CONF 80 (1997); BELLE Collaboration, Phys. Rev. Lett. **86**, 3228 (2001); BaBar Collaboration, Phys. Rev. Lett. **92**, 142002 (2004); FOCUS Collaboration, Phys. Lett. B **556**, 7 (2003); V.A. Kostelecký, Phys. Rev. Lett. **80**, 1818 (1998); Phys. Rev. D **61**, 016002 (2000); Phys. Rev. D **64**, 076001 (2001).
8. V.W. Hughes *et al.*, Phys. Rev. Lett. **87**, 111804 (2001); R. Bluhm *et al.*, Phys. Rev. Lett. **84**, 1098 (2000).
9. LSND Collaboration, Phys. Rev. D **72**, 076004 (2005); M.D. Messier (SK), in V.A. Kostelecký, ed., *CPT and Lorentz Symmetry III*, World Scientific, Singapore, 2005; B.J. Rebel and S.F. Mufson (MINOS), *ibid.*; V.A. Kostelecký and M. Mewes, Phys. Rev. D **69**, 016005 (2004); T. Katori *et al.*, Phys. Rev. D **74**, 105009 (2006); V. Barger *et al.*, Phys. Lett. B **653**, 267 (2007).
10. D.L. Anderson *et al.*, Phys. Rev. D **70**, 016001 (2004); E.O. Iltan, Mod. Phys. Lett. A **19**, 327 (2004).
11. Q.G. Bailey and V.A. Kostelecký, Phys. Rev. D **74**, 045001 (2006).
12. R. Bluhm and V.A. Kostelecký, Phys. Rev. D **71**, 065008 (2005); V.A. Kostelecký and R. Potting, Gen. Rel. Grav. **37**, 1675 (2005).
13. C.M. Will, *Theory and Experiment in Gravitational Physics,* Cambridge University Press, Cambridge, England, 1993.
14. K. Nordtvedt, Phys. Rev. **169**, 1017 (1968); C.M. Will, Astrophys. J. **163**, 611 (1971); Astrophys. J. **169**, 125 (1971).
15. J. Mueller *et al.*, gr-qc/0509114; J.G. Williams *et al.*, Phys. Rev. Lett. **93** 261101 (2004).
16. J. Battat, these proceedings.
17. J.G. Williams *et al.*, Int. J. Mod. Phys. D **13**, 567 (2004).
18. L.I. Schiff, Proc. Nat. Acad. Sci. **46**, 871 (1960); C. Lämmerzahl *et al.*, eds., *Gyros, Clocks, Interferometers ...: Testing Relativistic Gravity in Space,* Springer, Berlin, 2001.
19. G.M. Keiser *et al.*, in V.A. Kostelecký, ed., *CPT and Lorentz Symmetry III*, World Scientific, Singapore, 2005; J.M. Overduin, these proceedings.
20. I.H. Stairs, Living Rev. Rel. **6**, 5 (2003).
21. N. Wex and M. Kramer, astro-ph/0706.2382.
22. J.H. Taylor and J.M. Weisberg, Astrophys. J. **345**, 434 (1989); astro-ph/0211217.
23. E.G. Adelberger *et al.*, Phys. Rev. D **70**, 042004 (2004); J.C. Long *et al.*, Nature **421**, 922 (2003); S.J. Smullin *et al.*, Phys. Rev. D **72**, 122001 (2005).
24. W.H. Jensen, S.M. Lewis, and J.C. Long, these proceedings.
25. H. Müller *et al.*, these proceedings.

PREFERRED FRAME EFFECTS IN RELATIVISTIC BINARY PULSARS

M. KRAMER AND N. WEX
University of Manchester
Jodrell Bank Centre for Astrophysics
Alan-Turing Building
Manchester M13 9PL, U.K.
E-mail: Michael.Kramer@manchester.ac.uk

We summarize our recently presented consistent, theory-independent methodology to measure preferred-frame effects (PFEs) in binary pulsars that exhibit a high rate of periastron advance. In these systems the existence of a preferred frame for gravity leads to an observable characteristic signature in the timing data. A newly developed PFE timing model can be used to either measure or constrain the parameters related to a violation of the local Lorentz invariance of gravity in the strong internal fields of neutron stars. In particular, in the presence of PFEs we expect a set of the new timing parameters to have a unique relationship that can be measured and tested incontrovertibly. A combination of several suitable systems in a *PFE antenna array* provides full sensitivity to possible violations of local Lorentz invariance in strong gravitational fields in all directions of the sky.

1. Introduction

Some theories of gravity predict that the Universe's global matter distribution selects a preferred rest frame for local gravitational physics. In such theories the outcome of gravitational experiments depends on the motion of the laboratory with respect to this preferred frame. In particular, theories in which gravity is partially mediated by a vector field or a second tensor field are known to exhibit such preferred-frame effects whose strength is determined by cosmological matching parameters.

In the post-Newtonian limit, such preferred-frame effects are described by two phenomenological parameters, the parameterised post-Newtonian (PPN) parameters α_1 and α_2. In general relativity (GR) these two parameters are zero. Solar system experiments have already tightly constrained these two PPN parameters.[1,2] On the other hand, limits obtained in the weak gravitational fields of the solar system cannot rule out effects that only become significant in strong gravitational fields. For the well-motivated scalar-tensor theories, this has been demonstrated in a series of papers by Damour and Esposito-Farèse.[3,4,5,6] In order to describe preferred-frame effects in binary pulsars, Damour and Esposito-Farèse[7] introduced a non-boost-invariant Lagrangian that includes two strong field parameters $\hat{\alpha}_1$ and $\hat{\alpha}_2$, capable of accounting for strong-field contributions of a local Lorentz invariance of gravity. In the weak-field limit these two strong-field parameter are equal to α_1 and α_2, respectively.

In a recent paper Bailey and Kostelecký[8] study the violation of local Lorentz invariance within an effective field theory called the Standard-Model Extension (SME). Besides a number of solar system tests, they use their formalism to investigate the measurability of SME preferred-frame effects in binary pulsars. However, as already pointed out by Bailey and Kostelecký, their formalism does not account for strong field effects.

Here, we summarise a new method for testing preferred-frame effects related to strong gravitational fields, which uses binary pulsars with a high rate of periastron advance. Our method[9] provides independent and simultaneous tests for both the strong-field correspondent of α_1 and the strong-field correspondence of α_2. While previous tests,[7,10] by their probabilistic considerations, can only provide upper limits for preferred-frame effects, the method introduced in this paper is even capable of detecting a preferred frame in the universe, if it exists.

2. Testing Preferred-Frame Effects in Binary Pulsars

We derived the equations of motion from a general formalism[1] which includes strong field contributions using generic strong-field parameters.[9] As we do not impose boost-invariance, the resulting equations of motion contain terms that depend on the motion of the binary system with respect to a preferred frame. We showed that the rate of relativistic precession of periastron $\dot{\omega}$, the orbital eccentricity e, and the orbital period P_b should change in a characteristic way over time if a binary system moves relative to a preferred frame of reference. As these changes should become apparent in the timing data, we can – assuming a direction and velocity w for the preferred frame – use the data to constrain or even measure the strong field parameters which determine the strength of the preferred-frame effects.

Without making detailed assumptions on the strong-field properties of the underlying theory of gravity, most binary pulsars will not allow the determination of the gravitating masses and the orbital inclination i as required.[9] Indeed, most measured Post-Keplerian parameters, cannot be used as they are expected to have strong-field contributions, which are not quantified within a theory-independent framework. Moreover, one can precisely measure post-Keplerian parameters only in relativistic binary pulsars. In such cases, the preferred-frame effects cannot be described by a simple linear-in-time expressions. Fitting for first order time derivatives, therefore, cannot be used. This is true for any binary pulsar system where due to a relativistic orbital precession, the longitude of periastron has changed significantly since the time of its discovery, so that the time derivatives of the orbital parameters are trigonometric functions of the changing longitude of periastron, $\omega(t)$. Hence, the PFEs impose periodic changes on the orbital parameters themselves, with frequencies $\dot{\omega}$ and $2\dot{\omega}$ which need to be incooperated into the timing model to test for PFEs.

3. A PFE Timing Model

We extended the standard Damour & Deruelle (DD) timing model[11,12] to include the periodic orbital changes to allow us to fit for the amplitudes of these preferred-frame effects.[9] As we explain further below, the best test-case for our purpose is the Double Pulsar system where the orbital inclination, i, is very close to 90 degrees. In this case, changes in the projected semi-major axis are small and can be neglected. Denoting $\tilde{\chi}_0$ as the angle between the ascending node and the projection of $\mathbf{w}$ into the orbital plane, $\mathbf{w}_\perp$ (see Fig. 1), the changes in the remaining orbital parameters are then given by

$$\Delta\omega^{(w)}(T) = \eta_1^{(\omega)} \cos(\omega_L - \tilde{\chi}_0) - \eta_2^{(\omega)} \sin 2(\omega_L - \tilde{\chi}_0) \,, \tag{1}$$

where

$$\omega_L = \omega_0 + \dot{\omega}^{(1)}(T - T_0) \,, \tag{2}$$

and

$$\eta_1^{(\omega)} = \frac{\hat{\mathsf{Q}}_1}{12\hat{\mathsf{P}}}\left(\frac{1}{e} - eF_e^2\right)\left(\frac{w}{v_0}\right)\sin\psi \,, \quad \eta_2^{(\omega)} = \frac{\hat{\mathsf{Q}}_2}{12\hat{\mathsf{P}}}F_e^2\left(\frac{w}{v_0}\right)^2 \sin^2\psi \,. \tag{3}$$

as well as

$$\Delta e^{(w)}(T) = \eta_1^{(e)} \sin(\omega_L - \tilde{\chi}_0) + \eta_2^{(e)} \cos 2(\omega_L - \tilde{\chi}_0) \,. \tag{4}$$

where

$$\eta_1^{(e)} = \frac{\hat{\mathsf{Q}}_1}{6\hat{\mathsf{P}}}(1+e^2)F_e\left(\frac{w}{v_0}\right)\sin\psi \,, \quad \eta_2^{(e)} = \frac{\hat{\mathsf{Q}}_2}{12\hat{\mathsf{P}}}\frac{e(1+e^2)F_e}{\sqrt{1-e^2}}\left(\frac{w}{v_0}\right)^2 \sin^2\psi \,. \tag{5}$$

Finally,

$$\Delta M^{(w)}(T) = -\eta_1^{(M)} \cos(\omega_L - \tilde{\chi}_0) + \eta_2^{(M)} \sin 2(\omega_L - \tilde{\chi}_0) , \tag{6}$$

where

$$\eta_1^{(M)} = \eta_1^{(\omega)} \sqrt{1-e^2} + \frac{\hat{\mathsf{Q}}'_1}{\hat{\mathsf{P}}} e F_e \sqrt{1-e^2} \left(\frac{w}{v_0}\right)^2 \sin^2\psi , \tag{7}$$

$$\eta_2^{(M)} = \eta_2^{(\omega)} \sqrt{1-e^2} + \frac{\hat{\mathsf{Q}}'_2}{6\hat{\mathsf{P}}} \left(F_e \sqrt{1-e^2} - \tfrac{1}{2}\right) \left(\frac{w}{v_0}\right)^2 \sin^2\psi , \tag{8}$$

while v_0 is the velocity of the system and

$$F_e = \frac{1}{1+\sqrt{1-e^2}} . \tag{9}$$

With the timing data we can determine the amplitudes $\eta_{1,2}^{(w,e,\omega,M)}$ which are related to the strong-field parameters that describe PFE effects, i.e., $\hat{\mathsf{P}}, \hat{\mathsf{Q}}_1, \hat{\mathsf{Q}}_2, \hat{\mathsf{Q}}'_1, \hat{\mathsf{Q}}'_2$ which themselves depend on the chosen theory of gravity and of the structure of each body. They account for contributions of the highly relativistic interior of the neutron stars to the binary dynamics. In particular, it is assumed that these coefficients are independent of the interbody distance in the binary system. See Wex and Kramer[9] for details.

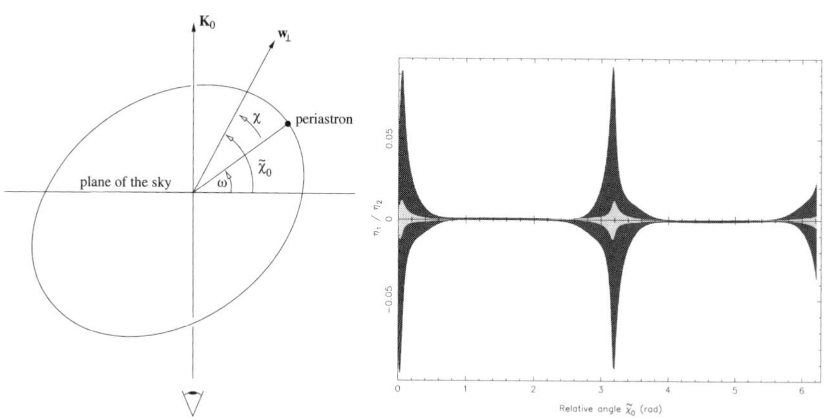

Figure 1. Left: definition of angles in the PFE timing model. Right: limits on the existence of preferred-frame effects present in the orbital motion of the Double Pulsar as measured by the parameters $\eta_1^{(\omega)}$ (dark grey) and $\eta_2^{(\omega)}$ (light grey) as introduced in the PFE timing model.

If the longitude of periastron has advanced for a considerable amount ($> \pi$) since the discovery of a binary pulsar, the timing data will allow to fit for all of these new timing parameters. If the binary motion shows the effect of a preferred-frame in the Universe, all five independent PFE amplitudes, $\eta_1^{(\omega)}$, $\eta_2^{(\omega)}$, $\eta_1^{(M)}$, $\eta_2^{(M)}$, and $\tilde{\chi}_0$ can be determined. One can even decide to fit additionally for $\eta_1^{(e)}$ and $\eta_2^{(e)}$, i.e., perform a fit simultaneously for these parameters and for $\eta_1^{(\omega)}$ and $\eta_2^{(\omega)}$. The existence of a preferred frame could then be confirmed by verifying the expected relationship between the PFE amplitudes in ω and e. The ratios of these amplitudes are unique in the sense that they are a function of e only and therefore independent of any strong-field parameters, and of the magnitude and direction of $\mathbf{w}$. For the parameters of the Double Pulsar (see below) we would expect to measure

$$\eta_1^{(\omega)}/\eta_1^{(e)} = 11.262\,, \tag{10}$$

$$\eta_2^{(\omega)}/\eta_2^{(e)} = 5.642 \tag{11}$$

if a preferred frame exists. In other words, they serve as a unique fingerprint for identifying PFE effects, providing incontrovertibly tests for their existence.

In the absence of any preferred-frame effect, we expect none of the PFE timing parameters to have significant values. In particular, the magnitude of PFE amplitudes should be consistent with zero, and the angle $\tilde{\chi}_0$ is undefined. In this case, one can obtain upper and lower limits for $\eta_1^{(\omega)}$, $\eta_2^{(\omega)}$, $\eta_1^{(M)}$, and $\eta_2^{(M)}$. A clear separation of the preferred-frame contribution ($\eta_1^{(\omega)}$, $\eta_2^{(\omega)}$) from the other relativistic terms ($\dot{\omega}^{(1)}$) in the precession of periastron is possible and will improve with time due to increasing coverage of ω-space.

4. The Double Pulsar

The PSR J0737−3039 binary pulsar system[13,14] is not only the only Double Pulsar system known, but it also provides us with the so far most stringent test of GR in the strong field regime.[15] Its large periastron advance makes it also a unique testbed to actually measure a possible violation of the Lorentz invariance for strong gravitational fields. This is greatly helped by the fortunate orientation of the binary orbit in space which allows us to observe the Double Pulsar under a nearly perfect edge-on geometry. The precise measurements of a Shapiro delay allows the determination of $\sin i$ and, combined with the measurement of the mass ratio, a theory-independent determination of both mass parameters.[9]

After applying our timing model, we find that the current data span is insufficient, so that we can only derive upper limits for the PFE amplitudes. Fig. 1 illustrates the 95% confidence for $\eta_1^{(\omega)}$ and $\eta_2^{(\omega)}$, consistent with the non-existence of preferred frames. In principle, we would also aim to fit for $\eta_1^{(M)}$ and $\eta_2^{(M)}$ but currently the periastron has not advanced far enough to separate these PFE amplitudes from other orbital parameters, in particular the decrease in the orbital period due to the emission of gravitational waves. With longer time-spans and more coverage of the periastron angle, such a fit will be possible. At the moment, we can translate the current timing results to obtain quantitative limits within the generalised EIH formalism.[9] The derived limits are currently not very tight, but are in fact several orders of magnitude worse than the limits for the corresponding weak-field PPN parameters α_1 and α_2. However, these limits are qualitatively different as they probe strong-field effects, and they will also improve considerably during the next couple of years. By 2013 the periastron will have advanced by nearly π since the pulsar discovery, which should also allow a separate measurement of $\eta_1^{(M)}$ and $\eta_2^{(M)}$ with comparable precision. This will provide us with measurements for the strong-field parameters $\hat{Q}_1$, $\hat{Q}_2$, $\hat{Q}'_1$, and $\hat{Q}'_2$.

5. Summary and Conclusions

We have developed a consistent methodology to measure preferred-frame effects (PFE) related to the strong internal gravitational fields in relativistic binary pulsars. We were able to show that in relativistic binary systems with a high rate of advance of periastron, a preferred frame will cause distinctive periodic changes in the orbital elements of the binary system that we can describe with a newly developed PFE timing model.

For inclination angles $i \simeq 90°$, the PFE amplitudes for changes in the longitude of periastron and the eccentricity are related by a factor which is independent of strong field parameters and parameters related to the preferred frame. This can be seen as a unique "fingerprint" of a preferred frame.

In the future, further binary systems with a variety of orientations in space will be discovered. In particular, with the Square-Kilometre Array (SKA) we expect to find about 100 relativistic double neutron star systems.[17] Some of these new systems will also be suitable sources for tests of local Lorentz-invariance of gravity, so that their observations can be combined to derive further constraints, as also pointed out by Bailey and Kostelecký.[8] Indeed, the sensitivity of additional systems to different directions in the sky will differ from that of the Double Pulsar, so that we can construct a dense network of antennae for studies of preferred-frame effects, a *PFE Antenna Array*. This results in a smooth and high sensitivity toward preferred-frame effects over the whole sky.

Acknowledgments

MK is grateful to the conference organizers for their support.

References

1. C. M. Will, *Theory and Experiment in Gravitational Physics*, Cambridge University Press, Cambridge, 1993.
2. C. M. Will, *Living Reviews in Relativity* **9**, 3 (2006).
3. T. Damour, and G. Esposito-Farèse, Class. Quantum Grav. **9**, 2093 (1992).
4. T. Damour, and G. Esposito-Farese, Phys. Rev. Lett. **70**, 2220 (1993).
5. T. Damour, and G. Esposito-Farèse, Phys. Rev. D **53**, 5541 (1996).
6. T. Damour, and G. Esposito-Farese, Phys. Rev. D **54**, 1474 (1996).
7. T. Damour, and G. Esposito-Farèse, Phys. Rev. D **46**, 4128 (1992b).
8. Q. G. Bailey, and V. A. Kostelecký, Phys. Rev. D **74**, 045001 (2006).
9. N. Wex, and M. Kramer, MNRAS **380**, 455 (2007).
10. N. Wex, in M. Kramer, N. Wex, and R. Wielebinski, eds., *Pulsar Astronomy - 2000 and Beyond, IAU Colloquium 177*, Astronomical Society of the Pacific, San Francisco, 2000, p. 113.
11. T. Damour, and N. Deruelle, Ann. Inst. H. Poincaré (Physique Théorique) **43**, 107 (1985).
12. T. Damour, and N. Deruelle, Ann. Inst. H. Poincaré (Physique Théorique) **44**, 263 (1986).
13. M. Burgay, *et al.*, Nature **426**, 531 (2003).
14. A. G. Lyne, *et al.*, Science **303**, 1153 (2004).
15. M. Kramer, *et al.*, Science **314**, 97 (2006).
16. J. M. Weisberg, and J. H. Taylor, Ap. J. **576**, 942 (2002).
17. M. Kramer, *et al.*, New Astronomy Review **48**, 993 (2004).

APOLLO: NEXT GENERATION LUNAR LASER RANGING

T. W. MURPHY, JR.* AND E. L. MICHELSEN
University of California, San Diego, CASS/0424
9500 Gilman Drive, La Jolla, CA 92093-0424, U.S.A.
**E-mail:* tmurphy@physics.ucsd.edu

J. B. R. BATTAT AND C. W. STUBBS
Harvard University, Physics Department
18 Hammond Street, Cambridge, MA 02138, U.S.A.

E. G. ADELBERGER AND H. E. SWANSON
University of Washington, Physics Department
Box 351560, Seattle, WA 98195-1560, U.S.A.

C. D. HOYLE
Humboldt State University, Physics and Astronomy Department
One Harpst Street, Arcata, CA 95521-8299, U.S.A.

R. J. MCMILLAN
Apache Point Observatory
2001 Apache Point Road, Sunspot, NM 88349-0059, U.S.A.

APOLLO (the Apache Point Observatory Lunar Laser-ranging Operation) is a new effort in lunar laser ranging that uses the Apollo-landed retroreflector arrays to perform tests of gravitational physics. APOLLO achieved its first range return in October, 2005, and began its science campaign the following spring. The strong signal (up to 2500 photons in a ten minute period) translates to one-millimeter random range uncertainty, constituting at least an order-of-magnitude gain over previous stations. One-millimeter range precision will translate into order-of-magnitude gains in our ability to test the weak and strong equivalence principles, the time rate of change of Newton's gravitational constant, the phenomenon of gravitomagnetism, the inverse-square law, and test new regimes of gravity and/or Lorentz violation.

1. Overview

Lunar Laser Ranging (LLR) has a long history[1] of providing many of our strongest tests of gravity.[2,3] LLR currently provides *the best* tests of the following gravitational parameters, at the indicated levels of precision:

- Strong Equivalence Principle (SEP) to 4×10^{-4};
- Weak Equivalence Principle (WEP) to 10^{-13} (tied with laboratory experiments in terms of $\Delta a/a$, but representing a single, sub-optimal pairing of materials);
- Time-rate-of-change of the gravitational constant ($\dot{G}/G$) to $< 10^{-12}$ per year;
- Gravitomagnetism (basis of frame dragging) to 0.1%;[4]
- Geodetic precession to 0.6%;
- Best test of $1/r^2$ to 10^{-10} times the strength of gravity at $\sim 10^8$ m length scales.

LLR thus far has not seen deviations from the expectations of general relativity.

The state-of-the-art in 2005 was 2 cm range precision, usually accomplished in an observing period lasting a few tens of minutes, and collecting 5–50 photons of returned laser energy. Typical performances of the two routine LLR stations in France (OCA) and Texas (MLRS) have been a return rate of 0.01 and 0.002 photons per pulse, respectively, to the larger Apollo 15 array. At 10 Hz pulse repetition rate, this corresponds to one photon every 10 and 50 seconds, respectively.

The LLR error budget is typically dominated by uncertainty associated with the tilt of the retroreflector array normal relative to the line of sight. These tilts—up to about 7° in each axis—are caused by "optical" librations of the moon. Even if the tilt is known precisely, the range measurement is spread temporally, with a peak-to-peak uncertainty in the ballpark of $a \tan 6° \approx 0.1a$, where a is the array dimension of roughly one meter. In a root-mean-square sense, the resulting 30–50 mm range uncertainty can be averaged to 1 mm uncertainty by gathering 900–2500 photons. This number is well outside the grasp of the OCA or MLRS stations.

A new lunar ranging apparatus, APOLLO[5] (the Apache Point Observatory Lunar Laser-ranging Operation), has begun operation in southern New Mexico on a mountain-top at an elevation of 2780 m. Using a 3.5 m telescope aperture and taking advantage of good atmospheric image quality ("seeing"), APOLLO is capable of receiving multiple photons per pulse.

2. APOLLO Project Status

We achieved our first unambiguous range results in October 2005, and began collecting science-quality data in April 2006. A steady-state campaign, with routine hour-long blocks of time distributed through the lunar month, began October 2006.

At this time, some of APOLLO's initial accomplishments are:

- As many as 2500 photons in a 500 second period
- Peak rates of ~0.6 photons per pulse, over 30 second intervals
- Range with ease at full moon (initial October 2005 acquisition at full phase)
- As many as 9 lunar return photons detected in a single pulse
- In strong runs, ~50% of photons arrive in multi-photon groups—even when the average rate is 0.25 photons per pulse

A sample run is shown in Fig. 1. The full-width at half-max (FWHM) of the lunar return is approximately 500 ps, corresponding to a root-mean-square (RMS) around 215 ps. Light makes a one-millimeter (one-way) round-trip in 6.7 ps, so that the per-photon RMS is about 32 mm. One would therefore require 1024 photons to achieve a centroid measure at the one-millimeter level, which is provided in this example of 1087 photons. Formally, this ~8 minute run presented a 7.8 ps, or 1.2 mm RMS uncertainty.

Figure 1 also hints at an important aspect of the *differential* nature of the APOLLO measurement. Namely, a local corner cube mounted at the exit aperture of the telescope intercepts a small portion of the outgoing laser beam, sending a light pulse back to the receiver. After attenuating the local return to the single-photon level, we are able to measure the departure (fiducial) time of the laser pulse using *the same* optical/detector/electronics system that is used for the lunar return. In this way, comparison of the lunar measurement to the interleaved fiducial measurement provides a differential measurement of the round-trip time from the local corner cube to the remote lunar array. The fiducial return is shown at the bottom of Fig. 1, and differs from the lunar return only by background and the tilt-induced temporal response of the lunar reflector array.

The performance of APOLLO can also be checked by repeated measurements to the same reflector on a given night. Subtracting the (imperfect) prediction from the measurements, one may plot residuals as a function of time, as in Figs. 2 and 3. The imperfect prediction results in an offset and

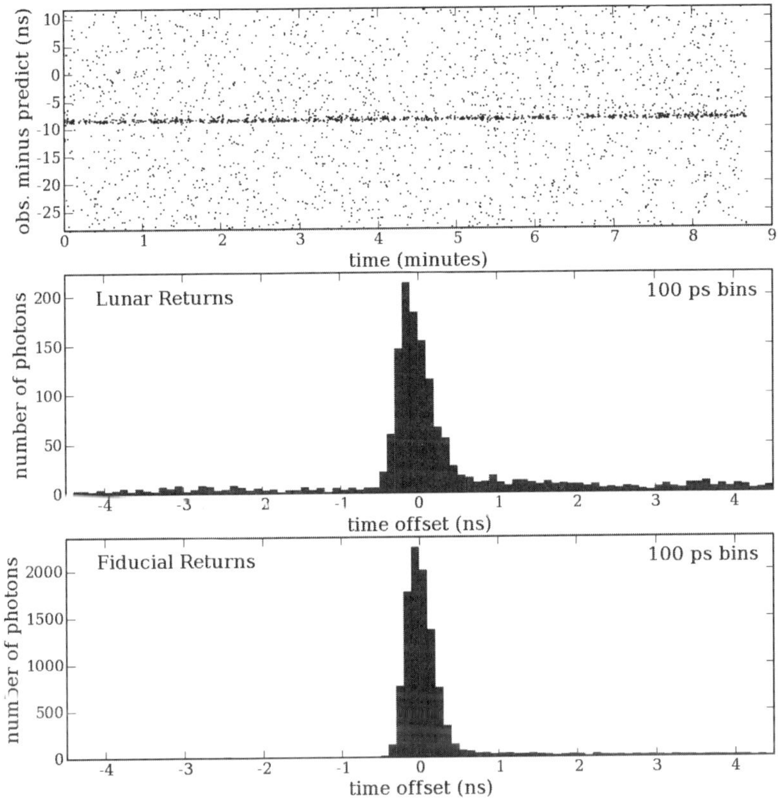

Figure 1. Sample return from the Apollo 11 array on 03 December 2006, in which 1087 photons were collected in a 10000 shot run. The reflector return is clearly visible against the background. The width is consistent with the temporal spread due to the tilted reflector array.

slope, but little or no quadratic component on these timescales. Note that the errors—estimated from the return width and photon number—appear to be consistent with deviations from a straight-line fit.

Initial integration of APOLLO data points into the LLR model at the Jet Propulsion Laboratory (JPL) indicates that the APOLLO points are perfectly consistent with the 37 year archive of lunar range measurements, producing post-fit residuals in the neighborhood of 6–8 mm. Though we are

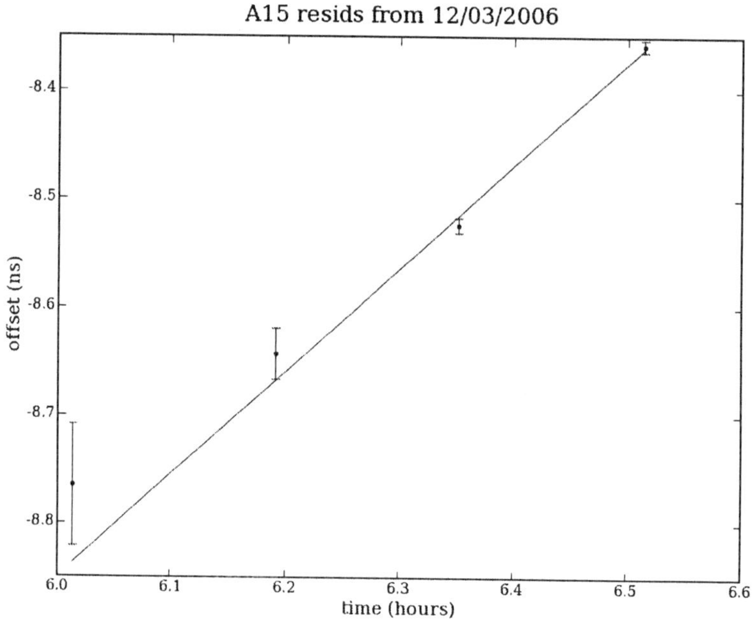

Figure 2. Sample residuals from a series of range measurements to the Apollo 15 reflector, representing a good night that ultimately delivered high returns. Each run represents 10000 laser shots, delivering 16, 132, 1287, and 1996 return photons, respectively, for estimated error bars corresponding to 8.4, 3.5, 1.1, and 0.9 mm. Collectively, this constitutes a 0.7 mm range measurement for the night. An approximate predicted range is subtracted, resulting in some residual offset and slope. The fact that the error bars are roughly consistent with the best-fit line indicates that the error estimates are approximately correct.

already falling into the sub-centimeter regime, we suspect that a significant part of the remaining residuals reflect the need for model improvements rather than data shortcomings (otherwise would imply a mis-estimate of typical error bars by a factor of three). But this is not unexpected: new data at higher precision than previously available are bound to require model improvements. The interesting question, then, is whether the model is capable of following the data with general relativity as the gravity model.

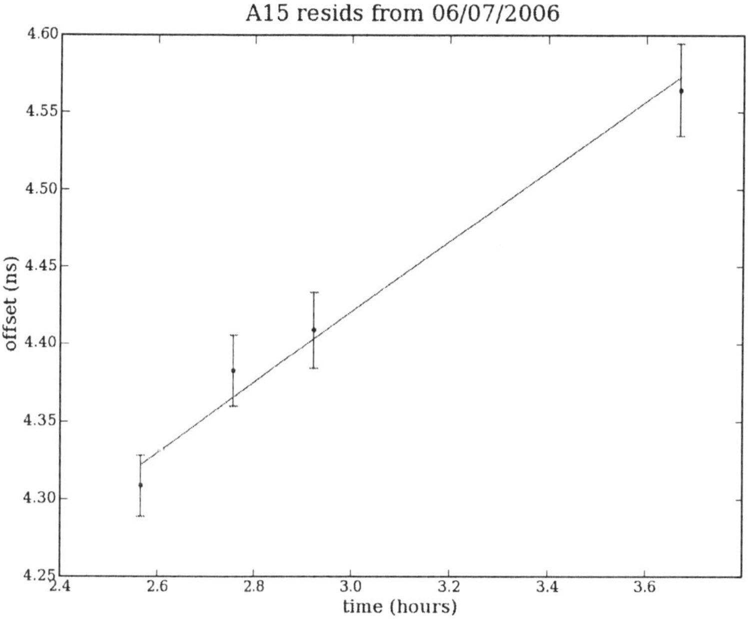

Figure 3. Sample residuals from a series of range measurements to the Apollo 15 reflector, on a mediocre night of modest returns. The four runs represent 5000, 5000, 8300, and 10000 laser shots, delivering 229, 241, 197, and 152 return photons, respectively, for estimated error bars corresponding to 3.0, 3.4, 3.7, and 4.5 mm. Collectively, this constitutes a 1.8 mm range measurement for the night. See the caption of Fig. 2 for more information.

3. Conclusion

APOLLO has initiated a data campaign unlike anything in the history of lunar ranging. Early work on the 2.7 meter telescope at McDonald Observatory approached single-photon-per-pulse performance, but at 0.3 Hz repetition rate and 3 ns pulse width. The APOLLO return rate is at least two orders-of-magnitude higher than that of other currently operating LLR stations (helped partly by a higher repetition rate), leading to millimeter-level range accuracies. It is expected that shortly APOLLO data will begin to impose new and meaningful constraints on fundamental physics parameters. For more on the relation of LLR to the Standard-Model Extension (SME) parameters, see the contribution to this conference by James Battat.[6]

References

1. J. O. Dickey, P. L. Bender, J. E. Faller, X. X. Newhall, R. L. Ricklefs, J. G. Ries, P. J. Shelus, C. Veillet, A. L. Whipple, J. R. Wiant, J. G. Williams, and C. F. Yoder, Science **265**, 482 (1994).
2. J. G. Williams, X. X. Newhall, and J. O. Dickey, Physical Review D **53**, 6730 (1996).
3. J. G. Williams, S. G. Turyshev, and D. H. Boggs, Physical Review Letters **93**, 261101 (2004).
4. T. W. Murphy, Jr., K. Nordtvedt, and S. G. Turyshev, Physical Review Letters **98**, 071102 (2007).
5. T. W. Murphy, Jr., E. G. Adelberger, J. B. R. Battat, L. N. Carey, C. D. Hoyle, P. LeBlanc, E. L. Michelsen, K. Nordtvedt, A. E. Orin, J. D. Strasburg, C. W. Stubbs, H. E. Swanson, and E. Williams, arXiV:0710.0890.
6. J. B. R. Battat, J. F. Chandler, and C. W. Stubbs, these proceedings; arXiV/0710.0701.

CONSTRAINTS ON LORENTZ VIOLATION FROM GRAVITY PROBE B

J.M. OVERDUIN
Gravity Probe B, Hansen Experimental Physics Laboratory
Stanford University, Stanford, CA 94305, U.S.A.
E-mail: overduin@relgyro.stanford.edu

After some historical remarks I summarize the current status of the Gravity Probe B mission and outline the implications of preliminary data for extensions of general relativity involving Lorentz violation.

The successful launch of Gravity Probe B on 20 April 2004 came exactly 100 years after the earliest published accounts of frame-dragging experiments, by August Föppl in Munich in 1904.[1] Föppl, working as he was before general (and for that matter special) relativity, was investigating the possibility of a coupling between the spin of the Earth and that of a pair of heavy flywheels whose rotation axis could be aligned along either lines of latitude or longitude. He was probably inspired by earlier experiments of countrymen Immanuel and Benedict Friedlaender (1896) involving torsion balances in the vicinity of spinning millstones, and by the writings of Ernst Mach, who famously speculated in 1883 that water in a spinning bucket might not exhibit the effects of centrifugal force "if the sides of the vessel increased in thickness and mass until they were ultimately several leagues thick." A sufficiently massive bucket, in other words, might carry the local inertial frame of the water around with it. Mach's Principle, as this idea came to be known, has proved to be of limited scientific value. Nevertheless Gravity Probe B (GPB) can be seen as a modern-day realization of Mach's proposal with an Earth-sized bucket and the role of water played by orbiting gyroscopes more than a million times more sensitive than the best inertial navigation gyros on Earth.

It is curious that Albert Einstein, whose theory of general relativity was originally inspired by Mach's principle, and whose earlier attempts at a gravitational theory all exhibited frame-dragging effects, did not himself work out the Machian implications of his new theory. That was left to

Hans Thirring and Josef Lense (1918), after whom the general relativistic frame-dragging effect is often named. (In a nice reversal of the usual course of events, Thirring had wanted to build an improved Föppl-type experiment and only reluctantly settled for doing the theoretical calculation after he was unable to obtain funding.[1]) The terms "frame-dragging" and "Lense-Thirring" are sometimes used interchangeably with "gravitomagnetic," based on the close analogy between Maxwell's equations and a subset of Einstein's field equations in the low-velocity, weak-field limit.[2] This identification can be misleading, however, as the distinction between gravitomagnetic and gravitoelectric is frame-dependent, and other phenomena besides frame-dragging are at least partly gravito-electromagnetic.[a] An example is the geodetic effect, which involves the transport of angular momentum through a gravitational field and was already studied two years before the Lense-Thirring effect by Willem de Sitter (1916). He showed that the Earth-Moon system would precess in the field of the Sun, an effect now called the solar geodetic effect (although "heliodetic" might be more descriptive). De Sitter's calculation was extended to rotating test bodies by Jan Schouten (1918) and Adriaan Fokker (1920), and the solar geodetic effect is now sometimes referred to as the Fokker-de Sitter effect. The idea of looking for these effects on Earth with a laboratory gyroscope was briefly considered in the 1930s by P.M.S. Blackett, who discarded it as impractical.[3] Technological progress during World War II, however, brought the problem back into the realm of possibility. An advertisement for a new "cryogenic gyroscope" in December 1959 prompted Leonard Schiff to revisit some earlier calculations involving tests of Mach's Principle and led to his elegant re-derivation of both the geodetic and frame-dragging effects in the form now known as the Schiff formula:

$$\vec{\Omega}_{\rm GR} = \vec{\Omega}_{\rm geo} + \vec{\Omega}_{\rm f-d} = \frac{3GM}{2c^2r^3}\left(\vec{r}\times\vec{v}\right) + \frac{GI}{c^2r^3}\left[\frac{3\vec{r}}{r^2}\left(\vec{S}\cdot\vec{r}\right) - \vec{S}\right], \qquad (1)$$

where M, I and $\vec{S}$ refer to the mass, moment of inertia and angular momentum of the central body and r and $\vec{v}$ are the orbital radius and instantaneous velocity of the gyroscope. In a nice example of scientific synchronicity, essentially the same results were arrived at independently months earlier by

[a]Indeed gravito-electromagnetism has already been observed to high precision, albeit indirectly, since solar system ephemerides are now accurate enough to require general relativistic corrections. See for example Tom Murphy's contribution to these proceedings. Gravito-electromagnetism as a subject considerably predates frame-dragging, going back at least to Michael Faraday's experiments with gravitational induction beginning in 1849.

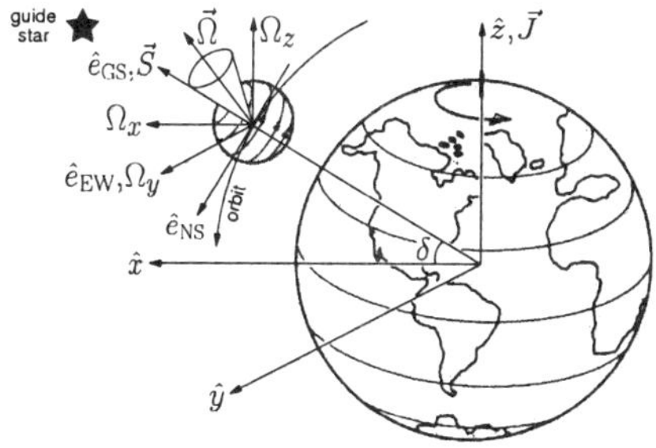

Figure 1. Geometry

George Pugh, a researcher at the Pentagon who also contributed the ingenious suggestion of shielding an orbiting gyroscope from non-gravitational disturbances inside a drag-free satellite.

Frame-dragging arises due to a spin-spin interaction between the gyroscope and rotating Earth, analogous to the interaction of a magnetic dipole with a magnetic field. In a polar orbit 642 km above the Earth, it causes a gyroscope's spin axis to precess in the east-west direction by 39 milliarcsec/yr, an angle so tiny that it is equivalent to the angular width of the planet Pluto as seen from Earth. The geodetic effect arises partly as a spin-orbit interaction between the spin of the gyroscope and the "mass current" of the moving Earth in the gyro rest frame. This is the analog of Thomas precession in electromagnetism. The spin-orbit interaction accounts for one-third of the total geodetic precession; the other two-thirds are not gravito-electromagnetic but arise due to space curvature around the massive Earth (the "missing inch").[2] In a 642 km polar orbit, the geodetic effect causes a gyroscope's spin axis to precess in the north-south direction by 6606 milliarcsec/yr, an angle comparable to that subtended by the planet Mercury as seen from Earth. It was against this theoretical background, and in full knowledge of the daunting challenge posed by these small numbers, that GPB was born one sunny afternoon in the Stanford University swimming pool where Schiff met low-temperature physicist Bill Fairbank and gyroscope specialist Bob Cannon. The program received its first NASA funding in March 1964.

In concept the experiment is simplicity itself: a gyroscope, a readout mechanism to monitor its spin axis, and a telescope to compare this axis with the line of sight to a distant guide star. In practice GPB grew into one of the most complex experiments ever flown, requiring at least a dozen new technologies that did not exist when it was conceived. Among these are the

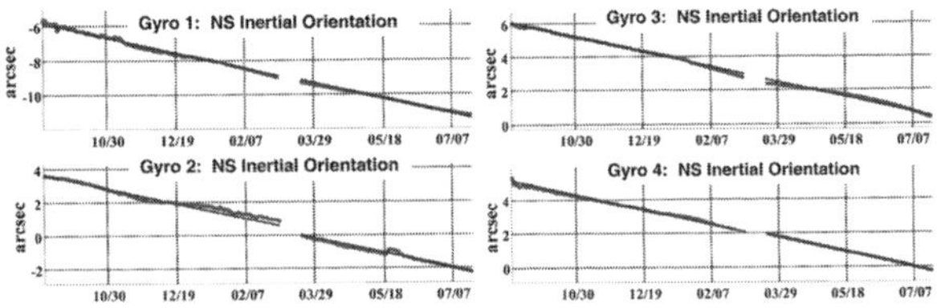

Figure 2. Seeing general relativity directly

world's roundest and most homogeneous gyroscope rotors and a suspension system operating to levitate and maintain them within microns of their housings over a dynamic range of eleven orders of magnitude in force. A novel readout scheme based on the superconducting London moment was developed using ultra-sensitive SQUID magnetometers. Expandable nested lead shields were employed to reduce the ambient magnetic field. New techniques were invented to spin up the gyros, reduce vacuum pressure and remove charge buildup on the rotors. Perturbing forces were suppressed by a drag-free control system whereby any one of the gyroscopes could be isolated as an inertial "plumb line;" the rest of the spacecraft was made to follow its motion by means of helium boiloff vented through a revolutionary porous plug and specially designed thrusters. The resolution of the onboard telescope, fastened to the gyro assembly by a novel quartz bonding technique, was enhanced by means of a beam splitter and image dividers. Non-inertial motions of the guide star, IM Pegasi, were compensated by the use of long-baseline radio interferometry to monitor its position relative to distant background quasars.

The initial orbit checkout phase of the mission, which lasted until 27 August 2004, has been discussed in detail by Mac Keiser at the previous meeting in this series.[4] The science phase which followed lasted until 14 August 2005 (or 353 days, just under the original goal of one full year). The final post-flight calibration phase then continued for a further 46 days until there was no longer enough liquid helium to maintain the experiment at cryogenic temperatures. Figure 2 shows one year's worth of preliminary north-south data. The predicted geodetic effect is already seen in the unprocessed gyro data to an accuracy level of 1%. Table 1 shows the general relativistic predictions in both the north-south (NS) and east-west (EW) directions compared to preliminary GPB results[5] (all figures in milliarcsec/yr). These numbers should be regarded as preliminary. As might be expected in an experiment that goes six or more orders of magnitude

Table 1. Preliminary GPB results (initial year-long 4-gyro average)[5]

	(Terrestrial) geodetic	Solar geodetic	Guide star proper motion	Net predicted (Ω_{GR})	Observed (Ω_{obsd})
NS	-6606	+7	$+28 \pm 1$	-6571 ± 1	-6578 ± 9 (1σ)
EW	-39	-16	-20 ± 1	-75 ± 1	-87 ± 9 (1σ)

in gyro drift rate beyond anything that has gone before, two unexpected factors have cropped up to complicate the data analysis. First, it became apparent during the science phase of the mission that there were variations in the polhode rate of the gyros (polhode motion had been expected, but its period had not been expected to change appreciably over the mission lifetime, given characteristic gyro spin-down periods on the order of 10,000 years). It is critical to understand and model these polhode variations in order to combine the data from successive orbits and thereby reduce the noise as far as possible within the limits inherent in the SQUID readout system (roughly 1 milliarcsec in 5 hrs). Second, misalignment torques (torques proportional to the angle between the spacecraft roll axis and the line of sight to the guide star) appeared during the post-flight calibration phase that were larger than expected. These classical torques must also be understood and modeled, because they can mimic the desired relativity signal. Both factors have been traced to larger-than-anticipated electrostatic patches on the gyroscopes. The misalignment torques are due to interactions between these patches and similar ones on the gyro housings, and the time-varying polhode periods are caused by the fact that these interactions extract energy from the spinning rotors, albeit at a rate that would be imperceptible were the rotors not so perfect. (The energy loss amounts to $\sim 10^{-13}$ W, four orders of magnitude below the heat radiation of $\sim 10^{-9}$ W from a gyro at 2.2 K.) A current worst-case upper bound on systematic error due to both factors together is 97 milliarcsec/yr in both directions.[5] However this will go down significantly. In expectation of the unexpected, GPB was designed to take various kinds of "redundant" data, and these are now proving their worth. In particular trapped-flux modeling is allowing the data analysis team to reconstruct the behavior of the gyro rotors in real time. This, in combination with more sophisticated method of data analysis, is expected to yield accuracies close to those originally envisioned. Final results are to be announced in December 2007.

The GPB data put new constraints on any possible departures from general relativity. Of particular interest for many at this meeting is the pure-gravity sector of the Standard-Model Extension (SME), a generalization of general relativity designed to incorporate violations of Lorentz symmetry.[6] The action of the SME contains besides the usual Einstein-Hilbert term a new tensor field $s^{\mu\nu}$ which couples to the traceless part of the Ricci tensor and induces spontaneous violations of local Lorentz invariance, acquiring vacuum expectation values $\bar{s}^{\mu\nu}$ which are assumed to be constant in the Minkowski limit. There are nine independent coefficients in $\bar{s}^{\mu\nu}$, of which three are of relevance to GPB because they give rise to an anomalous precession $\Delta\vec{\Omega}$ which adds to the general-relativistic terms in Eq. (1). In Earth-centered inertial coordinates ($\hat{x}, \hat{y}, \hat{z}$; Fig. 1) $\Delta\vec{\Omega}$ has the following components:

$$\begin{aligned} \Delta\Omega_x &= \tfrac{5}{4}\tilde{I}(-\tfrac{3}{5})\bar{s}^{jk}, \\ \Delta\Omega_y &= \tfrac{9}{8}\tilde{I}(-\tfrac{1}{3})\bar{s}^{tt} + \tfrac{1}{8}\tilde{I}(9)\bar{s}^{jk}, \\ \Delta\Omega_z &= \tfrac{5}{4}\tilde{I}(-\tfrac{3}{5})\bar{s}^{zk}, \end{aligned} \tag{2}$$

where $\tilde{I}(\beta) \equiv (1+\beta I/Mr^2)(GM/c^2r)^{3/2}(c/r) = 4404(1+0.272\beta)$ milliarcsec/yr at $r = 7027$ km. GPB constraints on the SME coefficents $\bar{s}^{\mu\nu}$ take the form $|\Delta\vec{\Omega}(\bar{s}^{\mu\nu})| < |\vec{\Omega}_{\rm obsd} - \vec{\Omega}_{\rm GR}|$. Apart from its intrinsic interest as a framework for Lorentz violation, the SME is so far unique in allowing for theoretical precessions *greater than*, as well as less than those predicted by general relativity.

Transforming to the coordinates ($\hat{e}_{\rm GS}, \hat{e}_{\rm NS}, \hat{e}_{\rm EW}$) used in the GPB data analysis (Fig. 1), we obtain

$$\begin{pmatrix} \Delta\Omega_{\rm GS} \\ \Delta\Omega_{\rm NS} \\ \Delta\Omega_{\rm EW} \end{pmatrix} = \begin{pmatrix} c_2\cos\delta\sin\alpha & c_1\cos\delta\cos\alpha + c_3\cos\delta\sin\alpha & c_1\sin\delta \\ -c_2\sin\delta\sin\alpha & -c_1\sin\delta\cos\alpha - c_3\sin\delta\sin\alpha & c_1\cos\delta \\ -c_2\cos\alpha & c_1\sin\alpha - c_3\cos\alpha & 0 \end{pmatrix} \begin{pmatrix} \bar{s}^{tt} \\ \bar{s}^{jk} \\ \bar{s}^{zk} \end{pmatrix}, \tag{3}$$

where $(c_1, c_2, c_3) = (4607, 4505, 1898)$ in milliarcsec/yr and $\delta = 16.84°$ and $\alpha = 343.3°$ are the declination and right ascension of the guide star in J2000.0 coordinates. In principle, inverting Eq. (3) and inserting experimental bounds on the quantities $\Delta\Omega$ yields upper bounds on $\bar{s}^{\mu\nu}$. Unfortunately, we have experimental limits only on $\Delta\Omega_{\rm NS}$ and $\Delta\Omega_{\rm EW}$. GPB is designed in order to maximize sensitivity to the geodetic and frame-dragging effects; of necessity this means minimizing sensitivity to precession along the perpendicular direction $\hat{e}_{\rm GS}$. In principle such a sensitivity could arise during periods when gyro spin $\vec{S}$ does not, in fact, point exactly at the guide star. In practice, however, the experiment is designed to keep the two aligned within milliarcseconds, and data are used only when this condition is met. At present, therefore, we have only two independent constraints on three coefficients. Adopting the 1σ values from Table 1, $|\Delta\Omega_{\rm NS}| = \Omega_{\rm NS,obsd} - \Omega_{\rm NS,GR} < 17$ milliarcsec/yr and $|\Delta\Omega_{\rm EW}| = \Omega_{\rm EW,obsd} - \Omega_{\rm EW,GR} < 22$ milliarcsec/yr. Given that the elements of the transformation matrix in Eq. (3) are of order ~ 2000 milliarcsec/yr, it follows that if the $\bar{s}^{\mu\nu}$ coefficients are of comparable magnitude, then they are at most of order $\sim 10^{-2}$.

In principle, it may be possible to break the degeneracy in Eq. (3) by using additional information; namely, limits on the absolute magnitude of the gyro spin $\vec{S}$ and spin-down rate $d\vec{S}/dt$, which are related to precession through the equation of motion $d\vec{S}/dt = \vec{\Omega} \times \vec{S}$. This might provide an independent constraint on $|\vec{\Omega}|$ and hence on $\Omega_{\rm GS}$. Such a procedure would be especially motivated by the prospect of obtaining an unambiguous upper limit on the time-time component $\bar{s}^{tt}$ of the Lorentz-violating field, which is difficult to constrain in any other way. In practice, however, any limits obtained in this way are likely to be weak because the precision of the new information is of order $\sim$ nHz. While this would be impressive in any other context, a nHz spin rate translates into a precession of order $\sim 10^7$ milliarcsec/yr, a million times larger than the GPB limits on $\Delta\Omega_{\rm NS}$ and $\Delta\Omega_{\rm EW}$. This issue will be the subject of further study.

Acknowledgments

Thanks go to Ron Adler, Quentin Bailey, Francis Everitt, Hans Fahr, Bob Kahn, Mac Keiser, Alan Kostelecký, Barry Muhlfelder, Holger Müller, Tom Murphy, Alex Silbergleit, Martin Tajmar and Bob Wagoner for valuable discussions.

References

1. J.B. Barbour and H. Pfister, eds., *Mach's Principle: From Newton's Bucket to Quantum Gravity*, Birkhäuser, Boston, 1995.
2. K.S. Thorne in J.D. Fairbank *et al.*, eds., *Near Zero*, W.H. Freeman, New York, 1988, p. 573.
3. C.W.F. Everitt in J.D. Fairbank *et al.*, eds., *Near Zero*, W.H. Freeman, New York, 1988, p. 587.
4. G.M. Keiser *et al.*, in V.A. Kostelecký, ed., *CPT and Lorentz Symmetry III*, World Scientific, Singapore, 2005, p. 115.
5. C.W.F. Everitt, presentation at the 18th Meeting on General Relativity and Gravitation (GRG18), Sydney, Australia, July 2007.
6. Q.G. Bailey and V.A. Kostelecký, Phys. Rev. D74 (2006) 045001.

PRELIMINARY RESULTS FROM A TEST OF CPT AND LORENTZ SYMMETRY USING A K-^{3}HE CO-MAGNETOMETER

T. W. KORNACK, G. VASILAKIS AND M. V. ROMALIS

Physics Department, Princeton University, Princeton, NJ 08544, U.S.A.

Lorentz and CPT symmetry has been tested using a K-^{3}He co-magnetometer which is sensitive to anomalous fields that couple to electron and nuclear spins differently than a normal magnetic field. Preliminary results obtained from data over the course of 15 months provide upper limits on the coupling energy of a CPT-violating field to neutron spin, $\tilde{b}^n < 1.4 \times 10^{-31}$ GeV, to proton spin, $\tilde{b}^p < 2.2 \times 10^{-30}$ GeV, and to electron spin, $\tilde{b}^e < 1.0 \times 10^{-28}$ GeV.

We have been operating a co-magnetometer using K and ^{3}He spins that is sensitive to the differential coupling of anomalous, CPT-violating fields to electron and nuclear spins. Very high sensitivity is achieved by operating with sufficiently high K density and near zero magnetic field to suppress K spin-exchange relaxation.[1] Strong spin-exchange interactions between polarized K atoms and ^{3}He allows the magnetometer to be operated in a self-compensating mode where the effects of external magnetic fields are passively cancelled.[2] The co-magnetometer remains sensitive to anomalous fields such as a vector Lorentz and CPT violating spin-coupling field b introduced by Kostelecký.[3,4] The co-magnetometer's insensitivity to magnetic

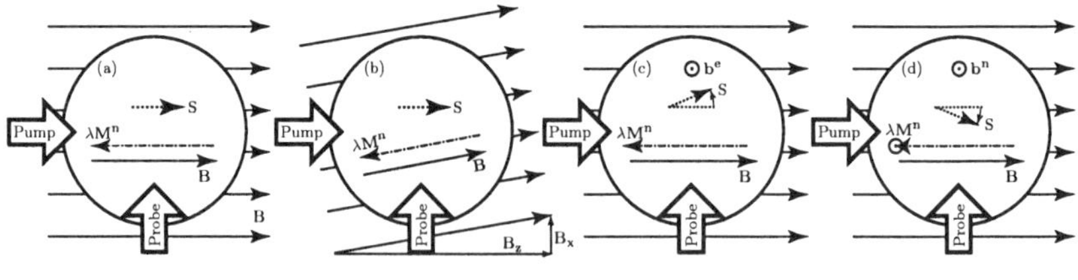

Figure 1. (a) Cell containing polarized K spin S and ^{3}He magnetization M^n. (b) ^{3}He moves to cancel applied field B in steady state. (c) Anomalous field b^e couples exclusively to S and generates a signal. (d) Anomalous field b^n couples exclusively to $I_{^3He}$; the resulting imbalanced field causes S to rotate and generate a signal.

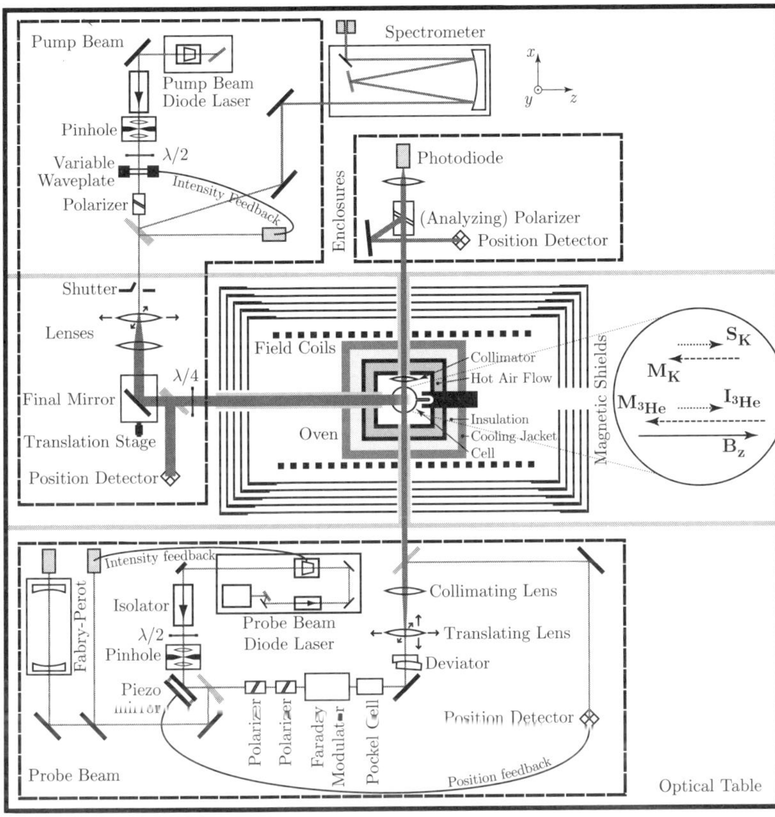

Figure 2. Schematic of the experiment depicting pump and probe lasers in a typical configuration.

fields and its sensitivity to anomalous fields is depicted in Fig. 1.

The co-magnetometer is implemented using an optical pump-probe setup shown in Fig. 2. The K and ^{3}He are held in a 2.5 cm diameter aluminosilicate glass cell, heated by forced hot air to achieve 4×10^{13} cm^{-3} density K vapor at 175°C. Circularly polarized light on the K D_1 resonance polarizes the K spins. The ~ 7 atm ^{3}He is polarized by spin-exchange with the polarized K atoms. External field coils cancel the magnetic field experienced by the K atoms. The probe beam measures the component of K polarization parallel to the beam using optical rotation; modulation of the probe beam enables high sensitivity to optical rotation. The apparatus additionally features wavelength, intensity and pointing feedback on the pump and probe beams.

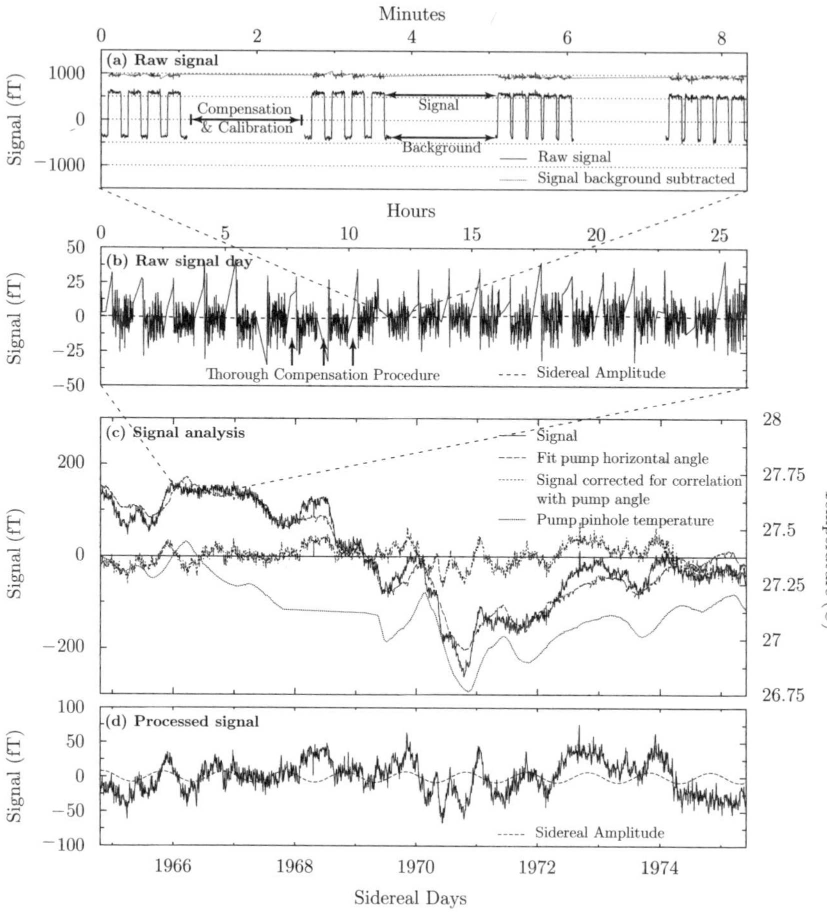

Figure 3. Typical data. (a) Raw signal showing background subtraction to remove short timescale drift. Gaps are for calibration and zeroing routines. (b) Raw signal showing hourly gaps for more thorough zeroing routines. (c) Typical signal analysis demonstrating temperature-driven systematic effects in the raw signal; here, the pump beam position is subtracted to give the uncorrelated signal trace. (d) The amplitude of the uncorrelated signal trace is analyzed to obtain the amplitude of the sidereal variation.

The operation of the co-magnetometer is illustrated in Fig. 3. The signal is measured continuously with interruptions for background measurements and various automated calibration and alignment procedures. Every 10 s, background measurements are performed to remove systematic drift due to mechanical instability of the optical table (Fig. 3a). Every minute or so, the field experienced by the K atoms is cancelled and the angle between the pump and probe beams is adjusted to maintain orthogonality. Every hour, a more thorough procedure zeroes out all fields, lightshifts and other misalignments (Fig. 3c). Greater detail on the apparatus and the operational details can be found in Ref. 5.

There are four systematic errors that can be explicitly removed from the signal: (1) any correlation between the signal and another, separate measurement, (2) purely linear or, more rarely, second-order polynomial drifts, (3) sudden jumps in the signal associated with mode hops of the lasers, and (4) temporary excursions from a nominal trend. Sources of systematic noise commonly include drift in the angle and position of the pump and probe beams and the temperature of certain components. Figure 3c shows an example of correlation between the signal and the pump beam position. In these data, the optical elements used for the pump beam were moving with changes in temperature. The pump position is scaled to fit the signal using a nonlinear least-squares method and the resulting correlated trace is subtracted from the signal to obtain the corrected signal in Fig. 3d.

After correlations and systematic drifts have been removed, the resulting signal is analyzed to search for the presence of an anomalous field. The signal due to an anomalous field would appear as a sinusoidal variation in the signal with a period of exactly one sidereal day. The sidereal time in this article is referenced in sidereal days using the standard J2000 epoch.

Fourier analysis is a reliable method for obtaining the sidereal amplitude of highly noise-compromised data. We seek the in-phase and out-of-phase Fourier components A_x and A_y with sidereal frequency,

$$\kappa S = A_x \cos(\Omega_\oplus t) + A_y \sin(\Omega_\oplus t) + \text{noise} \tag{1}$$

where κS is the calibrated signal in fT and $\Omega_\oplus = 1/\text{sd}$ is the rotation rate of the Earth. To obtain values for A_x and A_y, the signal data are multiplied by sine and cosine references with unit amplitude and periods of 1 sidereal day. The multiplied data are averaged over an integral number of periods and multiplied by 2 to give the amplitudes A_x and A_y.

From all the signal data recorded over 15 months, one must obtain a single pair of sidereal amplitudes A_x and A_y. Long-term data are presented in Fig. 4. There are several analytical steps to obtain these data. (1) The raw timeseries are broken down into a set of 1 sidereal day segments of continuous data. (2) Since the noise in the signal is dominated by slow drift, an rms of the data in a single day, A^i_{rms} is used as an estimate of the uncertainty σ_i for the purpose of relative weighting of different days. If the signal had only white noise, the uncertainty in the amplitude would

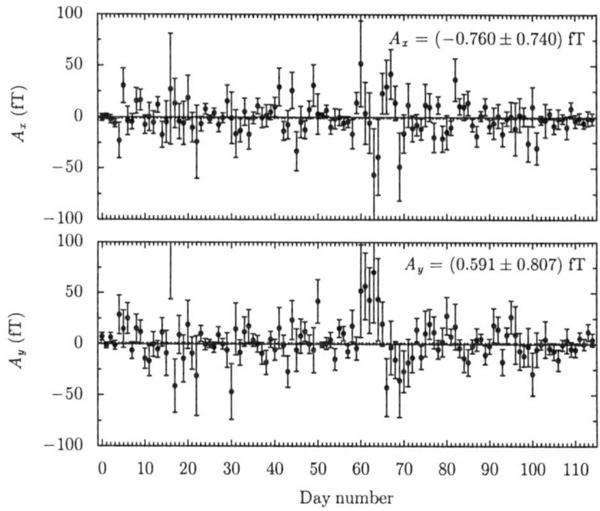

Figure 4. Summary of sidereal amplitudes of each day. These days were selected by limiting the noise level of each day to $A_{rms} < 100$ fT. The error bars set by A_{rms} were scaled by a factor of 0.7 to be consistent with the scatter of the data. The indicated results for A_x and A_y are weighted averages of these data.

be equal to $A^i_{rms}/\sqrt{N_{pts}}$ where N_{pts} is the number of points in the signal. (3) Without introducing any bias or strongly affecting the results, spurious data can be filtered out by setting an upper limit $A^i_{rms} < 100$ fT on the rms of the signal. (4) The sidereal amplitudes $A^i_{x,y}$ of the data in each day are obtained. (5) All the days of data with their amplitudes and uncertainties are combined using a weighted average to obtain a single pair of sidereal amplitudes A_x and A_y for the entire data set. (6) Since the uncertainties σ_i are not accurately known and are used only for relative weighting, the error bars can be uniformly scaled to provide $\chi^2 = 1$ so that the scatter in the repeated measurements is consistent with the uncertainty. We find that $\sigma_i = 0.7A^i_{rms}$ is consistent with the scatter of the data shown in Fig. 4. A histogram of the amplitudes is shown in Fig. 5.

The weighted average of these data gives the amplitudes

$$A_x = (-0.76 \pm 0.74)\ \text{fT}$$
$$A_y = (+0.59 \pm 0.81)\ \text{fT}. \tag{2}$$

To express this measurement in customary terms of energy-scale coupling to individual spins, one must carefully consider the components of the co-magnetometer system. The co-magnetometer signal S is sensitive to the

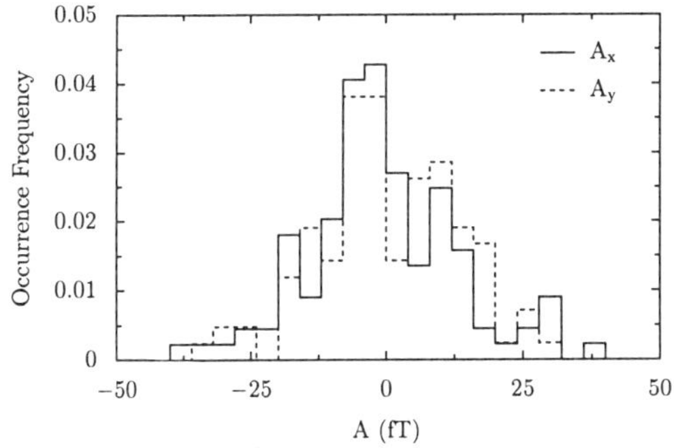

Figure 5. Distribution of the sidereal amplitudes of each day of data. These data reflect the amplitudes found in Fig. 4.

difference between the ^{3}He nuclear coupling and the K electron coupling to a field B:[2]

$$A = \kappa S = B_{^3\mathrm{He}} - B_{\mathrm{K}}. \tag{3}$$

Since the nuclear spin structure of ^{3}He is relatively well known, one can consider the proton as well as the neutron spin polarization in ^{3}He. To convert magnetic field measurements into an energy scale we equate the magnetic field interaction with a coupling to an anomalous $\tilde{b}$ field:

$$\mu_{^3\mathrm{He}} B_{^3\mathrm{He}} \langle \sigma_{^3\mathrm{He}} \rangle = \tilde{b}^n \langle \sigma_n \rangle + \tilde{b}^p \langle \sigma_p \rangle \tag{4}$$

$$\mu_{\mathrm{K}} B_{\mathrm{K}} \langle \sigma_{\mathrm{K}} \rangle = \tilde{b}^e \langle \sigma_e \rangle + \tilde{b}^p \langle \sigma_p \rangle. \tag{5}$$

For ^{3}He, several nuclear structure calculations[6,7,8,9] indicate that the spin polarization of the neutron in ^{3}He is $\langle \sigma_n \rangle / \langle \sigma_{^3\mathrm{He}} \rangle = 0.87 \pm 0.02$ and the spin polarization contributed by 2 protons is $\langle \sigma_p \rangle / \langle \sigma_{^3\mathrm{He}} \rangle = -0.05 \pm 0.01$. For the K the magnetic moment is dominated by the moment of the electron $\mu_{\mathrm{K}} = \mu_B$. Even though there is some contribution from proton polarization in the K nucleus, it is not significant because $\mu_{\mathrm{K}} \gg \mu_{^3\mathrm{He}}$.

The sidereal amplitudes measured by the co-magnetometer can be expressed in terms of spin coupling to anomalous fields referenced with respect to the celestial sphere. One must rotate in the x-y plane to correct for the Local Sidereal Time (LST) and divide by cos(40.3449°) to account for the latitude of the laboratory to obtain:

$$\begin{aligned} \tilde{b}^n_X &= (-3.7 \pm 8.1) \times 10^{-32}\ \mathrm{GeV} & \tilde{b}^n_Y &= (-9.0 \pm 7.5) \times 10^{-32}\ \mathrm{GeV} \\ \tilde{b}^p_X &= (+6.0 \pm 1.3) \times 10^{-31}\ \mathrm{GeV} & \tilde{b}^p_Y &= (+1.5 \pm 1.2) \times 10^{-31}\ \mathrm{GeV} \\ \tilde{b}^e_X &= (+2.8 \pm 6.1) \times 10^{-29}\ \mathrm{GeV} & \tilde{b}^e_Y &= (+6.8 \pm 6.1) \times 10^{-29}\ \mathrm{GeV} \end{aligned} \tag{6}$$

where the celestial axes $\hat{X}$ and $\hat{Y}$ intersect with the experiment's longitude at midnight and noon LST, respectively. These couplings are dominated by uncertainty, indicating that there is no anomalous effect at this level of sensitivity.

The following 1σ bounds can be set on the absolute value of $\tilde{b} = \sqrt{\tilde{b}_X^2 + \tilde{b}_Y^2}$:

$$
\begin{array}{ll}
\text{This experiment} & \text{(Previous measurements)} \\
\tilde{b}^n < 1.4 \times 10^{-31}\ \text{GeV} & (\tilde{b}^n < 1.1 \times 10^{-31}\ \text{GeV},\ {}^3\text{He-}{}^{129}\text{Xe maser}^{10}) \\
\tilde{b}^p < 2.2 \times 10^{-30}\ \text{GeV} & (\tilde{b}^p < 1.8 \times 10^{-27}\ \text{GeV, H maser}^{11}) \\
\tilde{b}^e < 1.0 \times 10^{-28}\ \text{GeV} & (\tilde{b}^e < 3.5 \times 10^{-31}\ \text{GeV, torsional pendulum}^{12}).
\end{array}
\quad (7)
$$

The limit on the neutron spin coupling is comparable to the best previous experiment.[10] The limit on the proton coupling is a factor of 10^3 better than the limit from the H maser,[11] although it is not independent of the neutron measurement. However, the combination of the He-Xe maser and our limits constrain the proton contribution at the level of a few 10^{-30} GeV. The recent electron spin results from the University of Washington[12] are much more sensitive and effectively eliminate possible contribution of electron coupling in our system. These results should be considered preliminary due to the relatively low sample size and the high systematic noise found in the majority of the data.

Significant improvements in the sensitivity of the co-magnetometer can be achieved by reducing systematic noise. Although the co-magnetometer can achieve 1 fT sensitivity in about 10 seconds, the poor long-term stability of the experiment provided 1 fT sensitivity after a year of data. A promising second generation of this experiment, CPT-II, has been designed from the ground up to speed the measurement time by rotating the experiment while reducing temperature sensitivity and mechanical instability.[13] In the mean time, the stationary apparatus is being used to to search for nuclear spin-dependent forces using a polarized ^{3}He source which can be reversed by adiabatic fast passage every few seconds.

Ongoing improvements to the experiment have reduced the noise of the co-magnetometer by a factor of around 5, mainly in the frequency region of 0.1–1 Hz. The present noise spectrum of the co-magnetometer is shown in Fig 6. The modulation scheme was changed from a Faraday rotator to a photo-elastic modulator which was found to give better performance for frequencies less than 1 Hz. The forced hot air oven heater was re-

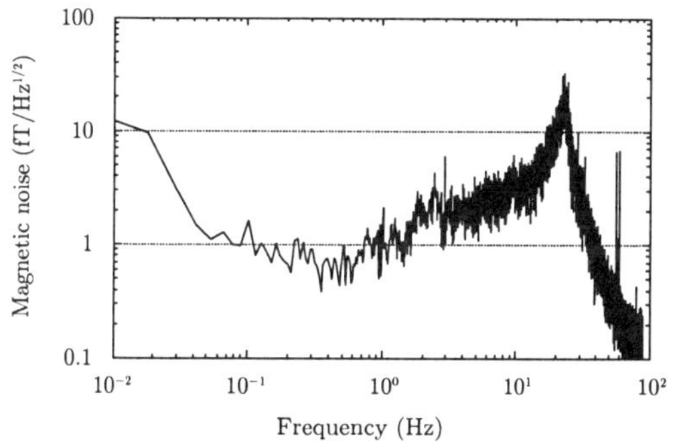

Figure 6. Performance of the improved co-magnetometer apparatus.

placed by a high frequency ac electric heater with twisted pair resistive wire. Greater laser stability and reliability is achieved by using DFB lasers for both pump and probe beams. The pump beam is amplified using a tapered amplifier with an intensity feedback. In order to minimize the effects of air currents and thermal expansion, the pump beam is coupled to a polarization-maintaining fiber after the amplifier to minimize the free-space path length.

References

1. J. C. Allred *et al.*, Phys. Rev. Lett. **89**, 130801 (2002).
2. T. W. Kornack, M. V. Romalis, Phys. Rev. Lett. **89**, 253002 (2002).
3. T. W. Kornack and M. V. Romalis, in V. A. Kostelecký, ed., *CPT and Lorentz Symmetry III*, World Scientific, Singapore, 2005.
4. V. A. Kostelecký and C. D. Lane, Phys. Rev. D **60**, 116010 (1998).
5. T. W. Kornack, dissertation, Princeton University, 2005.
6. J. L. Friar *et al.*, Phys. Rev. C **42**, 2310 (1990).
7. F. Bissey *et al.*, Phys. Rev. C **64**, 024004 (2001).
8. F. Bissey *et al.*, Phys. Rev. C **65**, 064317 (2002).
9. O. A. Rondon, Phys. Rev. C **60**, 035201 (1999).
10. D. Bear *et al.*, Phys. Rev. Lett. **89**, 209902 (2002).
11. D. F. Phillips *et al.*, Phys. Rev. D **63**, 111101 (2001).
12. B. R. Heckel *et al.*, Phys. Rev. Lett. **97**, 021603 (2006).
13. S. J. Smullin *et al.*, these proceedings.

CONSTRAINING QUANTUM GRAVITY WITH GLAST

F. KUEHN, R. HUGHES AND B. WINER
REPRESENTING THE GLAST LAT COLLABORATION
Physics Department, Ohio State University
191 W. Woodruff Ave.
Columbus, OH 43210, U.S.A.
E-mail: kuehn@mps.ohio-state.edu

We present results of ongoing work to understand GLAST's ability to constrain the scale of quantum gravity (QG). Using simulated gamma ray bursts with high energy emission, we find that if the scale of QG is 10^{19} GeV we are able to measure the effect to better than 10%. Our ability to distinguish the measured effect from a source intrinsic one is contingent on the discovery of multiple gamma ray bursts at different redshifts.

1. Introduction

1.1. *The GLAST experiment*

The Gamma Ray Large Area Telescope (GLAST) is a space based observatory consisting of two scientific instruments: the Large Area Telescope[1] (LAT) and the GLAST Burst Monitor[2] (GBM). The LAT is a pair conversion telescope with sensitivity in the energy range 20 MeV to greater than 300 GeV. The GBM is an array of scintillation detectors with sensitivity in the range 8 keV to 30 MeV. The scientific mission encompasses the discovery and monitoring of Gamma Ray sources in the universe.

1.2. *Gamma Ray Bursts and Quantum Gravity*

Gamma Ray Bursts (GRBs) are the brightest point sources of gamma rays in the universe. GRBs are essentially pulses of γ-rays spanning up to seven orders of magnitude in energy, and are typically visible in gamma ray telescopes for short periods of time. GRBs are classified into two basic types: short and long. Short bursts tend to last less than two seconds, whereas long ones can last up to hundreds of seconds. Optical follow-up observations have associated these outbursts with exploding stars,[3] or coalescing compact objects for long and short bursts,[4] respectively. These observations also show that GRBs are cosmological.

It is the large energy range, observed pulse structure, and cosmological distances that have led to the idea that GRBs can be used to test some observable consequences of QG, or Lorentz invariance violation (LIV).[5] Energy dependent modification to the speed of light result in photon arrival time delays for members of the same pulse. Specifically,

$$E_{QG} \sim \frac{L}{c}\frac{\Delta E}{\Delta t} \tag{1}$$

where Δt is the arrival time difference between two photons with energy splitting $\Delta E = E_2 - E_1$, L is the distance to the burst, c is the low energy (observed) speed of light, and E_{QG} is the energy scale of new physics, assumed to be the Planck scale $\sim 10^{19}$ GeV.

2. LAT Onboard detection and followup observations

Knowledge of the distance to the GRB is required for the association of time delays with a scale of QG. The distance is determined via optical followup observations which determine the redshift. To facilitate this program, the LAT is equipped with onboard software able to identify bright bursts onboard.[6] The software provides a position on the sky for the event and sends a rapid alert message via the GCN network.

GRBs are rapidly fading objects. It is easier to identify them in other wavebands if alert messages are sent soon after the initial burst. As the regular downlink intervals for data from the LAT are roughly spaced in three hour intervals, this system short circuits the process and will allow a better chance for ground identification.

3. Theoretical motivation

Motivation for testing the constancy of the speed of light comes from two general sources. Firstly, testing fundamental assumptions is a basic requirement of science. Secondly, there are numerous theoretical ideas related to QG in which the speed of light is not a constant.[7]

The notion of LIV is intimately tied to the idea that c may not be a constant. Attempts to understand and classify LIV are well documented elsewhere[8] and in these proceedings. However, there may be cases where Lorentz symmetry is maintained and yet the speed of light is not a constant. The concept of Doubly Special Relativity is one example.[9] Such notions provide strong motivation for the use of astrophysical tests because, in this case, a violation of the constancy of the speed of light would not show experimental signatures in Earth based LIV measurements since Lorentz symmetry is not broken.

The basis of the experiment to be performed here has been described elsewhere,[5] and is as follows. The speed of light is assumed to take the form

$$v = c\left(1 \pm \xi \frac{E}{E_{QG}}\right) \tag{2}$$

where ξ is an arbitrary constant usually taken to be one, E the energy of the photon, and the sign is indeterminate. This leads to an energy depend time delay in the arrival times of photons observed on Earth, yet emitted at the same time:

$$\Delta T = \xi \frac{\Delta E}{E_{QG}} \times \frac{L}{c} = \xi \frac{\Delta E}{E_{QG}} \times H_0^{-1} \int dz/h(z) \tag{3}$$

where ΔE is the energy between both photons, H_0^{-1} is the Hubble constant and the second term on the right shows the distance to the burst, L, as a function of cosmology. The ideal signature comes from a burst with short well defined pulses. It would show a delay in the arrival times of the photons from the pulses that is proportional to the distance to the burst and proportional to the measured energy of the γ-ray.

4. Current bounds

This measurement has been performed using γ-rays from GRBs at lower energies than those that will be observed by the LAT.[10] The data has come from experiments such as BATSE, BeppoSAX, HETE, INTEGRAL, and SWIFT. These all have sensitivity to energies lower than the lowest γ-rays observable by the LAT. A similar time delay measurement has been made using γ-rays from AGN by Whipple[11] and HESS. The GRB measurements have yielded bounds on E_{QG} (assuming $\xi = 1$) of $\sim 10^{16}$ GeV, while the AGN photons have pushed the limit up to $\sim 10^{17}$ GeV.

5. Current LAT sensitivity estimate

We present an analysis in progress to perform this measurement using future LAT data. As such, we are exploring the sensitivity of the LAT to the "QG" effect. We use simulated LAT data of bursts. These bursts are generated with spectral indices statistically equivalent to those BATSE detected, but with spectra extrapolated to LAT energies.[12]

5.1. *Analysis methodology*

The observable effect is a time delay in photons from the same pulse, but of different energies. We choose the sign of the effect to be such that higher energy photons travel more slowly than lower energy ones. In this scenario photons never travel faster than the "speed of light" and possible causality issues are avoided.

To fit for the effect we plot the arrival times versus the measured energies. For each pulse in a burst, we fit a straight line to the data and measure the slope. Equation 3 shows that the slope is related to the scale of QG. We bin all the data below a certain energy (in this case 550 MeV) into one point. It is allowed to float in the fit. We rely on the higher energy events for a possible signal using the lower energy data as a starting point for the fit.

Figure 1 is an example of this process. Here, the simulated burst consist of a single pulse. It was placed at a distance corresponding to $z \sim 1$, using the standard ΛCDM cosmology ($h = 0.7$, $\Omega_\Lambda = 0.7$, $\Omega_m = 0.3$, $\kappa = 0$). The

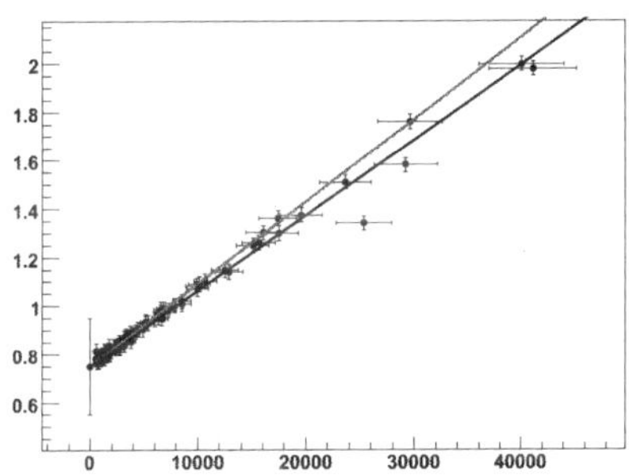

Figure 1. Arrival time (s) versus measured energy (MeV). The grey line represents the slope of the true effect. The black line shows the fit. The slope of this fit is consistent with the true value at the 1σ level.

slope of the true effect is shown by the grey line and the fit to the data is shown by the black line. When using bursts with multiple pulses, we fit all pulses simultaneously. This results in one estimated E_{QG} per burst. Fitting a few bursts in our sample with well defined pulses, we are able to determine E_{QG} to the 10% level for an input scale of 10^{19} GeV.

The determination of this effect will rely upon measurements of different bursts at different redshift. This will allow us to search for effects that are correlated with the distance and disentangle inherent source effects that could cause energy dependent delays in one particular burst. We are in the process of developing the combined burst analysis.

5.2. *Other GLAST LAT analysis*

The LAT collaboration has another analysis method with similar goals.[13] This algorithm is based on maximally cancelling any blurring effects on the entire GRB profile, by finding the amount of reverse-dispersion that optimally sharpens the profile. The procedure requires a model of the QG delay as a function of energy, but does not make use of a pulse or time-domain model of any kind, nor binning in either time or energy. This method is sensitive to finding $E_{QG} \sim 10^{19}$ GeV.

6. Discussion

The LAT has the potential to provide the most sensitive astrophysical test of QG using GRBs. This is due to a combination of the probed (high) energy regime, accurate timing information, and observable time structures/pulses associated with GRBs. If the effect is non-existing, bounds on the scale of QG will be placed at energies well above the expected scale of 10^{19} GeV. On the other hand, if the effect exists, the LAT is poised to discover it with good statistical certainty. Discovery is contingent on the use of multiple bursts at multiple redshifts to confirm that the effect is not intrinsic to a single source.

References

1. P.F. Michelson *et al.*, AIP Conf. Proc. **921**, 8 (2007).
2. G.G. Lichti *et al.*, AIP Conf. Proc. **906,** 89 (2007).
3. A.S. Fruchter *et al.*, Nature, **411**, 463 (2006).
4. D.B. Fox, *et al.*, Nature, **437**,845 (2005).
5. G. Amelino-Camelia, *et al.*, Nature, **393**, 763 (1998).
6. F. Kuehn, *et al.*, AIP Conf. Proc. **921**, 556 (2007).
7. D. Mattingly, gr-qc/0502097.
8. D. Colladay and V.A. Kostelecký, Phys. Rev. D **58**, 1160002 (1998); see also these proceedings.
9. J. Magueijo, and L. Smolin, gr-qc/0207085.
10. S. Boggs *et al.*, Ap. J., **611**, L77 (2004); J. Ellis *et al.*, Astron. Astrophys., **402**, 409 (2003); J. Bolmont *et al.*, astro-ph/0603725; B.E. Schaefer, Phys. Rev. Lett., **82**, 25 (1999).
11. S.D. Biller *et al.*, Phys. Rev. Lett. **83**, 2108 (1999).
12. N. Omodei, *et al.*, AIP Conf. Proc. **906**, 1 (2007).
13. J.D. Scargle, J.P. Norris, J.T. Bonnell, astro-ph/0610571.

RENORMALIZATION OF LORENTZ VIOLATING THEORIES

DAMIANO ANSELMI

Dipartimento di Fisica "E. Fermi" and INFN, Sezione di Pisa
Largo Bruno Pontecorvo 3, I-56127 Pisa, Italy
E-mail: damiano.anselmi@df.unipi.it

Unitary, Lorentz violating renormalizable theories can be constructed with the help of higher space derivatives. Renormalizability is governed by a "weighted power counting" criterion ensuring that no higher time derivatives are generated. A classification of the models is given, and a number of explicit examples are presented.

The set of power-counting renormalizable quantum field theories is considerably small. It can be enarged relaxing unitarity, locality and/or Lorentz invariance. Here I consider the renormalizable theories that can be constructed with the help of higher space derivatives and contain no higher time derivatives. Lorentz invariance is violated, but unitarity and locality are preserved. Renormalizability follows from a modified power-counting criterion, which weights time and space differently. The set of consistent theories is still very restricted, yet considerably larger than the set of Lorentz invariant theories. Renormalizable models exist in arbitrary spacetime dimensions.

I concentrate on scalar theories.[1] The generalization to fermions is straightforward. Gauge theories can be included following the same lines,[2] while gravity deserves a separate study. Lorentz violating models with higher space derivatives can be useful to define the ultraviolet limit of theories that are otherwise non-renormalizable or to study the new Lorentz violating Standard-Model Extension[3] and effective field theories. They can have also applications to nuclear physics, condensed matter physics and the theory of critical phenomena.

I first describe the models where the d-dimensional spacetime manifold M_d is split into the product of a $\widehat{d}$-dimensional submanifold $M_{\widehat{d}}$, containing time and possibly some space coordinates, and a $\overline{d}$-dimensional space submanifold $M_{\overline{d}}$. I assume that Lorentz and rotational invariance hold in the subspaces $M_{\widehat{d}}$ and $M_{\overline{d}}$, respectively. In the final part of the paper I describe how the construction can be extended to the most general case.

Each momentum p is split into components $\widehat{p}$, which live in $M_{\widehat{d}}$, and $\overline{p}$, which live in $M_{\overline{d}}$: $p = (\widehat{p}, \overline{p})$. The spacetime index μ is split into hatted and barred indices: $\mu = (\widehat{\mu}, \overline{\mu})$. A polynomial $P_{k,n}(\widehat{p}, \overline{p})$ in the momenta is called weighted polynomial of degree k and weight $1/n$, where k is a multiple of $1/n$, if $P_{k,n}(\xi^n \widehat{p}, \xi \overline{p})$ is a polynomial of degree kn in ξ. Clearly, $P_{k_1,n}(\widehat{p}, \overline{p}) P_{k_2,n}(\widehat{p}, \overline{p}) = P_{k_1+k_2,n}(\widehat{p}, \overline{p})$. A polynomial $H_{k,n}(\widehat{p}, \overline{p})$ is called homogeneous weighted polynomial of degree k and weight $1/n$, if $H_{k,n}(\lambda \widehat{p}, \lambda^{1/n} \overline{p}) = \lambda^k H_{k,n}(\widehat{p}, \overline{p})$. It is immediate to prove that a weighted polynomial $P_{k,n}$ of degree k can be expressed as a linear combination of homogeneous weighted polynomials $H_{k',n}$ of degrees $k' \leq k$.

Consider the free scalar theory

$$\mathcal{L}_{\text{free}} = \frac{1}{2}(\widehat{\partial}\varphi)^2 + \frac{1}{2\Lambda_L^{2n-2}}(\overline{\partial}^n \varphi)^2, \tag{1}$$

where Λ_L is an energy scale. Up to total derivatives it is not necessary to specify how the $2n$ derivatives $\overline{\partial}$ contract among themselves. The theory (1) is invariant under the weighted rescaling

$$\hat{x} \to \hat{x}\ \mathrm{e}^{-\Omega}, \qquad \bar{x} \to \bar{x}\ \mathrm{e}^{-\Omega/n}, \qquad \varphi \to \varphi\ \mathrm{e}^{\Omega(\not{d}/2-1)}, \tag{2}$$

where Ω is a constant parameter. Indeed, both lagrangian terms scale with the factor $\not{d} \equiv \widehat{d} + \overline{d}/n$, which is compensated by the scaling factor of the integration measure $\mathrm{d}^{\not{d}}x$ of the action.

Now, consider an interacting theory whose propagator is defined by (1). The vertices can be classified according to their number N of φ-legs, plus an extra label α that distinguishes different derivative structures. Each vertex of type (N, α) defines a monomial in the momenta of the fields. Denote the weighted degree of such a monomial by $\delta_N^{(\alpha)}$. A vertex containing p_1 derivatives $\widehat{\partial}$, p_2 derivatives $\overline{\partial}$ and N φ-legs is symbolically written as $\left[\widehat{\partial}^{p_1}\overline{\partial}^{p_2}\varphi^N\right]_\alpha$ and its degree is $\delta_N^{(\alpha)} = p_1 + p_2/n$. For the purposes of renormalization, quadratic terms different from those of Eq. (1) can be treated perturbatively as if they were "interactions" (i.e. two-leg vertices).

Consider a Feynman graph G made of L loops, E external legs, I internal legs and $v_N^{(\alpha)}$ vertices of type (N, α). The integral associated with G has the form

$$\mathcal{I}_G(k) = \int \frac{\mathrm{d}^{L\widehat{d}}\widehat{p}}{(2\pi)^{L\widehat{d}}} \int \frac{\mathrm{d}^{L\overline{d}}\overline{p}}{(2\pi)^{L\overline{d}}} \prod_{i=1}^{I} \mathcal{P}^{(i)}_{-2,n}(p,k) \prod_{j=1}^{V} \mathcal{V}^{(j)}_{\delta_j,n}(p,k),$$

where p are the loop momenta, k are the external momenta, $\mathcal{P}^{(i)}_{-2,n}$ are the propagators, which have weighted degree -2, and $\mathcal{V}^{(j)}_{\delta_j,n}$ are the vertices, with weighted degrees δ_j. The integral measure $\mathrm{d}^{\hat{d}}\widehat{p}\,\mathrm{d}^{\bar{d}}\overline{p}$ is a weighted measure of degree $\not{d}$. Performing a rescaling $(\widehat{k},\overline{k}) \to (\lambda\widehat{k},\lambda^{1/n}\overline{k})$, accompanied by an analogous change of variables $(\widehat{p},\overline{p}) \to (\lambda\widehat{p},\lambda^{1/n}\overline{p})$, it is immediate to prove that $\mathcal{I}_G(k)$ is a weighted function of degree

$$\omega(G) = L\not{d} - 2I + \sum_{j=1}^{V}\delta_j = L\not{d} - 2I + \sum_{(N,\alpha)}\delta_N^{(\alpha)}v_N^{(\alpha)}.$$

By the locality of counterterms, once the subdivergences of G have been inductively subtracted away, the overall divergent part of G is a weighted polynomial of degree $\omega(G)$ in the external momenta. The usual relations $L = I - V + 1$ and $E + 2I = \sum_{(N,\alpha)} N v_N^{(\alpha)}$ allow us to write

$$\omega(G) = d(E) + \sum_{(N,\alpha)} v_N^{(\alpha)}\left[\delta_N^{(\alpha)} - d(N)\right], \qquad d(X) \equiv \not{d}\left(1 - \frac{X}{2}\right) + X .$$

The theory is *i*) renormalizable, if it contains all vertices with $\delta_N^{(\alpha)} \leq d(N)$, and only those: $\omega(G)$ does not increase when the number of vertices increases; *ii*) super-renormalizable, if it contains all vertices with $\delta_N^{(\alpha)} < d(N)$, and only those: $\omega(G)$ decreases when the number of vertices increases; *iii*) strictly-renormalizable, if it contains all vertices with $\delta_N^{(\alpha)} = d(N)$, and only those: $\omega(G)$ does not depend on $v_N^{(\alpha)}$; *iv*) non-renormalizable, if it contains some vertices with $\delta_N^{(\alpha)} > d(N)$: $\omega(G)$ increases when the number of those vertices increases. The vertices with $\delta_N^{(\alpha)} = d(N)$ are called "weighted marginal", those with $\delta_N^{(\alpha)} < d(N)$ are called "weighted relevant" and those with $\delta_N^{(\alpha)} > d(N)$ are called "weighted irrelevant".

By locality, $\delta_N^{(\alpha)}$ cannot be negative. Moreover, polynomiality demands that there must exist a bound $N_{\max}$ on the number of legs that the vertices can contain. It is easy to show that these requirements demand

$$\not{d} > 2 \tag{3}$$

and the bound is

$$N_{\max} = \left[\frac{2\not{d}}{\not{d} - 2}\right], \tag{4}$$

where $[x]$ denotes the integral part of x.

Thus, in a renormalizable theory the weighted degree of divergence $\omega(G)$ of a graph G satisfies

$$\omega(G) \leq d(E). \tag{5}$$

The inequality (3) ensures also that $\omega(G)$ decreases when the number of external legs increases. Finally, since the vertices that subtract the overall divergences of G are of type (E,α) with $\delta_E^{(\alpha)} = \omega(G)$, it is immediate to check that the lagrangian contains all needed vertices. Indeed, Eq. (5) coincides with the inequality satisfied by $\delta_E^{(\alpha)}$. This completes the proof of renormalizability.

Perturbative unitarity demands that no higher-time derivative be present, both in the kinetic part and in the vertices. A lagrangian term with higher time derivatives would have $\delta_N^{(\alpha)} \geq 2$ for $N > 2$ or $\delta_2^{(\alpha)} > 2$. This cannot happen in a renormalizable theory, because Eq. (3) and $\delta_N^{(\alpha)} \leq d(N)$ imply $\delta_N^{(\alpha)} \leq 2$ in general and $\delta_N^{(\alpha)} < 2$ for $N > 2$. In particular, true vertices ($N > 2$) cannot contain any $\widehat{\partial}$-derivative at all, because the invariance under the Lorentz and rotational symmetries of $M_{\widehat{d}}$ and $M_{\overline{d}}$ excludes terms containing an odd number of $\widehat{\partial}$'s or an odd number of $\overline{\partial}$'s. Similar conclusions apply to the counterterms, because of Eq. (5). Therefore, renormalization does not turn on higher time derivatives, as promised.

The strictly renormalizable models have the $\delta_N^{(\alpha)} = d(N)$. Their lagrangian has the form

$$\mathcal{L}_{(\widehat{d},\overline{d})} = \frac{1}{2}(\widehat{\partial}\varphi)^2 + \frac{1}{2\Lambda_L^{2n-2}}(\overline{\partial}^n\varphi)^2 + \sum_{(N,\alpha)} \frac{\lambda_{(N,\alpha)}}{N!\Lambda_L^{(n-1)(N+\widehat{d}-\widehat{d}N/2)}} \left[\overline{\partial}^{nd(N)}\varphi^N\right]_\alpha, \tag{6}$$

where $\lambda_{(N,\alpha)}$ are dimensionless couplings. Examples are the $d = 4$ $2(n+1)$-dimensional φ^4-theories

$$\mathcal{L}_{(2,2n)} = \frac{1}{2}(\widehat{\partial}\varphi)^2 + \frac{1}{2\Lambda_L^{2(n-1)}}(\overline{\partial}^n\varphi)^2 + \frac{\lambda}{4!\Lambda_L^{2(n-1)}}\varphi^4 \tag{7}$$

and the $d = 3$, $(2+n)$-dimensional φ^6-theories, whose lagrangian is

$$\mathcal{L}_{(2,n)} = \frac{1}{2}(\widehat{\partial}\varphi)^2 + \frac{1}{2\Lambda_L^{2(n-1)}}(\overline{\partial}^n\varphi)^2 + \frac{\lambda_6}{6!\Lambda_L^{2(n-1)}}\varphi^6, \tag{8}$$

when n is odd and

$$\mathcal{L}_{(2,n)} = \frac{1}{2}(\widehat{\partial}\varphi)^2 + \frac{1}{2\Lambda_L^{2(n-1)}}(\overline{\partial}^n\varphi)^2 + \frac{1}{4!\Lambda_L^{2(n-1)}}\sum_\alpha \lambda_\alpha \left[\overline{\partial}^n\varphi^4\right]_\alpha + \frac{\lambda_6}{6!\Lambda_L^{2(n-1)}}\varphi^6, \tag{9}$$

when n is even. Four-dimensional models are (9) with $n = 2$ and

$$\mathcal{L}_{(1,3)} = \frac{1}{2}(\widehat{\partial}\varphi)^2 + \frac{1}{2\Lambda_L^2}(\overline{\Delta}\varphi)^2 + \frac{\lambda_6}{6!\Lambda_L^4}\varphi^4(\overline{\partial}\varphi)^2 + \frac{\lambda_{10}}{10!\Lambda_L^6}\varphi^{10}.$$

The weighted scale invariance (2) still holds at the classical level, but is anomalous at the quantum level. Non-homogeneous theories can be obtained from the homogeneous ones adding super-renormalizable terms, which are those that satisfy the strict inequality $\delta_N^{(\alpha)} < d(N)$.

Instead of splitting the spacetime manifold into two submanifolds, we can split it into the product of more submanifolds, eventually one for each coordinate. This analysis covers the most general case. It is still necessary to distinguish a $\widehat{d}$-dimensional submanifold $M_{\widehat{d}}$ containing time from the $\overline{d}_i$-dimensional space submanifolds $M_{\overline{d}_i}$, $i = 1, \ldots \ell$, so write $M_d = M_{\widehat{d}} \otimes \prod_{i=1}^{\ell} M_{\overline{d}_i}$.

Denote the space derivatives in the ith space subsector with $\overline{\partial}_i$ and assume that they have weights $1/n_i$. Then the kinetic term of the lagrangian reads

$$\mathcal{L}_{\rm kin} = \frac{1}{2}(\widehat{\partial}\varphi)^2 + \frac{1}{2}\varphi P_2(\overline{\partial}_i, \Lambda_L)\varphi,$$

where $P_2(\overline{\partial}_i, \Lambda_L)$ is the most general weighted homogeneous polynomial of degree 2 in the spatial derivatives, $P_2(\lambda^{1/n_i}\overline{\partial}_i, \Lambda_L) = \lambda^2 P_2(\overline{\partial}_i, \Lambda_L)$, invariant under rotations in the subspaces $M_{\overline{d}_i}$. The Λ_L-dependence is arranged so that P_2 has dimensionality 2. The previous analysis can be repeated straightforwardly. It is easy to verify that the weighted power-counting criterion works as before with

$$d\!\!\!/ = \widehat{d} + \sum_{i=1}^{\ell} \frac{\overline{d}_i}{n_i}.$$

References

1. D. Anselmi and M. Halat, arxiv:0707.2480.
2. D. Anselmi, in preparation.
3. D. Colladay and V.A. Kostelecký, Phys. Rev. D 58 (1998) 116002; T. Jacobson, S. Liberati and D. Mattingly, Ann. Phys. 321 (2006) 150; and other contributions to these proceedings.

RENORMALIZATION OF GAUGE THEORIES WITH LORENTZ VIOLATION

D. COLLADAY AND P. MCDONALD
Division of Natural Sciences
5800 Bay Shore Rd
Sarasota, FL 34243, U.S.A.
E-mail: colladay@ncf.edu

In this talk, the explicit one-loop renormalization of pure Yang-Mills with Lorentz violation is discussed. Multiplicative renormalization is used to demonstrate universality of the coupling constant as the various two-, three-, and four-point diagrams are consistent with the proposed renormalization scheme. Divergences can therefore be consistently cancelled and the one-loop beta functions can be derived yielding the running of the Lorentz-violating coupling constants.

1. Introduction

Pure Yang-Mills theories are remarkable as they are highly nonlinear, but remain renormalizable despite their mathematical complexity. Moreover, there has been much recent interest in theoretical and experimental aspects of possible Lorentz violation in nature.[1] In this talk, I will discuss the extension of the standard renormalization techniques to Yang-Mills theory in the presence of Lorentz-violating terms. The details of the calculations may be found in the literature.[2] Related work includes an analysis of one-loop renormalization in the QED sector,[3] an extension of a subset of terms to curved spacetime,[4] and effects of ghost sector Lorentz violation on Higgs through spontaneous symmetry breaking.[5] In addition, renormalization properties of scalar fields with Lorentz violation have recently been studied.[6] The presence of instantons in Yang-Mills theories with Lorentz violation has also been discussed in the literature.[7]

An overview of Yang-Mills with Lorentz violation is given in the Notations and Conventions section. The CPT-even and CPT-odd Lorentz-violating terms are considered independently in the following sections. The resulting beta functions and the running of the couplings are then discussed.

2. Notations and Conventions

The pure Yang-Mills Lagrangian is constructed in the conventional way using the curvature $F^{\mu\nu}$ and the connection A^μ. Incorporating Lorentz-violating terms that are both gauge invariant and power-counting renormalizable yields the following action

$$S_M(A) = -\frac{1}{2}\int d^4x Tr\left[F^{\mu\nu}F_{\mu\nu} + (k_F)_{\mu\nu\alpha\beta}F^{\mu\nu}F^{\alpha\beta}\right. \\ \left. + (k_{AF})^\kappa \epsilon_{\kappa\lambda\mu\nu}(A^\lambda F^{\mu\nu} - \frac{2}{3}igA^\lambda A^\mu A^\nu)\right] \quad . \quad (1)$$

The k_F terms are CPT-even while the k_{AF} terms are CPT-odd. Gauge-fixing terms including Faddeev-Popov ghost fields are also introduced in the standard way

$$S_g = -\int d^4x Tr\left[\frac{1}{\xi}(\partial^\mu A_\nu)^2 + \bar{c}(\partial^\mu D_\mu)c\right] \quad . \quad (2)$$

We work to lowest-order in the Lorentz-violating parameters and use dimensional regularization to regularize the integrals. This makes it possible to separate the one-loop contributions into two independent computations based on CPT symmetry since to lowest-order there is no mixing. To simplify the analysis, k_F is taken to be trace-free, although the more general case can be handled using more sophisticated functional determinant techniques.[8] Both the current techniques and the functional determinant approach yield consistent results.

Parameters δ_i^j are introduced in the counterterm Lagrangian such that i represents the parameter k_F or k_{AF}, while the superscript j represents the number of bosons flowing into the vertex. These parameters are then used to calculate the multiplicative renormalization factors Z_i for the Lorentz-violating couplings.

3. One-Loop Calculations

In the following section, the combination

$$\tilde{g}^2 = \frac{g^2}{(4\pi)^2}C_2(G)\Gamma\left(2-\frac{d}{2}\right) \quad (3)$$

appears often. The factor $C_2(G)$ is the quadratic Casimir invariant of the Lie algebra and $\Gamma(x)$ is the standard gamma function exhibiting the usual pole at $d = 4$. The standard values for the coupling constant renormalization factor

$$Z_g = 1 - \frac{11}{6}\tilde{g}^2 ,$$

and the wave function renormalization parameter

$$Z_3 = 1 - \left(\frac{5}{3} + \frac{1}{2}(1-\xi)\right)\tilde{g}^2 ,$$

are retained in the Lorentz-violating case since the Lorentz-preserving diagrams are unmodified. The parameter ξ is the conventional gauge fixing term. At one loop, the k_F terms yield only CPT-even radiative corrections, while the k_{AF} terms yield only CPT-odd corrections. This occurs because of the unique symmetry properties of these terms and the preservation of these symmetries at one-loop.

Calculation of the two-point function yields the following counterterms for these parameters:

$$\delta^{2g}_{k_F} = \left(4 + \frac{1}{2}(1-\xi)\right)\tilde{g}^2 , \tag{4}$$

and

$$\delta^{2g}_{k_{AF}} = \left(-2 + \frac{1}{2}(1-\xi)\right)\tilde{g}^2 . \tag{5}$$

These expressions fix the renormalization factors $(k_F)_{B\mu\nu\rho\sigma} = Z_{k_F}(k_F)_{\mu\nu\rho\sigma}$ and $(k_{AF})^{\kappa}_{B} = Z_{k_{AF}}(k_{AF})^{\kappa}$ with

$$Z_{k_F} = 1 + \frac{7}{3}\tilde{g}^2 \tag{6}$$

and

$$Z_{k_{AF}} = 1 - \frac{11}{3}\tilde{g}^2 . \tag{7}$$

To check the theoretical consistency of the multiplicative renormalization procedure, it is necessary to verify that the three- and four-point diagrams produce the expected results given the above renormalization conditions. Explicit calculation demonstrates that this is in fact the case, as the three- and four-point diagrams produce exactly the expected terms based on multiplicative renormalization.

4. Beta Functions

If the validity of the multiplicative renormalization scheme is assumed to apply to all-orders, it is possible to compute the beta functions that determine the running of the coupling constants. These beta functions are calculated as

$$\beta_g = -\frac{11g^3}{3(4\pi)^2} C_2(G) \ , \tag{8}$$

for the coupling g, the same as in the conventional case and,

$$\beta_{k_F} = \frac{14g^2}{3(4\pi)^2} C_2(G) k_F \ , \quad \beta_{k_{AF}} = -\frac{22}{3}\frac{g^2}{(4\pi)^2} C_2(G) k_{AF} \tag{9}$$

for the Lorentz-violating terms. The running is then calculated in terms of the parameter[a]

$$Q(\mu) = 1 + \frac{22g_0^2}{3(4\pi)^2} C_2(G) \ln \frac{\mu}{\mu_0} \ , \tag{10}$$

that determines the running of g as

$$g^2(\mu) = Q^{-1} g_0^2 \ . \tag{11}$$

The running of the Lorentz-violating parameters may then be expressed as

$$k_F = (k_F)_0 Q^{7/11} \ , \quad k_{AF} = (k_{AF})_0 Q^{-1} \ . \tag{12}$$

The result of this analysis is that the CPT-odd k_{AF} term runs in exactly the same way as the standard coupling g does. The k_{AF} terms run in the opposite direction, but with the same basic logarithmic growth as g.

5. Summary

It has been demonstrated through explicit one-loop calculations that pure Yang-Mills theory in the presence of Lorentz-violating terms is in fact renormalizable to lowest order in the parameters. All of the divergences can be consistently subtracted using multiplicative renormalization techniques as in the conventional case. Calculations indicate patterns that imply that this procedure may apply to all-orders, but this remains to be verified in future work. If the assumption of full renormalizabilty is made, it is possible to compute the beta functions and analyze the running of the Lorentz-violating parameters. The CPT odd k_{AF} term is found to run in exactly

[a] Parameters with a subscript 0 represent running couplings evaluated at the reference scale μ_0.

the same way as the conventional coupling constant g. The behavior of the k_F term is similar in functional behavior, but runs in the opposite sense. Another interesting fact is that it turns out to be unnecessary to modify the ghost sector to accommodate the Lorentz violating terms.

References

1. See, for example, V.A. Kostelecký, ed., *CPT and Lorentz Symmetry III*, World Scientific, Singapore, 2005; and these proceedings.
2. D. Colladay and P. McDonald, Phys. Rev. D **75**, 105002 (2007).
3. V. A. Kostelecký, C. Lane, and A. Pickering, Phys. Rev. D **65**, 056006 (2002); Phys. Rev. Lett. **91** 031801 (2003).
4. G. de Barredo-Peixoto and I. L. Shapiro, Phys. Lett. B **642** 153 (2006).
5. B. Altschul, Phys. Rev. D **73** 045004 (2006).
6. D. Anselmi and M. Halat, hep-th/0707.2480; D. Anselmi, these proceedings.
7. D. Colladay and P. McDonald, J. Math. Phys. **45** 3228 (2004).
8. D. Colladay and P. McDonald, in preparation.

A POEM-BASED TEST OF THE WEP USING A SOUNDING ROCKET (SR-POEM)

ROBERT D. REASENBERG AND JAMES D. PHILLIPS
Smithsonian Astrophysical Observatory
Harvard-Smithsonian Center for Astrophysics
60 Garden Street
Cambridge, MA 02138, U.S.A.
E-mail: reasenberg@cfa.harvard.edu

We are developing an instrument (SR-POEM) for detecting a possible violation of the weak equivalence principle (WEP) while on a free-fall trajectory provided by a sounding rocket. An SR-POEM experiment compares the rate of fall of two test substances with a single-flight accuracy of $\sigma(\eta) \leq 10^{-16}$. It has evolved from POEM,[1] a ground-based test of the WEP, whose design was inspired by the successful WEP test at JILA.[2]

In the SR-POEM instrument, there are four test mass assemblies (TMA), each about 0.5 kg. There are two with each kind of sample, arranged in a square on a plane perpendicular to the symmetry axis (z axis) of the payload and close to the CM of the free-flying assembly. The two TMA along a diagonal of the square have the same kind of sample. At a distance of perhaps 0.3 m along the z axis, there is a highly stable plate (e.g., ULE glass) that holds four of our tracking frequency laser gauges (TFG)[3] with the same in-plane spacing and orientation as the TMA. The TFG measure the distances from (concave, partially reflecting) mirrors on the stable plate to (flat) mirrors on the TMA. We assume that further development of the TFG will bring the precision to 0.1 pm/$\sqrt{Hz}$, which is comfortably above the photon counting statistics limit. These TFG are augmented to provide the error signals required to align the two-mirror cavities.[4] As in POEM, each TMA is surrounded by the plates of a multi-component capacitance gauge that can sense six degrees of freedom for each TMA and can apply force and torque electrostatically.

In operation, each TMA is continuously observed by the corresponding TFG. During a measurement, the TMA are free to translate, but their orientation is constrained by the suspension system (TMA-SS), which uses the capacitance gauge electrodes to apply torque. Position measurements made with the capacitance gauge just before and just after the WEP measurement will be used to determine the average transverse motion of each TMA and thus the differential Coriolis acceleration.

At the start of the WEP measurement phase of the flight, the z axis is aligned with the direction to the Earth and the TMA have been uncaged. The capacitance gauge is used to assess and then correct the position and motion of the TMA in all degrees of freedom. Then the translation portion of the TMA-SS is disabled and, following that, data taken by the TFG may be used for WEP measurements, each lasting 40 s nominally. There will be about eight such measurements during the science phase of the flight, with the instrument orientation inverted between successive measurements to cancel most systematic errors. During these rotational slews, the translation control of the TMA-SS will again be enabled.

When the TMA separate from the caging mechanism, they will acquire a potential that we expect to be of the order of 0.1 V. During the setup phase, before the payload has reached the altitude where the WEP measurements can be made, this potential will be measured by applying DC to the capacitance plates and neutralized iteratively using UV light and a bias potential on the plates. Residual charge, which we expect to cause 0.001 V scale potential and be stable, will have an effect on $\hat{\eta}$, the estimate of the WEP-violation parameter from the experiment, that will tend to cancel with the payload inversions. The variation of surface potential on the TMA and capacitance plates cannot be eliminated, although it can be made small ($\approx$ 1 mV rms) by choice of surface material (e.g., gold).[5] Any constant force in the measurement direction will cancel as a result of the payload inversions. Temporal variation of this force is expected to contribute systemic error well below the expected WEP measurement precision.[5]

In the short ($\approx 10^3$ s) period of free fall, it will not be possible for the instrument to reach thermal equilibrium. Yet, thermal stability is an essential aspect of the SR-POEM experiment. Our approach is to ensure long thermal time constants and, where necessary, to use materials with a low coefficient of thermal expansion (CTE). By flying around midnight, we avoid direct solar heating, which would exceed other thermal perturbations and be highly directional in spacecraft coordinates.

We seek to minimize two adverse thermal effects. First (direct), temperature change within the precision measurement system can result in a direct measurement error. Second (indirect), temperature change in any pat of the payload can result in a change of the mass distribution and therefore a change in the accelerations of the TMA. Since a local mass generally couples differently to each of the TMA, a shift in the location of local mass could mimic a WEP signature. However, a slow position drift of local mass will tend to produce an effect that will cancel when the instrument orientation is reversed.

The four TFG are mounted on a highly stable plate, which we assume here to be selected ULE glass with a CTE of $\alpha \approx 1 \times 10^{-8}/K$. We require that, during a 40 s measurement period, the change of shape of the plate not introduce more than 0.1 pm into the WEP measurement. However, simple (cylindrical) bending or (spherical) doming will not matter to the extent that the TMA are symmetrically mounted, because the WEP signal is found by subtracting the average measurement of one diagonal pair of TMA from that of the other. Similarly, only a change of shape that adds a quadratic temporal term to the required combination of TFG measurements can masquerade as an WEP violation. Because of the alternating orientations of the instrument for successive WEP measurements, the offending shape change would need to be correlated with the measurement cycles.

The precision instrument is mounted inside an aluminum vacuum chamber with a 1/4 inch wall. The chamber is mounted to the payload tube through a thermal isolating structure. The inside and outside of the chamber are gold plated giving those surfaces a low emissivity, $\epsilon = 0.02$. Radiative exchange between the payload tube ($\epsilon = 0.1$) and the chamber wall causes the chamber to approach the payload-tube temperature with a time constant of 1.5×10^5 s. Inside the chamber, the most critical component is the TFG plate. It is a ULE disk, 20 cm in diameter and 4 cm thick, with $\epsilon \approx 1$. The TFG plate is suspended from the ULE metering tube by three titanium flexures, each controlling two degrees of freedom. Radiative exchange between the chamber wall and the TFG plate causes the plate to approach the chamber wall temperature with a time constant of 5.5×10^5 s. The corresponding time constant for the much thinner (1 cm) metering tube is 1.4×10^5 s. These time constants were calculated without consideration of the beneficial effect of the mu-metal shield that is included between the chamber and the precision instrument. These long thermal time constants and the lack of significant thermal input on the time scale of the inversion cycle ($P_{inv} \approx 140$ s) suggest that the direct thermally induced systematic error will be well below the expected statistical uncertainty of the WEP estimate.

The most important heat source for changing the mass distribution in the neighborhood of the TMA is the radio transmitter, which is mounted on a shelf in an end section of the payload.[a] During the measurement phase of the flight, heat from the radio will raise the temperature of the most affected 1 meter section of the aluminum payload tube by about 6 K, causing the shelf to move away from the TMA by 0.14 mm. If the end section with the radio is properly balanced, it will look like an on-axis mass, and its motion will have no effect on the WEP signal. However, if we assume that the 36 kg radio section has a 5 cm mass offset (cf. 18 cm radius of the cylindrical payload), then the effect over the entire measurement period grows nearly linearly to $\Delta(\eta) \approx 1.4 \times 10^{-18}$. However, the imbalance is unlikely to be so large and there would be strong cancellation of this effect by the inversions. The indirect thermal effect is not a problem.

Near the Earth's surface, $dg/dh \approx 3 \times 10^{-7} g_0/m$ and $d^2g/dh^2 \approx 1.5 \times 10^{-13} g_0/m^2$, where h is altitude and $g_0 = g(h)$ is evaluated at $h = 0$. Thus, for the second order term to contribute an error in acceleration of $10^{-17}g$ would require one set of TMA to be displaced from the other vertically by about 1 cm. However, the positioning of the TMA at the start of a measurement will be uniform to better than 10 micron. Similarly, tilting the instrument by enough to displace the TMA of one along-diagonal pair from the other pair by ± 1 cm, yields an additional differential acceleration of about $10^{-17}g$. We expect such tilts to be at least 1000 fold smaller.

During a WEP measurement free-fall time Q, the TMA will move along the instrument z axis as a result of payload gravity, Earth's gravity gradient, and spacecraft acceleration due to low-thrust forces (drag, gas leaks, radiation pressure from a warm Earth, etc.). They will therefore explore an inhomogeneous gravity field. In 40 s, the Earth's gravity gradient acting on TMA that are 1 cm from the horizontal plane through the payload CM will displace those TMA by 24 micron or, with proper initial velocity, 6 micron.

Since the TMA are not located at (but are quite close to) the center of mass of the payload, they accelerate away from (or toward) that point

---

[a]During the atmospheric ascent, the outer skin of the rocket is heated by about 130 K because of friction with the air. This (17 inch diameter) insulated skin is ejected at about 100 s after launch. The effect on the (14 inch diameter) payload tube is to increase its temperature by about 1 K. Following skin ejection, the payload tube cools radiatively by about 4 mK/s. Neither of these temperature changes results in much perturbation at P_{inv}, and such a thermal signal would be attenuated by factors of $\approx$ 6,000 in the transfer to the vacuum chamber and then to the precision structure. Therefore, they do not bias the WEP measurement.

in proportion to their displacement from the CM multiplied by the Earth's gravity gradient tensor, plus a contribution from the payloads gravity. The gravity gradient tensor changes smoothly and predictably as the payload follows its free-fall trajectory; Earth's gravity field is precisely known, as will be the payload trajectory. For the payload close to its nominal (vertical) orientation, the gravity gradient produces very nearly the same TMA acceleration (in payload coordinates) independent of whether the instrument is nadir or zenith pointing. The differences, which come from the next term in the expansion of the Earth's field, are shown above to be negligible.

We envision a sequence of sounding rocket flights. With each successive flight, we would seek to use our flight experience to improve the measurement accuracy and would use a different pair of test substances so as to address the several components of matter that might contribute to a WEP violation. Long term, SR-POEM could evolve into an instrument for a non-cryogenic orbital mission or the technology could be merged with that being developed for STEP to yield a better cryogenic mission.

References

1. R.D. Reasenberg and J.D. Phillips, Int. J. Mod. Phys. D, in press.
2. T.M. Niebauer, M.P. McHugh, and J.E. Faller, Phys. Rev. Lett. **59**, 609 (1987).
3. J.D. Phillips and R.D. Reasenberg, Rev. Sci. Inst. **76**, 064501 (2005).
4. N.M. Sampas, Ph.D. thesis, U. Colorado, 1990; N.M. Sampas and D.Z. Anderson, Applied Optics **29**, 394 (1990).
5. N.A. Robertson *et al.*, Class. Quantum Grav. **23**, 2665 (2006).

TESTING ALTERNATIVE OSCILLATION SCENARIOS WITH ATMOSPHERIC NEUTRINOS USING AMANDA-II DATA FROM 2000 TO 2003

J. AHRENS

Institute of Physics, Mainz University
Staudinger Weg 7
D-55099 Mainz, Germany
E-mail: jens.ahrens@lycos.de

J. L. KELLEY

Department of Physics, University of Wisconsin
1150 University Avenue
Madison, WI 53706, U.S.A.
E-mail: jkelley@icecube.wisc.edu

FOR THE ICECUBE COLLABORATION

http://www.icecube.wisc.edu

The AMANDA-II neutrino telescope detects upward-going atmospheric muon neutrinos penetrating the Earth from the Northern Hemisphere via the Cherenkov light of neutrino-induced muons, allowing the reconstruction of the original neutrino direction. Due to the high energy threshold of about 50 GeV, the declination spectrum is minimally affected by mass-induced neutrino oscillations; however, alternative oscillation models predicting effects at high energy can be tested and constrained. Of particular interest are models that allow one to test Lorentz invariance and the equivalence principle. Using the AMANDA-II data from the years 2000 to 2003, a sample of 3401 candidate neutrino-induced events was selected. No indication for alternative oscillation effects was found. For maximal mixing angles, an upper limit is set on both the Lorentz violation parameter $\delta c/c$ and the equivalence principle violation parameter $2|\phi|\delta\gamma$ of 5.3×10^{-27} at the 90% confidence level.

1. Introduction and detector description

Cosmic ray particles entering the Earth's atmosphere generate a steady flux of secondary particles, including muons and neutrinos. High energy muons pass through the atmosphere and can penetrate several kilometers of ice and rock, while atmospheric neutrinos of energies only above roughly 40 TeV start to be absorbed in the Earth. Lower energy muon neutrinos penetrating the diameter of the Earth can oscillate into tau neutrinos. However, the oscillation maxima at 30 GeV[1] and below are beneath the AMANDA-II threshold. Departures from conventional mass-induced oscillations could emerge at higher neutrino energies due to relativity-violating effects (see below). Such mechanisms would distort the expected angular distribution and energy spectrum of atmospheric neutrinos and could be detectable by AMANDA-II.

The AMANDA-II neutrino telescope is embedded 1500 – 2000 m deep in the transparent and inert ice of the Antarctic ice sheet, close to the geographic South Pole. AMANDA-II consists of 677 optical modules (OMs) on 19 vertical strings, which are arranged in three approximately concentric circles of 60 m, 120 m and 200 m diameter. Muons produced in ν_μ-nucleon interactions can be directionally reconstructed by observing the Cherenkov radiation that propagates through the ice to the array of photosensors. To ensure that the observed muon is due to a neutrino interaction, the Earth is used as a filter against atmospheric muons, and only tracks from the Northern Hemisphere (declination $\delta > 0°$) are selected.

2. Phenomenology of standard and alternative neutrino oscillations

It is commonly accepted that standard (mass-induced) $\nu_\mu \rightarrow \nu_\tau$ oscillations[a] are responsible for the measured deficit of atmospheric muon neutrinos.[1] Atmospheric neutrino data can also be used to test non-standard oscillation mechanisms that lead to observable differences at higher neutrino energies. Various new physics scenarios can result in neutrino flavor mixing. Two of these scenarios, which can be described in a mathematically analogous way, have been tested in this analysis. The underlying theories assume small deviations from the principles of the theory of relativity and lead to measurable neutrino oscillations:

- In theories predicting violation of Lorentz invariance (VLI), a set of additional neutrino eigenstates with different maximal attainable velocities (MAV) c_n/c is introduced, violating special relativity.[2]

[a]In the regime of atmospheric neutrino oscillations, it suffices to consider a two-flavor system of eigenstates (ν_μ, ν_τ).

- In theories predicting violation of the weak equivalence principle (VEP), gravitational neutrino eigenstates are introduced which couple with distinct strengths γ_n to a gravitational potential ϕ, conflicting with the universal coupling assumed in general relativity.[3,4]

The main difference between these oscillation scenarios and standard oscillations is the linear energy dependence of the oscillation frequency, shifting observable oscillation effects into the energy range of AMANDA-II. For the sake of simplicity, we will focus on the VLI scenario. As both theories are mathematically equivalent, the results can be transferred to the VEP case by simply exchanging the relativity-violating oscillation parameters $\delta c/c \rightarrow 2|\phi|\delta\gamma$ and mixing angles $\Theta_c \rightarrow \Theta_\gamma$.

Combining standard and VLI oscillations, one obtains three systems of neutrino eigenstates (flavor, mass, and MAV eigenstates), resulting in a total of 5 oscillation parameters: the mass-squared difference Δm^2, two mixing angles Θ_m and Θ_c, the VLI parameter $\delta c/c$, and a complex phase η. Fixing $\Delta m^2 = 2.3 \times 10^{-3}\,\mathrm{eV}^2$ and $\Theta_m = 45°$, the survival probability may then be written as:

$$P(\nu_\mu \rightarrow \nu_\mu) = 1 - \sin^2 2\Theta \; \sin^2(\Omega \; L) \tag{1}$$

$$2\Theta = \arctan(s/t) \qquad \Omega = \sqrt{s^2 + t^2} \tag{2}$$

$$\begin{aligned} s &= 2.92 \times 10^{-3}\,|\,1/E_\nu + 8.70 \times 10^{20} \; \delta c/c \; \sin 2\Theta_c \; E_\nu \; \mathrm{e}^{i\eta}| \,, \\ t &= 2.54 \times 10^{18} \; \delta c/c \; \cos 2\Theta_c \; E_\nu \,. \end{aligned} \tag{3}$$

Here the muon neutrino path length L is expressed in km and the neutrino energy E_ν in GeV. For the given set of parameters, one can observe a significant effect within the analyzed energy range (100 GeV – 10 TeV) and declination range ($\delta \geq 20°$), for certain values of Θ_c and $\delta c/c$.

The VLI effects can also be viewed in terms of the Standard-Model Extension, an effective quantum field theory which adds all coordinate-independent renormalizable Lorentz-violating operators to the Standard Model.[5,6] In particular, the above oscillations correspond to non-zero and differing eigenvalues of c_L, the dimensionless coefficient of the Lorentz-violating Lagrangian term $\frac{1}{2}i(c_L)_{\mu\nu ab}\bar{L}_a\gamma^\mu \overleftrightarrow{D}^\nu L_b$.[7] Additionally, in this scenario only rotationally invariant (time) components of c_L are non-zero (so-called "fried chicken" models), and we consider only a two-flavor system. With these restrictions, $\delta c/c$ in eq. 3 is equivalent to the difference in eigenvalues $(c_a - c_b)$ of the 2×2 matrix c_L^{TT}.

3. Data selection and analysis method

The data used in this analysis are selected from 7.9×10^9 events recorded from 2000 to 2003, representing a total livetime of 807.2 days. Quality selection criteria based on various track reconstruction quantities are used to reduce contamination of the sample by mis-reconstructed atmospheric muons to a level of approximately 4% (see Ref. 8 for more details). The resulting number of selected neutrino candidate events is 3401. Reconstructed declination and the number of OMs triggered in an event (N_{ch}, an energy estimator) are used as observables to search for any deviations in the atmospheric neutrino spectrum that are dependent on flight length or neutrino energy. A full simulation chain, including neutrino absorption in the Earth, neutral current regeneration, muon propagation, and detector response is used to simulate the response of AMANDA-II to atmospheric neutrinos.[9,10] The expected atmospheric muon neutrino flux before oscillations is taken from Lipari.[11]

The analysis method uses a χ^2-test to compare the declination and N_{ch} distributions of data with Monte Carlo simulations including VLI oscillation effects. The systematic uncertainties affecting the Monte Carlo prediction are integrated into the χ^2 expression: uncertainties in the normalization due to detector response and theoretical models of atmospheric neutrino flux (30%); uncertainty due to the relative production rate between kaons and pions (6%); and uncertainty in the sensitivity of the OMs (11.5%). The optimal binning of the χ^2 expression is determined using toy Monte Carlo studies, and the exclusion regions for alternative oscillation effects are obtained by scanning through the oscillation parameter space.

4. Results and Outlook

The analysis of the final atmospheric neutrino sample finds no evidence for alternative oscillations, and a preliminary upper limit on the VLI parameter $\delta c/c$ is set of 5.3×10^{-27} at the 90% confidence level, for nearly maximal mixing angles $\Theta_c \approx \pm\pi/4$. The dependence on the unconstrained phase η is found to be small (see figure 1); the most conservative limit is obtained for $\cos\eta = 0$. The limit can also be interpreted in the context of VEP theories, leading to an upper limit of $2|\phi|\delta\gamma \leq 5.3 \times 10^{-27}$. This result is competitive with the limits obtained using data from Super-Kamiokande[12] and MACRO.[13] However, AMANDA is not sensitive to small mixing angles due to the systematic errors and its higher energy threshold.

A likelihood analysis of the 2000-2006 AMANDA-II data sample is

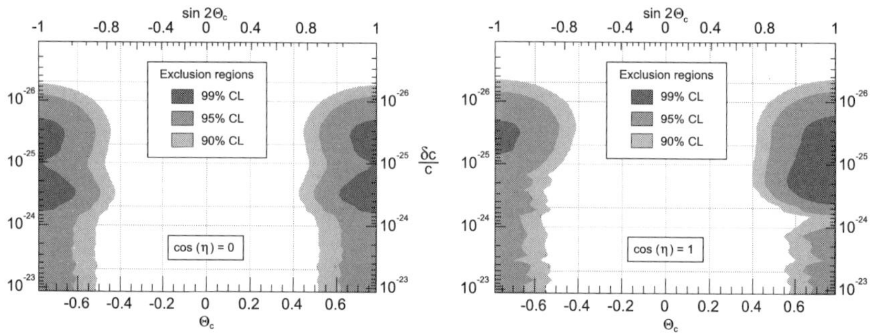

Figure 1. Shown are preliminary exclusion regions for VLI (VEP) oscillation effects, left for $\cos\eta = 0$, right for $\cos\eta = 1$.

in progress, with improved systematic error estimation and increased sensitivity.[14] This analysis will also extend the technique to search for evidence of quantum decoherence resulting from interaction of neutrinos with the background space-time foam.[15] The next-generation IceCube detector, when completed in 2010, will be able to extend the sensitivity to VLI effects by about one order of magnitude.[16]

References

1. Y. Ashie *et al.*, Phys. Rev. D **71**, 112005 (2005); E. Aliu *et al.*, Phys. Rev. Lett. **94**, 081802 (2005).
2. S. Coleman and S.L. Glashow, Phys. Lett. B **405**, 249 (1997).
3. M. Gasperini, Phys. Rev. D **38**, 2635 (1988).
4. A. Halprin and C.N. Leung, Phys. Rev. Lett. **67**, 14 (1991).
5. D. Colladay and V.A. Kostelecký, Phys. Rev. D **58**, 116002 (1998).
6. V.A. Kostelecký and M. Mewes, Phys. Rev. D **69**, 016005 (2004).
7. V.A. Kostelecký and M. Mewes, Phys. Rev. D **70**, 031902 (2004).
8. J. Ahrens and J.L. Kelley for the IceCube Collaboration, Proc. of the 30th Intl. Cosmic Ray Conf., Mérida, Mexico (2007).
9. G. C. Hill, Astropart. Phys. **6**, 215 (1997).
10. D. Chirkin and W. Rhode, hep-ph/0407075 (2004).
11. P. Lipari, Astropart. Phys. **1**, 195 (1993).
12. M.C. González-García and M. Maltoni, Phys. Rev. D **70**, 033010 (2004).
13. G. Battistoni *et al.*, Phys. Lett. B **615**, 14 (2005).
14. J.L. Kelley for the IceCube Collaboration, Proc. of the First Workshop on Exotic Physics with Neutrino Telescopes, astro-ph/0701333 (2007).
15. D. Morgan *et al.*, Astropart. Phys. **25**, 311 (2006).
16. M.C. González-García, F. Halzen, and M. Maltoni, Phys. Rev. D **71**, 093010 (2005).

MEASUREMENT OF THE GROUND-STATE HYPERFINE SPLITTING OF ANTIHYDROGEN

B. JUHÁSZ AND E. WIDMANN

Stefan Meyer Institute for Subatomic Physics
Boltzmanngasse 3
1090 Wien, Austria
E-mail: bertalan.juhasz@oeaw.ac.at

The ASACUSA collaboration at the Antiproton Decelerator at CERN is planning to measure the hyperfine splitting of the ground state of antihydrogen using an atomic beam line to test the CPT invariance. The spectrometer will consist of two sextupoles for spin selection and analysis, and a microwave cavity to flip the spin of the antihydrogen atoms. The antihydrogen atoms will be produced in either a radiofrequency Paul trap or in a cusp trap. Numerical simulations show that such an experiment is feasible if ~200 antihydrogen atoms per second can be produced in the ground state, and that an accuracy of better than 10^{-7} can be reached.

1. Introduction

The antihydrogen atom ($\overline{\mathrm{H}}$), which consists of an antiproton ($\overline{\mathrm{p}}$) and a positron, is the antimatter counterpart of hydrogen. Since it is the simplest neutral antimatter atom, it can be used to test the Charge-Parity-Time reversal (CPT) invariance. The 1S ground state of hydrogen (antihydrogen) is split due to the interaction between the electron (positron) spin and the proton (antiproton) spin, and can have quantum numbers $F = 0, 1$ (total spin), and $M = \{0\}, \{-1, 0, 1\}$ (projection of F). The states with quantum numbers $F = 0$ (singlet state) and $F = 1$ (triplet state) have different energies even at zero external magnetic field.

The ground-state hyperfine splitting (GS-HFS) frequency ν_{HF} in hydrogen (antihydrogen) is proportional in the leading order to the proton (antiproton) spin magnetic moment μ_p:

$$\nu_{\mathrm{HF}} \simeq \frac{16}{3}\left(\frac{m_p}{m_p + m_e}\right)^3 \frac{m_e}{m_p}\frac{\mu_p}{\mu_N}\frac{\alpha^2}{h} Ry \simeq 1.42 \text{ GHz}. \tag{1}$$

The antiproton spin magnetic moment $\mu_{\overline{\mathrm{p}}}$ is known to a precision of only 0.3%, therefore a measurement of ν_{HF} for antihydrogen (which has never been done before) with a precision of only 10^{-6} can already lead to an improvement of the value of $\mu_{\overline{p}}$ by three orders of magnitude.

Recently, the group of V.A. Kostelecký has developed an extension of the Standard Model of elementary particles that includes both CPT violating and Lorentz invariance violating terms in the Lagrangian of the quantum field theory.[1] These terms result in a correction to the energies of the hyperfine sublevels:

$$\begin{aligned} \Delta E^{\mathrm{H}}(m_J, m_I) = {} & a_0^e + a_0^p - c_{00}^e m_e - c_{00}^p m_p \\ & + (-b_3^e + d_{30}^e m_e + H_{12}^e) m_J/|m_J| \\ & + (-b_3^p + d_{30}^p m_p + H_{12}^p) m_I/|m_I|, \end{aligned} \tag{2}$$

where m_J and m_I are the characteristic spin quantum numbers of the electron (positron) and the proton (antiproton), respectively, at large external magnetic fields. The parameters a, d and H change sign for antihydrogen, therefore the GS-HFS frequencies in hydrogen and antihydrogen can be different.

An important feature of this theory is that the parameters a and b are not dimensionless; instead, they have a dimension of energy (or frequency). From this it follows that it is not the relative but the *absolute* precision of an experiment that matters when one is comparing matter and antimatter systems for CPT tests. Therefore the $\overline{\mathrm{H}}$ GS-HFS measurement can be competitive with the oft-quoted 'most sensitive CPT limit' of $|m_{K^0} - m_{\overline{K}^0}|/m_{K^0} < 10^{-18}$.[2] This is because the K^0-$\overline{K}^0$ measurement corresponds to only ~100 kHz on the absolute scale, while the $\overline{\mathrm{H}}$ GS-HFS measurement with a relative precision of only 10^{-6} would have an absolute precision of 1.42 GHz $\times 10^{-6}$ = 1.42 kHz.

2. Proposed experimental method

The most precise measurements on the ground-state hyperfine levels of hydrogen have been carried out with hydrogen masers. Such masers can keep the hydrogen atoms in the radiofrequency field for long enough time for precise spectroscopy, as the experimental resolution is inversely proportional to the interaction time with the field. Unfortunately, similar measurements with antihydrogen would require a maser cavity covered with antimatter coating, which is clearly impossible.

Antihydrogen atoms could be kept in the RF field by a confining magnetic field. However, the required fields would be too strong and would shift and broaden the energy levels too much.

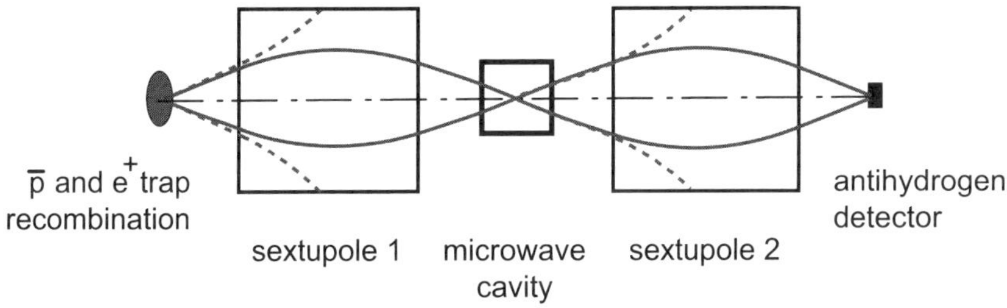

Figure 1. Schematic view of the proposed atomic beam spectrometer with the antihydrogen source, the two sextupole magnets and the microwave cavity. The trajectories drawn with solid lines represent $\overline{\mathrm{H}}$ atoms in low-field seeker states, while the dashed lines represent $\overline{\mathrm{H}}$ atoms in high-field seeker states.

Thus the ASACUSA collaboration at CERN's Antiproton Accelerator (AD) plans to use a method[3] similar to classical atomic beam (Stern-Gerlach type) experiments to measure the $\overline{\mathrm{H}}$ GS-HFS. This method has the advantage that it can work with relatively high-temperature (50-100 K) atoms, while trapping would require ultra-cold (< 1 K) atoms. A schematic view of the proposed atomic spectrometer line is shown in Fig. 1.

The antihydrogen atoms will be produced in either a two-frequency Paul trap or a cusp trap, which are currently under development.[3] Depending on the sign of their magnetic moment, the four hyperfine states of the ground-state $\overline{\mathrm{H}}$ atoms can be divided into two pairs (see Fig. 2): 'high-field seekers', which will move towards regions of higher magnetic field in the inhomogeneous field of a sextupole magnet and thus will be defocused, and 'low-field seekers', which will be attracted towards the sextuple axis and thus will be focused onto a microwave cavity. The focused atoms will then pass through a second sextupole, which will again focus them onto an antihydrogen detector. However, if a $\overline{\mathrm{H}}$ atom in e.g. the $(F, M) = (1, -1)$ low-field seeker state is converted into the $(0, 0)$ high-field seeker state by a microwave field of appropriate frequency in the cavity, then the atom will be defocused and will not reach the detector. Thus the on-resonance count rate in the detector will drop from the constant off-resonance count rate.

Simulations showed that unlike the Paul trap, the cusp trap could provide a partially polarized $\overline{\mathrm{H}}$ atoms due to the focussing/defocussing effect of the strong inhomogeneous magnetic field of such a trap. In this case, the first sextupole could be omitted from the setup.

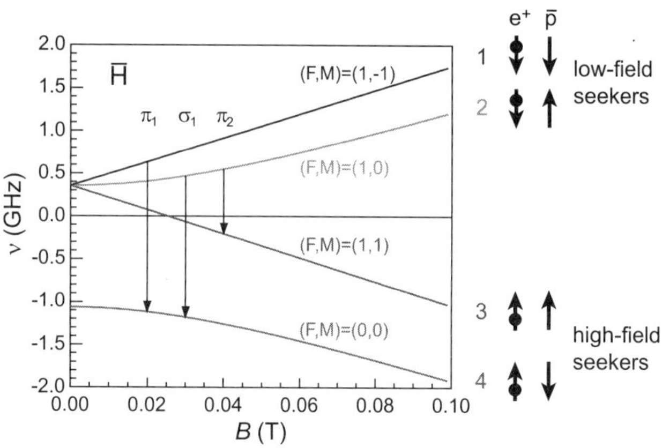

Figure 2. Frequencies (i.e. energies) of the four hyperfine states of antihydrogen as a function of the external magnetic field B. The transitions observable with the proposed method are also drawn.

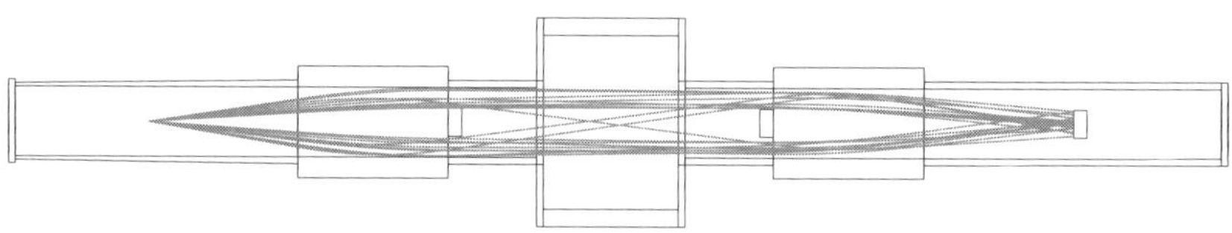

Figure 3. Simulated trajectories of the antihydrogen atoms (with microwave resonance off) in the spectrometer line. The total length from the source to the detector is ~150 cm.

3. Monte Carlo simulations

Simulations using the GEANT4 toolkit[4] have been carried out to estimate the expected count rates and experimental resolution. The preliminary results showed that a spectrometer line with two sextupoles, each with an internal diameter of 10 cm, an effective length of 25 cm, and a pole tip field of 3 Tesla (see Fig. 3) can have a total transmission efficiency of 2–5 $\times 10^{-4}$. Using the expected production rate of 200 $\overline{\mathrm{H}}$/s, the detection rate at the antihydrogen detector will be 2–5 $\overline{\mathrm{H}}$/min.

Note that in the setup in Fig. 3, most of the $\overline{\mathrm{H}}$ trajectories are parallel to the beam axis between the two sextupoles, and not focused onto the center of the microwave resonance cavity. Such a geometry results in a larger velocity acceptance of the spectrometer which increases the transmission efficiency drastically.

Figure 4 shows a simulated resonance scan of one of the transitions

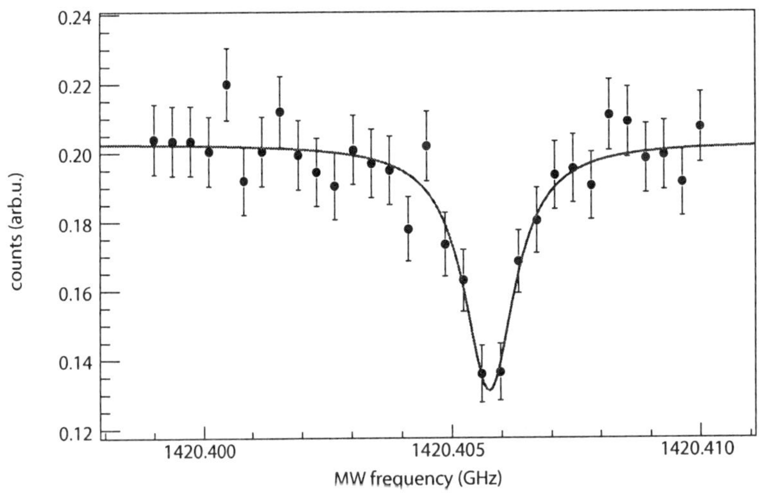

Figure 4. Simulated microwave resonance scan of one of the hyperfine transitions. The fitted line is an inverted Lorentzian function.

using the cusp trap and only one sextupole. A fit of a Lorentzian function to these data points reveals that the FHWM of the line is ~1.2 kHz, while the error of the center frequency is below 0.1 kHz. This corresponds to a relative error of $\sim 5 \times 10^{-8}$, thus we can safely assume that the measurement will have a precision better than 10^{-7}, or 10^{-21} GeV on the energy scale.

Acknowledgments

This work was supported by the Grant-in-Aid for Specially Promoted Research (15002005) of Monbukagakusho of Japan, the exchange visit programme between the Austrian and the Hungarian academies of science, and the Hungarian Scientific Research Fund (OTKA T046095).

References

1. R. Bluhm, V.A. Kostelecký, and N. Russell, Phys. Rev. Lett. **82**, 2254 (1999).
2. W.-M. Yao *et al.*, J. Phys. G **33**, 1 (2006).
3. ASACUSA proposal addendum, CERN/SPSC 2005-002, SPSC P-307 Add.1 (2005).
4. S. Agostinelli *et al.*, Nucl. Instrum. Methods A **506**, 250 (2003).

SPONTANEOUS LORENTZ VIOLATION AND BARYOGENESIS

JING SHU

Enrico Fermi Institute, Department of Physics
and Kavli Institute for Cosmological Physics
University of Chicago, 5640 S. Ellis Avenue, Chicago, IL 60637, U.S.A.
and
HEP Division, Argonne National Laboratory
9700 Cass Ave., Argonne, IL 60439, U.S.A.
E-mail: jshu@theory.uchicago.edu

In the presence of background fields that spontaneously violate Lorentz invariance, a matter-antimatter asymmetry can be generated even in thermal equilibrium. In this paper we systematically investigate models of this type, showing that either high-energy or electroweak versions of baryogenesis are possible, depending on the dynamics of the Lorentz-violating fields. We identify two scenarios of interest: baryogenesis from a weak-scale pseudo-Nambu-Goldstone boson with intermediate-scale baryon-number violation, and sphaleron-induced baryogenesis driven by a constant-magnitude vector with a late-time phase transition.

1. Introduction

The observed universe manifests a pronounced asymmetry between the number density of baryons n_b and antibaryons $n_{\bar{b}}$. However, the origin of the baryon number asymmetry remains a major puzzle for cosmology and particle physics. In a classic work, Sakharov argued that three conditions are necessary to dynamically generate a baryon asymmetry in an initially baryon-symmetric universe: (1) baryon number non-conserving interactions; (2) C and CP violation; (3) departure from thermal equilibrium. In deriving these conditions, the assumption is made that CPT is conserved. If Lorentz invariance is violated, then CPT is also violated, and one can generate baryons in thermal equilibrium. This idea is first implemented in the context of "spontaneous baryogenesis"[1] scenario and has subsequently been elaborated upon in various ways.[2,3,4,5,6,7]

In most of the previous studies, the effects of sphaleron transitions are not discussed or mistakenly believed to wash out the baryon asymmetry if

$B-L$ is zero. The role of sphaleron transitions in thermal equilibrium is to adjust different particle density distributions in a way that preserves $B-L$ to minimize the total free energy. In the presence of a Lorentz-violating background, the one particle free energy is modified and final net baryon number density should be quite different from the one without Lorentz-violating background. In this situation, sphaleron transitions will generate a nonzero $B+L$. Because of that, we reconsider previously studied models and construct new models.[8] We find the final net baryon number density largely depends on how the effective chemical potential ga_0 evolves with time. This provide us a general rule to categorize different models of baryogenesis via spontaneous Lorentz violation and understand them in a unified picture. We also identify two scenarios of potential interest. One is the case of a simple constant-magnitude timelike vector field coupled to J^{μ}_{B+L} where appropriate baryon asymmetry could be generated by electroweak sphalerons alone and the other is that of a derivatively-coupled pseudo-Nambu-Goldstone boson with a weak scale mass and $(B-L)$-violating interactions that freeze out at a Majorana neutrino mass scale of order 10^{10} GeV.

2. Baryogenesis in the presence of Lorentz violation

We consider the theory of a vector field A_μ with a nonzero vacuum expectation value (vev), coupled to a current J^μ in the matter fields which corresponds to some continuous global symmetry. The vector field gets a condensate $A_\mu = (a_0, 0, 0, 0)$ and that makes the interaction term $\mathcal{L}_{int} = gA_\mu J^\mu \to -ga_0 Q$, where Q is the conserved charge. Such an interaction term now acts like a "chemical potential" $\mu_b^0 = ga_0$ for the matter fields, which splits the free energy of particle and anti-particle. Because of this effect, there will be a non-zero baryon number density generated by baryon-number violating interactions in thermal equilibrium $n_B = n_b - n_{\bar{b}} = g_b T^3\left[\pi^2 \frac{\mu_b^0}{T} + \left(\frac{\mu_b^0}{T}\right)^3\right]/6\pi^2 \simeq g_b \mu_b^0 T^2/6 \sim \mu_b^0 T^2$, where g_b counts the internal degrees of freedom of the baryons. When the B-violating interactions mentioned above become ineffective ($\Gamma \leq H$), we get the final baryon asymmetry

$$\frac{n_B}{s} \sim \frac{\mu_b^0}{g_{*s} T_F} = \frac{ga_0}{g_{*s} T_F}, \tag{1}$$

with the entropy density $s = (2\pi^2/45) g_{*s} T^3$, where T_F is the temperature at which the baryon number production is frozen out.

With such a spontaneous Lorentz violation background, the energy dispersion relation is modified to $E = \sqrt{K_i^2 + m^2} \pm \mu^0$, where K_i is the momentum of the fermion and $\pm$ are for fermion and anti-fermion respectively. The net baryon number density now becomes $B^{(\mu)} = -\frac{2N}{13} T^2 (3\mu^0 + \frac{1}{N} \sum_{i=1}^{N} \mu_i^0)$, where the parameters μ and μ_i are the chemical potentials of the quarks and the ith lepton, respectively. If there is leptonic flavor violation in thermal equilibrium, one can write it in terms of $\mu_L^0 \equiv \frac{1}{N} \sum_{i=1}^{N} \mu_i^0$ and $\mu_B^0 = 3\mu^0$. The fact that $B \propto (\mu_B^0 + \mu_L^0) = 2\mu_{B+L}^0$ tells us a nonzero net baryon number density can be spontaneously generated through sphaleron transitions in thermal equilibrium in the presence of a nonzero time-like vector background coupled to J_{B+L} current.

As we know sphaleron transitions connect baryon and lepton number, we need to consider both the baryon number current and lepton number current that couple to the background field. It is convenient to rewrite B and L currents in terms of the $B + L$ and $B - L$ currents. From Eq. (1), we know that

$$\frac{n_{B-L}}{s} = \frac{\mu_-^0(T_-)}{g_{*s} T_-} = \frac{g_- a_0(T_-)}{g_{*s} T_-}, \qquad \frac{n_{B+L}}{s} = \frac{\mu_+^0(T_+)}{g_{*s} T_+} = \frac{g_+ a_0(T_+)}{g_{*s} T_+}, \tag{2}$$

where T_- and T_+ are the lowest freeze-out temperature for any interactions that could violate $B - L$ and $B + L$, respectively. T_+ is the sphaleron freeze-out temperature which is roughly 150 GeV. T_- ranges from TeV to GUT scale and is very model dependent. Notice that $T_- \ll T_+$, so whether $n_{B+L} \ll n_{B-L}$ or $n_{B+L} \gg n_{B-L}$ will only depend on whether $\mu^0(T)/T$ is an increasing or decreasing function with respect to $1/T$. The net baryon number $n_B = (n_{B+L} + n_{B-L})/2$, so we know that n_B is of the same order as $\max\{n_{B+L}, n_{B-L}\}$. From Eq. (2), we get

$$\frac{n_B}{s} = \begin{cases} \dfrac{n_{B+L}}{2s} \sim \dfrac{g_+ a_0(T_+)}{g_{*s} T_+} & \text{if } \dfrac{|\mu^0(T)|}{T} \text{ increases as a function of } 1/T, \\[2ex] \dfrac{n_{B-L}}{2s} \sim \dfrac{g_- a_0(T_-)}{g_{*s} T_-} & \text{if } \dfrac{|\mu^0(T)|}{T} \text{ decreases as a function of } 1/T. \end{cases}$$

3. Present-day constraints on Lorentz violation

In principle, there are no real experimental constraints as baryogenesis happens at the early time while all experiments are at present. Nevertheless, the highly constrained experimental bounds today suggest that the spontaneous Lorentz-violating background undergoes a phase transition if it doesn't decay away or roll to an extremely small value.

The direct constraints between baryon number current and Lorentz violating background field are coming from neutral meson mixing. Only the difference Δa_μ between the corresponding two a_μ coefficients is observable. The experimental constraint comes from the parameter $\Delta a_\mu = r_{q_1} a_\mu^{q_1} - r_{q_2} a_\mu^{q_2}$, where $a_\mu^{q_1}$, $a_\mu^{q_2}$ are Lorentz-violating coupling constants for the two valence quarks in the meson, and where the factors r_{q_1} and r_{q_2} allow for quark-binding or other normalization effects. Experiments studying neutral K-mesons have constrained two combinations of Δa for d and s quarks, with bounds in the Sun-centered frame of approximately

$$|\Delta a_0| \leq 10^{-20}\ \text{GeV} \tag{3}$$

by the KTeV Collaboration at Fermilab.[9,10] Other experiments with D mesons have constrained two combinations of Δa for the u and c quarks at about 10^{-15} GeV (FOCUS Collaboration, Fermilab).[9,11] There are even more constrained results from axial vector current and astrophysics, but the couplings in those constraints are essential to generate baryons.

4. Sources of Lorentz violation

We first consider that Lorentz-violating vector field A_μ has a constant expectation value in the vacuum (a_0 = constant). Our discussion could also be generalized to a ghost field and high rank-tensor condensate.[2] The vector condensation is achieved through a Mexican-hat potential[a] $V(A_\mu) = \frac{\mu^2}{2} A_\mu A^\mu + \frac{\lambda}{4}(A_\mu A^\mu)^2$. At the classical level, the timelike component for a vector field with minimal ground state energy is given by ${a_0}^2 = \frac{\mu^2}{\lambda}$. Since $\mu^0/T = g a_0/T$ is increasing as the universe expands (and T decreases), the relevant freeze-out temperature is thus $T_+ = 150$ GeV, due to sphaleron transitions. From Eq. (1) we obtain

$$\frac{n_B}{s} \sim \frac{g a_0}{g_{*s} T_+} \sim g a_0 (10^4\ \text{GeV})^{-1} \approx 10^{-10}\,. \tag{4}$$

Such a field condensate would seem to violate the experimental constraints discussed in Section 3. One way to accommodate the experimental limits is to imagine a phase transition for the A_μ field itself. We can replace the coefficient of the mass term μ^2 with $({\mu'}^2 - \alpha|\Phi|^2)$ in the potential, where Φ is the Higgs doublet. At high temperatures, the Higgs expectation value $\langle\Phi\rangle$ vanishes, and we get a non-zero vacuum expectation value of the vector

[a]Notice we use the minus metric here.

background field. At late times, $|\Phi|^2 = v^2$. If $\mu'^2 - \alpha v^2$ is negative, we get a zero vev for vector background field. So the Lorentz symmetry is restored.

Another possibility is that the Lorentz-violating background is time-dependent. A simple way to achieve this is to image A_μ is the gradient of a slowly-rolling scalar field ϕ. The chemical potential term a_0 is then given by $\dot{\phi}/f$. When one apply this to a quintessence field which has "tracking" behavior $\dot{\phi} \propto \sqrt{V(\phi)} \propto \sqrt{\rho_{\rm back}} \propto T^2$, we can see that the freeze-out temperature is determined by T_-, which is very model dependent. Similarly, one can consider the interaction between the derivative of the Ricci scalar curvature $\mathcal{R}$ and the baryon number current J^μ from the effective theory of gravity[7] $\frac{1}{M_*^2}\int d^4x\sqrt{-g}(\partial_\mu \mathcal{R})J^\mu$. The net baryon number density obtained is proportional to an even higher positive power of freeze-out temperature T_F, since $\dot{\mathcal{R}} \propto \dot{\rho}$, where ρ is the total energy density.

Finally we consider the Lorentz-violating background arises as the gradient of a pseudo-Nambu-Goldstone boson (PNGB) $\phi = f\theta$, where f is the scale of spontaneous symmetry breaking. A PNGB remains overdamped in its potential until its mass becomes comparable to the Hubble parameter, at which time it will roll to its minimum and begin to oscillate and decay like a massive particle. If we want to generate a baryon asymmetry of the right amplitude, then from $n_B/s \sim \dot{\phi}/f g_{*s} T_F = 10^{-10}$ and $g_{*s} \sim 100$ we require $\dot{\phi} = 10^{-8} f T_F$. The PNGB obeys the equation of motion $\ddot{\phi} + 3H\dot{\phi} + \frac{dV}{d\phi} = 0$ and we drop the term $\ddot{\phi}$ as we are interested in the rolling phase. For typical values $\phi \sim f$, we have $\dot{\phi} \sim H^{-1}\frac{dV}{d\phi} \sim m^2 f M_{pl}/T^2$. Thus, to achieve successful baryogenesis requires that the freeze-out temperature satisfy $\frac{T_F^3}{m^2} \sim 10^8 M_{pl}$. The ideal circumstance would be if freeze-out occured when the field had just begun to roll substantially, but not yet begun to oscillate. This corresponds to $H \sim m$, which implies $T_F^2 \sim m M_{pl}$. Comparing to the expression of $\dot{\phi}$ shows that PNGB baryogenesis works if the freeze-out temperature is at an intermediate scale $T_F \sim 10^{-8} M_{pl} \sim 10^{10}$ GeV and the PNGB mass is $m \sim T_F^2/M_{pl} \sim 100$ GeV.

5. Conclusion

We have investigated the possible origin of the observed baryon asymmetry in the presence of a coupling between a Lorentz-violating vector field and the baryon current, especially reconsidered the effects of sphaleron transitions. If a_0/T is increasing with time, then the final net baryon number density is determined by the freeze-out temperature $T_+ \approx 150$ GeV. For the opposite case, the final net baryon number density is determined by the freeze-out temperature T_- which is model-dependent. Most previous works[2,3,4,5,7] consider a slow rolling scalar field as the spontaneous Lorentz-violating background, and the absolute value of the effective chemical is decreasing with time, so the right net baryon number density generated depends on a high freeze-out temperature T_-. In order to obtain the right net baryon number density, the coupling times the time component of the background field ga_0 should be not too small, and experiment constraints at present suggest that we need some dynamical mechanism to decrease a_0. We first consider a constant Lorentz-violating background case and sphaleron transitions will be the main source to generate the baryon asymmetry. We can imagine that a phase transition occurs in between freeze-out and today. Our investigation of the PNGB scenario reveals that the most natural implementation of this idea requires PNGB with weak-scale masses (100 GeV) and $(B - L)$-violating interactions that freeze out at an intermediate scale of around 10^{10} GeV naturally from the decay of Majorana neutrinos. We therefore consider this scenario to be quite promising.

References

1. A. G. Cohen and D. B. Kaplan, Phys. Lett. **B 199**, 251 (1987); A. G. Cohen and D. B. Kaplan, Nucl. Phys. **B 308**, 913 (1988).
2. O. Bertolami, D. Colladay, V.A. Kostelecký, and R. Potting , Phys. Lett. **B 395**, 178, (1997).
3. A. De Felice, S. Nasri, and M. Trodden, Phys. Rev. **D 67**, 043509 (2003); M. Trodden, hep-ph/0302151.
4. Mingzhe Li, Xiulian Wang, Bo Feng and Xinmin Zhang, Phys. Rev. **D 65**, 103511 (2002); Mingzhe Li, Xinmin Zhang, Phys. Lett. **B 573** 20 (2003); Peihong Gu, Xiulian Wang, Xinmin Zhang, Phys.Rev. **D 68**, 087301 (2003); Xiao-Jun Bi, Pei-hong Gu, Xiu-lian Wang, Xin-min Zhang, hep-ph/0311022; Hong Li, Ming-zhe Li, Xin-min Zhang, hep-ph/0403281; Xiao-Jun Bi, Jian-Xiong Wang, Chao Zhang, Xin-min Zhang, hep-ph/0404263.
5. M. Yamaguchi, Phys. Rev. **D 68**, 063507 (2003); R. H. Brandenberger and M. Yamaguchi, Phys. Rev. **D 68**, 023505 (2003); F. Takahashi and M. Yamaguchi, Phys. Rev. **D 69**, 083506 (2004).
6. T. Chiba *et al.*, Phys. Rev. Lett. **92**, 011301 (2004).
7. H. Davoudiasl, R. Kitano, C. Kribs, H. Murayama, and P.J. Steinhardt, Phys. Rev. Lett. **B 93**, 201301 (2004).
8. S. M. Carroll and J. Shu, Phys. Rev. **D 73**, 103515 (2006).
9. V. A. Kostelecký, ed., *CPT and Lorentz Symmetry II*, Singapore, World Scientific, 2002.
10. KTeV Collaboration, A. Alavi-Harati *et al.*, Report EFI 99-25 (1999).
11. FOCUS Collaboration, J. Link *et al.*, Phys. Lett. **B 485**, 62 (2000); FOCUS Collaboration, J. Link *et al.*, Phys. Rev. Lett. **86**, 2955 (2001).

LORENTZ INVARIANCE TESTED WITH FAST OPTICAL ION CLOCKS IN A STORAGE RING

G. GWINNER
Department of Physics and Astronomy
University of Manitoba
Winnipeg, MB R3T 2N2, Canada
E-mail: gwinner@physics.umanitoba.ca

S. REINHARDT, G. SAATHOFF, D. SCHWALM, AND A. WOLF
Max-Planck-Institut für Kernphysik
69029 Heidelberg, Germany

G. HUBER, S. KARPUK, AND C. NOVOTNY
Institut für Physik
Universität Mainz
55099 Mainz, Germany

We have carried out an improved test of relativistic time dilation. The new method uses fast optical atomic clocks with large, but different Lorentz boosts, and combines advanced ion storage and cooling techniques with optical frequency counting using a frequency comb. $^7Li^+$ ions are prepared at 6.4% and 3.0% of the speed of light in a storage ring. Their time is read with an accuracy of 2×10^{-10} using laser saturation spectroscopy. The comparison of the Doppler shifts yields a time dilation measurement represented by a Mansouri-Sexl parameter $|\hat{\alpha}| \leq 8.4 \times 10^{-8}$, consistent with Special Relativity. In the context of the Lorentz and CPT violating Standard-Model Extension, we put the limit $\tilde{\kappa}_{tr} \leq 8.4 \times 10^{-8}$ in the photon sector. Measurements with circularly polarized light promise to put constraints on various untested parameters in the particle sector.

1. Introduction

The relativistic Doppler effect was already proposed as an experimental test of relativity by Einstein in 1907.[1] Time dilation leads to the *ether-independent* relativistic Doppler formula $\nu_0 = \nu_l \gamma(1 - \beta \cos\theta)$, where ν_l and ν_0 denote the frequencies in the laboratory reference frame of the observer and the particles' rest frame moving at velocity $v = \beta c$ with respect to

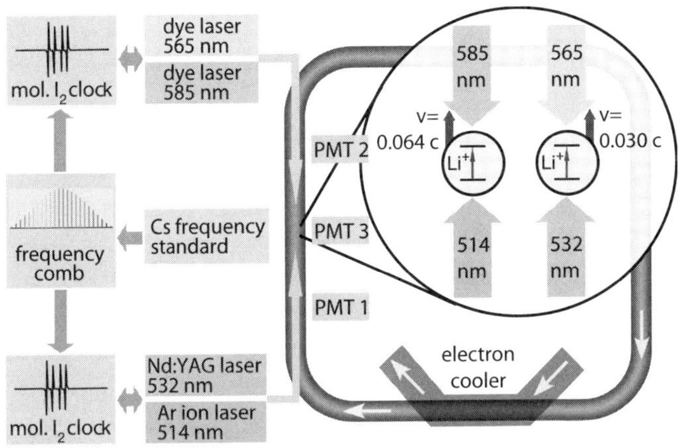

Figure 1. Schematic layout of the TSR. Li^+ ions circulate in the 55 m circumference ring. In the electron cooler, cold electrons are overlapped with the ions and provide cooling. The measurements at the two different velocities are carried out sequentially. In the experiment, the two lasers are coupled into the ring from the same side and are retro-reflected.

the observer, respectively; θ is the angle of observation with respect to the particles' movement as measured in the lab frame, and $\gamma = 1/\sqrt{1-v^2/c^2}$. We are using a modern version of the experiment by Ives and Stilwell,[2] where two laser beams, parallel and anti-parallel with the atomic motion, excite the same transition of rest-frame frequency ν_0. Within Special Relativity (SR), the lasers will have laboratory frequencies of $\nu_p = \nu_0/\gamma(1-\beta)$ and $\nu_a = \nu_0/\gamma(1+\beta)$, respectively, and the product of the two expressions yields

$$\nu_0^2 = \nu_a \nu_p. \tag{1}$$

It is common to parametrize possible deviations from SR using the Robertson-Mansouri-Sexl (RMS) test theory,[3] where deviations in the time dilation sector are quantified by the test parameter $\hat{\alpha}$, and we get

$$\frac{\nu_{\rm p}\nu_{\rm a}}{\nu_0^2} = 1 + 2\,\hat{\alpha}\,(\beta^2 + 2\,\vec{\beta}_{\rm lab}\cdot\vec{\beta}) + \mathcal{O}(c^{-4}), \tag{2}$$

where $c\vec{\beta}_{\rm lab}$ is the velocity of the lab against a preferred cosmic frame, which is generally taken to be the cosmic microwave background rest frame. The β^2 term used in our measurement allows to determine $\hat{\alpha}$ absolutely *without* having to rely on the precise knowledge of $\beta_{\rm lab}$ (for $\beta \gg \beta_{\rm lab}$).

2. Experiment

In our experiment at the Max Planck Institute for Nuclear Physics, $^7Li^+$ ions are accelerated by a tandem Van-de-Graaff accelerator and injected into the Test Storage Ring (TSR) shown in Fig. 1. The helium-like $^7Li^+$ exhibits the strong $2s\ ^3S_1 \to 2p\ ^3P_2$ transition at 548 nm in its metastable triplet spectrum. The 50 s vacuum lifetime of the metastable ground state $2s\ ^3S_1$ is quenched to 10-20 s due to collisions with the rest gas, still sufficiently long to perform laser spectroscopy. To extract time dilation from a measurement of the Doppler shifts at one ion velocity, the rest frame transition frequency needs to be known accurately. Since the best available measurement by Riis et al.[4] has an uncertainty of 400 kHz, which was the limiting factor in our previous time dilation measurement,[5] we set up a new experiment with ion beams at two different velocities, $\beta_1 = 0.030$ and $\beta_2 = 0.064$. The Doppler-shifted frequencies $\nu_a^{(1,2)}$, $\nu_p^{(1,2)}$ measured at β_1 and β_2 can be combined using Eq. (2) (neglecting the sidereal term) to

$$\frac{\nu_a^{(2)}\nu_p^{(2)}}{\nu_a^{(1)}\nu_p^{(1)}} = \frac{1+2\hat{\alpha}\beta_2^2}{1+2\hat{\alpha}\beta_1^2} \approx 1 + 2\hat{\alpha}(\beta_2^2 - \beta_1^2), \tag{3}$$

independent of the rest frame frequency. As $\beta_2^2 - \beta_1^2 \approx 0.8\beta_2^2$, the sensitivity is not significantly diminished.

The moving clocks are read using laser saturation spectroscopy. The laboratory frequencies ν_p and ν_a of the parallel and anti-parallel laser beams (with respect to the ion beam) must obey relation (2) for resonance, which is indicated by a dip in the fluorescence spectrum. Through permanent cooling of the ions by a cold electron beam, the ion beam's width shrinks to $\approx 250\ \mu$m, the divergence to $\approx 50\ \mu$rad, and the longitudinal momentum spread to $\delta p/p = 3.5 \times 10^{-5}$, leading to a Doppler width of the transition of about 2.5 GHz full-width half maximum. This broadening is overcome in saturation spectroscopy by selecting a narrow velocity class of the order of the natural linewidth; two lasers are overlapped parallel and anti-parallel with the ion beam, respectively, and excite the clock transition. If the laser beams are sufficiently intense, the fluorescence yields get saturated, i.e., they become almost independent of the intensity, whereas at low intensity they increase linearly. In this saturation regime, each ion contributes nearly the same amount of fluorescence when exposed to one or both lasers. In general, the lasers interact with ions of two different velocity classes which then both contribute separately to the total fluorescence. However in the case Eq. (2) holds, both lasers are resonant with the same velocity class β and the number of fluorescing ions is cut in half. The corresponding decrease of the total fluorescence is referred to as the Lamb dip. The co-propagating laser (a Nd:YAG laser at 532 nm for β_1 and an argon-ion laser at 514 nm for β_2) is fixed in frequency by locking it to a well-known iodine (I_2) line, whereas the counter-propagating light is generated by a tunable dye laser (at 565 nm and 585 nm for β_1 and β_2, respectively). The dye laser frequency is referenced to a second, I_2-stabilised dye laser by a tunable frequency-offset lock. The iodine lines for the dye laser are calibrated using an optical frequency comb.[6] All laser frequencies are known absolutely to 70 kHz during the whole experiment.

The mean velocity of the ion beam is adjusted for the fixed-frequency laser at $\nu_{\mathrm{p}}^{(1,2)}$ to select ions in the centre of the velocity distribution. The dye laser is scanned around $\nu_{\mathrm{a}}^{(1,2)}$ and the Lamb dip in the fluorescence is recorded with photomultipliers (PMT) from the side; its frequency is measured with respect to the I_2 clock in the laboratory frame.

The observed resonance widths are in accordance with the natural linewidth of the $2\,^3S \rightarrow 2\,^3P$ transition of 3.7 MHz, once the broadening mechanisms present in our experiment are accounted for.

3. Results

3.1. *Robertson-Mansouri-Sexl interpretation*

Taking all systematic errors into account, the transition frequencies ν_{a} and ν_{p} measured at $\beta_1 = 0.030$ and $\beta_2 = 0.064$ yield SR values for the rest frame frequency (Eq. (1)) of

$$\sqrt{\nu_{\mathrm{a}}^{(1)}\nu_{\mathrm{p}}^{(1)}} = (546\;466\;918\;577 \pm 108)\ \mathrm{kHz}$$
$$\sqrt{\nu_{\mathrm{a}}^{(2)}\nu_{\mathrm{p}}^{(2)}} = (546\;466\;918\;493 \pm 98)\ \mathrm{kHz},$$

respectively. From Eq. (3) follows a test parameter

$$\hat{\alpha} = (-4.8 \pm 8.4) \times 10^{-8}, \tag{4}$$

which is consistent with SR.[7]

3.2. *Standard-Model Extension*

In the leading order, the limit on $\hat{\alpha}$ in RMS theory is identical to the constraint on $\tilde{\kappa}_{\mathrm{tr}}$ in the photon sector of the Standard-Model Extension (SME),[8] hence we obtain

$$\tilde{\kappa}_{\mathrm{tr}} \leq 8.4 \times 10^{-8}. \tag{5}$$

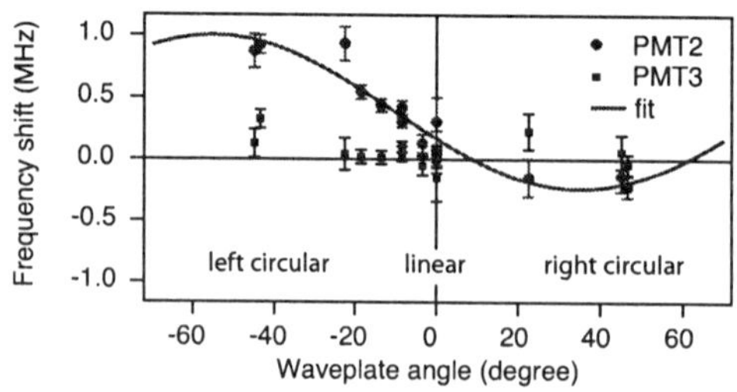

Figure 2. Frequency of the Lamb dip as a function of the polarization of the laser light for two fluorescence detectors.

In the particle sector, limits on the quadrupole parameters $\tilde{c}_Q, \tilde{c}_{TJ}$ for the proton and the electron had been derived by Lane[9] from our previous measurement.[5] Since then, much stronger limits have been measured[10] and deduced from astronomical observations.[11] In the present experiment, we have taken additional data with circularly polarized light, as shown in Fig. 2. At the position of detector PMT2, a residual magnetic field causes a polarization dependence of the Lamb dip frequency via the Zeeman effect. The absence of magnetic fields at PMT3 leads to a polarization independence of the resonance as observed in that detector. Non-vanishing SME dipole parameters would cause a polarization dependence of the Lamb dip, hence the PMT 3 data can constrain them. Preliminary investigations show that a Doppler experiment with circularly polarized light is sensitive to the so far untested parameters $\tilde{b}_Z, \tilde{d}_{ZX}, \tilde{d}_Z, \tilde{g}_{DZ}, \tilde{g}_{IJ}, \tilde{g}_Q, \tilde{g}_{TJ}$, even if individual m_F levels cannot be resolved in the measurement.

References

1. A. Einstein, Ann. Phys. **328**, 197 (1907).
2. H. E. Ives and G. R. Stilwell, J. Opt. Soc. Am. **28**, 215 (1938).
3. H. P. Robertson, Rev. Mod. Phys. **21**, 378 (1949); R. Mansouri and R. U. Sexl, Gen. Rel. Grav. **8**, 809 (1977).
4. E. Riis *et al.*, Phys. Rev. A **49**, 207 (1994).
5. G. Saathoff *et al.*, Phys. Rev. Lett. **91**, 190403 (2003).
6. Th. Udem *et al.*, Nature **416**, 233 (2002); S. Reinhardt *et al.*, Opt. Commun. **261**, 282 (2006).
7. S. Reinhardt *et al.*, accepted for publication in Nature Physics.
8. M. Hohensee *et al.*, Phys. Rev. D **75**, 049902 (2007); P. L. Stanwix *et al.*, Phys. Rev. D **74**, 081101 (2006).
9. C. D. Lane, Phys. Rev. D **72**, 016005 (2005).
10. P. Wolf *et al.*, Phys. Rev. Lett. **96**, 060801 (2006).
11. B. Altschul, Phys. Rev. Lett. **96**, 201101 (2006).

BOSE-EINSTEIN CONDENSATES AND LORENTZ VIOLATION

D. COLLADAY AND P. MCDONALD
New College of Florida
Sarasota, FL 34243, U.S.A.
E-mail: mcdonald@ncf.edu

We investigate Bose-Einstein condensates as a potential testing ground for detection of Lorentz violation. For certain optically trapped condensates our calculations, performed in the context of the Standard-Model Extension, yield good potential bounds for some momentum independent SME parameters coupled to spin.

1. Introduction

The recent past has witnessed remarkable progress on the production of Bose-Einstein condensates (BEC). Using a variety of trapping techniques it is now possible to produce reasonably large condensates using a number of atomic species in an array of different spin states. The fragility of these large, coherent quantum mechanical systems make them difficult to create and maintain. Changing perspective, this fragility can be viewed as a sensitivity to environment which might be used to detect small variations in background fields. The purpose of this paper is to develop this idea in the context of Lorentz violation (LV).

The study of Lorentz violation has undergone rapid growth in the last fifteen years. There is a well-developed theoretical framework for studying Lorentz violation, the Standard-Model Extension (SME),[1] in which spontaneous symmetry breaking is used to generate couplings between standard fields and vacuum expectations of tensor fields which parametrize symmetry violations. Using the SME, a framework for studying the statistical mechanics and thermodynamics of systems in the presence of Lorentz violation has been deveolped.[2] We use this framework to bound expected effects of symmetry breaking on a variety of condensates. This work is closely related to previous results.[3]

2. Calculations for Particles Without Interactions

The Hamiltonian for noninteracting spin-0 bosons in a box is:[2]

$$H = \frac{p^2}{2m} + A + C_j \frac{p_j}{m} + F_{jk} \frac{p_j p_k}{2m} \tag{1}$$

where the coefficients A, C, and F govern the effective Lorentz violation. Using the Hamiltonian (1) and a previously established framework for statistical mechanics calculations in the presence of Lorentz violation, we can compute the expected number of particles in excited states and the expected energy.[2] The results of the calculation indicate that the effect of Lorentz violation can be absorbed into an effective mass.

To study magnetically trapped gases, we add to the Hamiltonian H of (1) a harmonic potential,

$$V_{\text{trap}} = \frac{1}{2} m \sum_i (\omega_i x_i)^2 .$$

Using the framework cited above we can compute first order corrections to the ground state.[3] For ground state energy we get

$$E_{GS} = A + \frac{1}{4} \hbar \sum_i F_{ii} \omega_i \tag{2}$$

while for the ground state wave function we get (using the momentum representation and neglecting a phase),

$$\phi_{GS}(p) = \left(\frac{1}{p \hbar m \omega_{ho}} \right)^{\frac{3}{4}} \left(1 + \frac{1}{8} \text{Tr}(F) \right) \exp \left(- \frac{1}{2m\hbar} \eta(p, F) \right) \tag{3}$$

where $\omega_{ho} = (\omega_1 \omega_2 \omega_3)^{\frac{1}{3}}$ and $\eta(p, F) = \sum_i \frac{p_i^2}{\omega_i} + \sum_{i,j} F_{ij} \frac{p_i p_j}{\omega_i + \omega_j}$. These results suggest a possible experiment: we create a condensate, remove the trapping potential, measure the velocity distribution and compare to the perturbed ground state distribution. Unfortunately, the current shape sensitivity is about 1%, far to rough to produce interesting results.[5]

In addition to the ground state for spin-0 bosons, we can explicitly compute the expected number of particles in excited states and the critical temperature. This leads to an expression for the expected number of particles in the ground state:

$$\frac{N_0}{N} = 1 - \left(\frac{T}{T_c} \right)^3 . \tag{4}$$

This implies that Lorentz violation has no effect on ground state occupancy.

We can treat noninteracting spin polarized hydrogen similarly. Working in center of mass coordinates, we write the relevant piece of the Lorentz violating Hamiltonian as

$$H_{LV} = \sum_{e,p} A^{(r)} + B_j^{(r)}\sigma_{(r)}^j + (C_j^{(r)} + D_{jk}^{(r)}\sigma_{(r)}^k)\frac{P_j}{M} + (F_{jk}^{(r)} + G_{jkl}^{(r)}\sigma_{(r)}^l)\frac{P_jP_k}{2M}$$

where M is the mass of the atom and the superscripts $r = e$ and $r = p$ correspond to the Lorentz violating parameters associated to the electron and proton. To trap hydrogen magnetically, it must be in the triplet spin configuration. Taking the configuration to be $|F, m_F\rangle = |1, 1\rangle$, the calculation of ground state corrections is then straightforward:

$$E_{GS} = \sum_{e,p}\left[A + B_3\frac{1}{4}\hbar\sum_i(F_{ii} + G_{ii3})\omega_i\right]. \tag{5}$$

For the configuration $|F, m_F\rangle = |1, -1\rangle$, ground state corrections are obtained by flipping the signs of the spin coupling B and G.

To treat more complicated atoms (^{7}Li, ^{23}Na, and ^{87}Rb), we can proceed as we did with hydrogen, introducing terms in the Hamiltonian for each constituent part. The complexity of the resulting Hamiltonian precludes treatment by direct computation. To proceed we note that the atoms of interest share a common structure: each has nuclear spin of $3/2$ and one valence electron (in particular, they can be trapped in either spin state 1 or spin state 2). This suggests that we adopt a Schmidt model to continue the analysis: we concentrate the nuclear spin in the odd nucleon (i.e., a proton) and proceed as we did with hydrogen, obtaining explicit corrections to the ground state which depend on the spin couplings as in Eq. (5).

3. Interactions

In the absence of Lorentz violation the standard approach to accounting for two-body interactions in BEC involves using a Born approximation at low energy. More precisely, one introduces a scattering length

$$a = -(m/4\pi\hbar^2)\int U(r)dr \tag{6}$$

where $U(r)$ is the interatomic potential. One then replaces $U(r)$ with the effective potential

$$V(r) = a(4\pi\hbar^2/m)\int U(r)dr. \tag{7}$$

This leads to a modified Hamiltonian, and, upon expanding about the condensate wave function, the Gross-Piatevski equation for the condensate wave function. One then employs mean field methods to analyze condensate behavior.

In the presence of Lorentz violation, we proceed in similar fashion, introducing a scattering length as in Eq. (6) and an effective two-body potential as in Eq. (7).[a] Replacing the interatomic potential by the effective potential gives a modified Hamiltonian which then yields a modified Gross-Piatevski equation for the condensate wave function Φ:

$$i\hbar\frac{\partial}{\partial t}\Phi = \left[-\frac{\hbar^2}{2m}\nabla^2 + H_{LV} + V_{\text{trap}} + \frac{4\pi\hbar^2 a}{m}|\Phi|^2\right]\Phi \qquad (8)$$

where H_{LV} is the Lorentz violating portion of the Hamiltonian.

To proceed with the analysis, we employ the Thomas-Fermi approximation and assume that the kinetic terms are small with respect to interactions over the condensate. This allows us to neglect the kinetic terms in the Hamiltonian, as well as all momentum dependent contributions in the Lorentz violating Hamiltonian. Assuming a strong magnetic trap, the terms in H_{LV} which contribute are then the momentum independent spin couplings. Under these assumptions, direct calculation allows us to conclude that Lorentz violating effects do not affect bulk condensate properties.[3]

4. Optical Traps and Sensitive Tests

Optical trapping produces condensates exhibiting a superposition of various spin states. To analyze such condensates we assume the Thomas-Fermi approximation and introduce spin dependent scattering lengths to obtain an energy

$$K = \int drn(r)\left[H_{LV} + V_{\text{trap}} + c_0 n/2 + c_2 n\langle \mathbf{F}\rangle^2/2 + E_{ze}\right]$$

where n is condensate density, $\mathbf{F}$ is the total spin, c_i define combinations of total spin-0 and total spin-2 scattering states, and E_{ze} is the Zeeman energy contributed by stray magnetic background fields. Assuming well-shielded experiments and setting the proton angular momentum $l = 1$, we can minimize the energy functional. Our results indicate that the condensate that forms depends on the sign of the coefficient c_2. In particular, when $c_2 < 0$

[a]For BEC, the scattering length is very sensitive to the structure of the interatomic potential. This effect may be experimentally detectable.[4]

the condensate exhibits collective polarization with a spin that couples to the momentum independent spin coupling in the LV Hamiltonian. The effect, should it occur, mimics the effect of a constant external magnetic field. This LV effect can be differentiated from the effect of a stray field: rotation of the condensate experiment will tend to produce stray fields with time dependent direction, while the LV effect will have constant direction.

Our computation provides a means of probing for Lorentz violation using BEC. Given the above discussion of the effect of LV on spin polarized condensates, the torque applied by a LV background field on a condensate cloud should be comparable to the inertia at some reasonable angular acceleration. Assuming the radius of the cloud to be $\sim$ 1 micron, $m \sim 100$ GeV, and the angular acceleration to be $\sim 10^{-19} m^{-2}$, we obtain an estimated sensitivity of $|\mathbf{B}| \sim 10^{-29}$ GeV, comparable to the best CPT and Lorentz tests to date.[6]

5. Summary

Lorentz violation does not affect many thermodynamic properties of cold gases. For noninteracting condensates LV perturbs the ground state wave function, though experimental sensitivity probably precludes seeing an effect. For optically trapped condensates with multiple spin components LV can mimic a background magnetic field. In this case, the effect grows linearly with the size of the cloud and may be detectable.

References

1. D. Colladay and V.A. Kostelecký, Phys. Rev. D **55**, 6760 (1997); Phys. Rev. D **58**, 116002 (1998).
2. D. Colladay and P. McDonald, Phys. Rev. D **70**, 125007 (2004).
3. D. Colladay and P. McDonald, Phys. Rev. D **73**, 105006 (2006).
4. D. Comparat, private communication (2006).
5. L. V. Hau *et al.*, Phys. Rev. A **58**, R54 (1998).
6. See, for example, V.A. Kostelecký, ed., *CPT and Lorentz Symmetry III*, World Scientific, Singapore, 2005; and these proceedings.

EMBEDDING GEOMETRY AND DECOMPOSITION OF GRAVITY

T. E. CLARK, S. T. LOVE, CHI XIONG

Department of Physics, Purdue University
West Lafayette, Indiana 47907-2036, U.S.A.
Email: clark@physics.purdue.edu, loves@physics.purdue.edu, xiong@purdue.edu

T. TER VELDHUIS

Department of Physics & Astronomy, Macalester College
Saint Paul, Minnesota 55105-1899, U.S.A.
Email: terveldhuis@macalester.edu

We decompose higher dimensional gravity and study the physics of extra vector(s) in a brane-bulk world scenario as well as their relations to the embedding geometry. The possibility of 4D spontaneous Lorentz violation due to the extra vector(s) is discussed.

The original goal of Kaluza-Klein theory[1] is to unify gauge fields and gravity, while in the present paper we use it for studying the higher dimensional geometry and the decomposition of gravity. We obtain *extra* gauge field(s), which are not to be identified with the photon or other known gauge bosons in 4D. The higher dimensional Kaluza-Klein metric will lead to a 4D graviton (spin 2), new gauge bosons (spin 1) and the dilaton (or radion, spin 0). We start by considering the zero mode of 5D Kaluza-Klein metric and the corresponding 4D effective theory ($\mu = 0, 1, 2, 3; M = 0, 1, 2, 3, 5$):

$$G_{MN} = \rho^{-\frac{1}{3}} \begin{pmatrix} g_{\mu\nu} + \rho A_\mu A_\nu & \rho A_\mu \\ \rho A_\nu & \rho \end{pmatrix}. \tag{1}$$

Compactified on a circle with radius r, the 4D effective action is[1]

$$\begin{aligned} S_G &= -\frac{1}{2k_5^2} \int d^4x dy \ \ e^{(5)} R^{(5)} \qquad\qquad (k_5^2 = 2\pi r k_4^2) \\ &= -\frac{1}{2k_4^2} \int d^4x \ \ e^{(4)} [R^{(4)} + \frac{1}{4} \rho F_{\mu\nu} F^{\mu\nu} + \frac{1}{6\rho^2} \partial^\mu \rho \partial_\mu \rho] \end{aligned} \tag{2}$$

where $F_{\mu\nu} = \partial_\nu A_\mu - \partial_\mu A_\nu$ and k_5, k_4 are the 5D and 4D gravitational constants respectively. We then add a 3-brane in the 5D bulk which breaks the 5D translation and Lorentz symmetry spontaneously. The position of the brane is described by the embedding function $Y^M = Y^M(x^\mu)$ and x^μ are the coordinates on the brane. The brane action is the Nambu-Goto action built from the induced metric $h_{\mu\nu}$[2]

$$S_{\text{brane}} = -f^4 \int d^4x \sqrt{det\ h_{\mu\nu}}, \qquad h_{\mu\nu} = G_{MN} \partial_\mu Y^M \partial_\nu Y^N. \tag{3}$$

In static gauge, we choose $Y^\mu = x^\mu, Y^5 = \phi(x)$ and the Goldstone boson ϕ describes the brane fluctuation for a 5D spacetime with non-dynamical gravity. When we consider a curved spacetime metric as in Eq. (1) and compactify the 5D theory on a circle, an extra vector field appears in the induced metric $h_{\mu\nu}$ after the field ϕ is "eaten" by A_μ (set $\rho = 1$)

$$h_{\mu\nu} = g_{\mu\nu} + (A_\mu + \partial_\mu\phi)(A_\nu + \partial_\nu\phi) = g_{\mu\nu} + X_\mu X_\nu. \tag{4}$$

If we work in the Einstein frame and rescale the vector X_μ to get the normalized Maxwell term, the mass of X_μ is found to be $m_X^2 = f^4/M_{(4)}^2$. For $f \sim TeV$, this mass is very small. However, this is not true in general. The dilaton(s) from the higher dimension(s) can change the mass by a "dilaton factor". To see how this dilaton factor could change the mass of the brane gauge field, we use a toy model based on a two-step dimensional reduction from $(4+\delta)$D to 5D and then to 4D: $M_4 \times S^1 \times E \longmapsto M_4 \times S^1 \longmapsto M_4$ and $\hat{G}_{\hat{M}\hat{N}} = \text{diag}\{G_{MN}(x^\mu, y), \quad e^{-2\sigma(x^\mu, y)}\gamma_{ij}(z^k)\}$. After changing to the Einstein frame and rescaling the (induced) metric, we find that the mass of the brane gauge field X_μ changes by a "warped" factor, i.e., $m_X^2 = f^4\ e^{\frac{4}{3}\sigma\delta}/M_{(4)}^2$, so it may get exponentially increased to a large value. It is X_μ rather than ϕ which are the physical degrees of freedom relevant to the study of brane phenomenology. The non-derivative coupling of X_μ to the Standard Model can be obtained from the brane action including the Standard Model fields

$$S_{X-SM} = \int d^4x \frac{m_X^2}{F_X^4} T_{SM}^{\mu\nu} X_\mu X_\nu, \quad T_{SM}^{\mu\nu} \equiv \frac{\delta S_{brane}}{\delta h_{\mu\nu}}|_{h_{\mu\nu}=\eta_{\mu\nu}}. \tag{5}$$

Therefore the 4D effective action of the brane gauge field is (without derivative couplings)

$$S_{X-SM} = \int d^4x [L_{SM} - \frac{1}{4} F_{\mu\nu}F^{\mu\nu} + \frac{1}{2} m_X^2 X^\mu X_\mu + \frac{m_X^2}{F_X^4} T_{SM}^{\mu\nu} X_\mu X_\nu]. \tag{6}$$

Note that we only keep the leading order terms and higher order terms like $\sim (X^\mu X_\mu)^2$ are neglected.

Now we consider the derivative coupling of X_μ to matter and gravity by studying the embedding geometry of this brane-bulk world scenario.[3,4] We start with the embedding frame ($e_\mu = Y^M_{\ ,\ \mu}\ \partial_M,\ n_i$) which satisfies the following conditions (n_i are the normal vectors to the brane)

$$G_{MN}\partial_\mu Y^M \partial_\nu Y^N = h_{\mu\nu}, \quad G_{MN}\partial_\mu Y^M n_i^N = 0, \quad G_{MN} n_i^M n_j^N = \delta_{ij}. \quad (7)$$

The covariant derivatives of the embedding frame basis are

$$\nabla_\mu e_\nu = \Gamma^\lambda_{\mu\nu} e_\lambda - K^i_{\mu\nu} n_i \ , \quad \nabla_\mu n^i = K^i_{\mu\nu} e^\nu + B^{ij}_\mu n_j. \quad (8)$$

The extrinsic curvature $K^i_{\mu\nu}$, the d-dimensional connection $\Gamma^\lambda_{\mu\nu}$, and the twist potential B^{ij}_μ are introduced through the above Gauss-Weingarten equations.[4] They can be expressed in terms of the embedding frame basis and their covariant derivatives (note that the twist potential vanishes in co-dimension one case)

$$K^i_{\mu\nu} = -n^i_M \nabla_\mu (Y^M_{,\nu}) \ , \ \Gamma^\lambda_{\mu\nu} = h^{\lambda\sigma} Y^M_{,\sigma} G_{MN} \nabla_\mu (Y^M_{,\nu}) \ , \ B_{\mu ij} = n_{Mj} \nabla_\mu n^M_i . \quad (9)$$

Based on the Gauss-Weingarten equations, it is straightforward to deduce their integrability conditions, the Gauss-Codazzi-Ricci equations,[4] which relate the higher dimensional Riemann curvature tensor to its lower dimensional components, plus the extrinsic curvature and the twist potential as

$$\begin{aligned}
\hat{R}_{KLMN}\ e^K{}_\rho e^L{}_\sigma e^M{}_\mu e^N{}_\nu &= R_{\rho\sigma\mu\nu} + K^i_{\mu\sigma} K_{\nu\rho\ i} - K^i_{\nu\sigma} K_{\mu\rho\ i} \\
\hat{R}_{KLMN}\ n^{Ki} e^L{}_\sigma e^M{}_\mu e^N{}_\nu &= \tilde{\nabla}_\nu K^i_{\mu\sigma} - \tilde{\nabla}_\mu K^i_{\nu\sigma} \\
\hat{R}_{KLMN}\ n^{Ki} n^{Lj} e^M{}_\mu e^N{}_\nu &= F^{ij}_{\mu\nu} + K^i_{\nu\tau} K^{\tau\ j}_\mu - K^i_{\mu\tau} K^{\tau\ j}_\nu . \quad (10)
\end{aligned}$$

Here $\tilde{\nabla}_\mu K^i_{\nu\sigma} = \nabla_\mu K^i_{\nu\sigma} - B^{ij}_\mu K_{\nu\sigma}$, $F^{ij}_{\mu\nu} = \partial_\nu B^{ij}_\mu - \partial_\mu B^{ij}_\nu + B^{ik}_\mu B^j_{\nu k} - B^{ik}_\nu B^j_{\mu k}$. Note that in some literature, the "induced" metric is defined as $P_{MN} \equiv G_{MN} - n^i_M n_{Ni}$ and $P^N_M \equiv \delta^N_M - n^i_M n^N_i$ is the operator for projecting higher-dimensional fields onto the brane. On the other hand $\tilde{P}^N_M \equiv e_M{}^\nu e^N{}_\nu$ can also be viewed as a projector and the two can be identified as $P^N_M = \tilde{P}^N_M$. For simplicity, we consider a 3-brane embedded in a 5D spacetime so the twist potential vanishes and last equation of Eq. (10) becomes trivial.[3] One then can derive the Einstein equations on the brane (which slightly differ from Ref. 3 since we use the $\tilde{P}^N_M$ as the projectors)

$$\begin{aligned}
R_{\mu\nu} - \frac{1}{2} R h_{\mu\nu} + E_{\mu\nu} - Q_{\mu\nu} &= \frac{2}{3} k_5^2 [\hat{T}_{MN} e^M{}_\mu e^N{}_\nu + (\hat{T}_{MN} n^M n^N - \frac{1}{4}\hat{T}) h_{\mu\nu}] \\
\nabla_\tau K^\tau_\mu - \nabla_\mu K &= k_5^2\ n^M e^N{}_\mu \hat{T}_{MN} \quad (11)
\end{aligned}$$

where $E_{\mu\nu} = \hat{C}_{KLMN}\, n^K n^M e^L{}_\mu e^N{}_\nu$ ($\hat{C}_{KLMN}$ is the 5D Weyl tensor) and $Q_{\mu\nu} = (K_{\mu\nu}K - K_{\mu\tau}K^\tau_\nu) - \frac{1}{2}h_{\mu\nu}(K^2 - K_{\sigma\tau}K^{\sigma\tau})$. Focusing on the "global" case where the 5D spacetime is *flat*, and working in static gauge $Y^\mu = x^\mu$, $Y^5 = \phi(x)$ (noting that $e^M_\mu = \partial_\mu Y^M$ is *not* the 4D vierbein) gives

$$\begin{aligned}
&e^\nu{}_\mu = \delta^\nu{}_\mu \ , \qquad e^5{}_\mu = \partial_\mu\phi \ , \qquad E_{\mu\nu} = 0 \\
&n_\mu = -\partial^\mu\phi/\sqrt{1+\partial_\tau\phi\,\partial^\tau\phi} \ , \quad n_5 = 1/\sqrt{1+\partial_\tau\phi\,\partial^\tau\phi} \\
&h_{\mu\nu} = \eta_{\mu\nu} + \partial_\mu\phi\,\partial_\nu\phi \ , \quad h^{\mu\nu} = \eta^{\mu\nu} - \partial^\mu\phi\,\partial^\nu\phi/(1+\partial_\tau\phi\,\partial^\tau\phi) \\
&K_{\mu\nu} = \partial_\mu\partial_\nu\phi/\sqrt{1+\partial_\tau\phi\,\partial^\tau\phi} \\
&R_{\rho\sigma\mu\nu} = (\partial_\mu\partial_\rho\phi\,\partial_\nu\partial_\sigma\phi - \partial_\mu\partial_\sigma\phi\,\partial_\nu\partial_\rho\phi)/(1+\partial_\tau\phi\,\partial^\tau\phi).
\end{aligned} \tag{12}$$

Equations (11) are consistency equations for ϕ and its derivatives. Since the extra dimensional translation is spontaneously broken, the dynamics of the Nambu-Goldstone field ϕ is given by the conservation of the broken symmetry current $\partial_M \hat{T}^{M5} = F_\phi^2 \Box\phi + ... = 0$. In order to solve these equations one has to specify the stress energy tensor $\hat{T}_{MN}$. One may convert the equations for ϕ to an action which includes the leading couplings like $\partial_\mu\phi\,\partial_\nu\phi\, T^{\mu\nu}_{SM}$ plus other higher order derivative terms.

The "local" case, which corresponds to a general *curved* 5D spacetime, is more complicated and one will obtain a set of differential equations for spin-2 (4D graviton), spin-1 (4D vector) and spin-0 (4D dilaton). If we only consider the vector part, the Higgs mechanism happens again and one can simply make the replacement $\partial_\mu\phi \to \partial_\mu\phi + A_\mu \to X_\mu$ in Eq. (11). One must be careful, however, since $\partial_\mu\partial_\nu\phi$ has the ambiguity of being replaced by $\partial_\mu X_\nu$ or by $\partial_\nu X_\mu$. One also has to figure out the relation between n_μ and X_μ. The phenomenology of X_μ has been studied in Ref. 6 by considering its couplings to the Standard Model fields, based on the 4D coset approach and the method of non-linear realizations which included derivative coupling terms.

As we mentioned earlier, terms like $\sqrt{-g}\ X^\mu X_\mu, \sqrt{-g}\ (X^\mu X_\mu)^2$ and $\sqrt{-g}\ \Lambda$ are allowed in the action. With the particular parameters and signs they can give a potential $\lambda(X^\mu X_\mu - \eta^2)^2$ which suggests a non-vanishing vacuum expectation value for X^μ and hence the 4D Lorentz symmetry is spontaneously broken.[7] That is, the extra vector X_μ from the higher dimensional spacetime metric triggers a further spontaneous Lorentz symmetry breaking in 4D. For its couplings to gravity and to the Standard Model, the results found in Refs. 5, 6 are similar to what has been discussed in Ref. 9, which used a "gauged ghost condensation" to build models with spontaneous Lorentz symmetry breaking. Nevertheless, one should notice that the stability of the vacuum is assumed in the non-linear realization approach but becomes quite subtle in the (gauged) ghost condensation scenario[8,9] which we leave to further investigation.

Acknowledgments

The work of TEC, STL and CX was supported in part by the U.S. Department of Energy under grant DE-FG02-91ER40681 (Task B). The work of TtV was supported in part by a Cottrell Award from the Research Corporation.

References

1. See, for example, T. Appelquist, A. Chodos and P. Freund, eds. *Modern Kaluza-Klein Theories*, Addison-Wesley, Menlo Park, 1987; J.M. Overduin and P.S. Wesson, gr-qc/9805018.
2. R. Sundrum, Phys. Rev. D, **59**, 085009 (1999); N. Arkani-Hamed, S. Dimopoulos and G. Dvali, Phys. Rev. D **59**, 086004(1999); T. Kugo and K. Yoshioka, Nucl. Phys. B **594**, 301 (2001).
3. T. Shiromizu, K. Maeda and M. Sasaki, Phys. Rev. D **62**, 024012 (2000).
4. See, for example, L.P. Eisenhart, *Riemannian Geometry*, Princeton University Press, Princeton, 1966; Y. Aminov, *The Geometry of Submanifolds*, Gordon and Breach, London, 2001; M.D. Maia, E.M. Monte, Phys. Lett. A 297(2002); M. Heydari-Fard, M. Shirazi, S. Jalalzadeh and H.R. Sepangi, Phys. Lett. B **640**, 1 (2006).
5. T.E. Clark, S.T. Love, M. Nitta and T. ter Veldhuis, Phys. Rev. D **72**, 085014 (2005); S.T. Love, J. Phys. A **40**, 7049 (2007); T.E. Clark, S.T. Love, M. Nitta, T. ter Veldhuis and C. Xiong, Phys. Rev. D **75**, 065028 (2007); T.E. Clark, S.T. Love, M. Nitta, T. ter Veldhuis and C. Xiong, hep-th/0703179.
6. T. Clark, S. Love, T. ter Veldhuis and C. Xiong, in preparation.
7. See, for example, V.A. Kostelecký, ed., *CPT and Lorentz Symmetry III*, World Scientific, Singapore, 2005; R. Bluhm and V.A. Kostelecký, Phys. Rev. D **71**, 065008(2005); R. Bluhm, hep-ph/0506054.
8. N. Arkani-Hamed, H. Cheng, M. Luty and S. Mukohyama, JHEP **05** 074 (2004).
9. H. Cheng, M. Luty, S. Mukohyama and J. Thaler, JHEP **05** 076 (2006).

THE FIRST LUNAR LASER RANGING CONSTRAINTS ON GRAVITY SECTOR SME PARAMETERS

J.B.R. BATTAT, J.F. CHANDLER AND C.W. STUBBS
Harvard-Smithsonian Center for Astrophysics
60 Garden Street,
Cambridge, MA 02138, U.S.A.
E-mail: jbattat@cfa.harvard.edu

We present the first constraints on pure-gravity sector Standard-Model Extension (SME) parameters using Lunar Laser Ranging (LLR). LLR measures the round trip travel time of light between the Earth and the Moon. With 34+ years of LLR data, we have constrained six independent linear combinations of SME parameters at the level of 10^{-6} to 10^{-11}. There is no evidence for Lorentz violation in the LLR dataset.

1. Introduction and Motivation

Two of us (C.W.S. and J.B.R.B.) are members of the Apache Point Observatory Lunar Laser-ranging Operation (APOLLO), a next-generation LLR facility, capable of millimeter-precision lunar range measurements (see the article by T.W. Murphy in these proceedings). The APOLLO project was motivated by the realization that an order-of-magnitude improvement in fundamental physics constraints (e.g., equivalence principle, gravitomagnetism, gravitational $1/r^2$ law and $\dot{G}$, to name a few) could be achieved with straightforward improvements to the standard LLR apparatus.

With the recent description of the pure-gravity sector of the SME,[1] we learned that LLR can also provide incisive constraints on Lorentz Violation. The predicted LLR observable under Lorentz Violation is a periodic perturbation to the Earth-Moon range with the leading order effects occuring at four distinct frequencies: 2ω, ω, $2\omega - \omega_0$ and $\Omega_\oplus$. Here ω is the lunar orbital (sidereal) frequency, ω_0 is the anomalistic lunar orbital frequency and $\Omega_\oplus$ is the mean Earth orbital (sidereal) frequency. Although the APOLLO program is still in the data collection phase, there are more than three decades of freely-available archival LLR data on a public archive.[2] In this article, I present the SME parameter constraints that result from our analysis of archival LLR data. These are the first LLR-based constraints on pure-gravity SME parameters.

2. The LLR Dataset and Analysis Software

LLR measures the time of flight of photons between a telescope on the Earth and corner cubes on the lunar surface.[3] LLR data is typically presented in "normal points" which are typically generated from a few to a hundred lunar signal photons collected over a span of 1 to 5 minutes. Our analysis makes use of archival data from September 1969 through December 2003.

In the analysis of LLR normal points, a set of residual ranges is computed by subtracting the model's predicted range from the observed range. The range sensitivity with respect to each model parameter (the partial derivatives) is also computed at the time of each normal point. The residuals and the partial derivatives are then used to compute optimal model parameter values via a weighted linear least-squares fit. For our analysis, we used the Planetary Ephemeris Program (PEP),[5] which is currently maintained by one of us (J.F.C.). To our knowledge, it is the only publicly available LLR analysis software.

Typically, the lunar range model is formulated in the parametrized post-Newtonian (PPN) framework,[4] which permits model-independent constraints on metric theories of gravity. At present, no ephemeris models explicitly incorporate SME parameters. You can, however, think of these models as implicitly including the SME parameters but with values pegged at zero (i.e., no SME perturbation to the lunar orbit). It is therefore only necessary to compute, by hand, the partial derivative of range with respect to each SME model parameter (see Table 1). The analysis code can then provide SME parameter adjustments simultaneously with the other model parameters. A covariance matrix including the correlations between the SME parameters and all other model parameters is also produced.

Table 1. SME parameter partial derivatives. Symbols used here are explained in Ref. 1.

SME Parameter	Partial Derivative of Lunar Range with Respect to SME Parameter
$\bar{s}^{11} - \bar{s}^{22}$	$-\frac{r_0}{12}\cos\left(2\omega t + 2\theta\right) - \frac{\omega e r_0}{16(\omega-\omega_0)}\cos\left[\left(2\omega - \omega_0\right)t + 2\theta\right]$
$\bar{s}^{12}$	$-\frac{r_0}{6}\sin\left(2\omega t + 2\theta\right) - \frac{\omega e r_0}{8(\omega-\omega_0)}\sin\left[\left(2\omega - \omega_0\right)t + 2\theta\right]$
$\bar{s}^{02}$	$-\frac{\omega(\delta m) v_0 r_0}{M(\omega-\omega_0)}\cos\left(\omega t + \theta\right)$
$\bar{s}^{01}$	$\frac{\omega(\delta m) v_0 r_0}{M(\omega-\omega_0)}\sin\left(\omega t + \theta\right)$
$\bar{s}_{\Omega_\oplus,c}$	$V_\oplus r_0 \left(\frac{b_1}{b_2}\right)\cos\left(\Omega_\oplus t\right)$
$\bar{s}_{\Omega_\oplus,s}$	$V_\oplus r_0 \left(\frac{b_1}{b_2}\right)\sin\left(\Omega_\oplus t\right)$

The main drawback to this approach is that one cannot perform an iterative analysis in which one takes the best-fit model parameter values and uses them to re-integrate the equations of motion to refine the model parameter values. We accept this limitation because the non-SME model parameters have been highly refined through iterative solutions over the past several decades and so the solution sits firmly in the linear regime already. Furthermore, the addition of the SME parameters preserves the linearity because the lunar range is strictly linear in the SME parameters (see Table 2 of Ref. 1), so no iteration is necessary.

3. Systematic Errors

The solar system is complex. Predictions of the lunar range rely on models of planetary and asteroid positions, gravitational harmonics of the Sun, Earth and Moon and various relativistic and non-gravitational effects (to name a few). Solar system models have many hundreds of parameters that influence the Earth-Moon range time. There are strong correlations between model parameters. As a result, solutions will suffer from systematic errors in model parameter estimates that can dominate the formal errors reported by the least-squares analysis. In this work, we account for the underestimation of model parameter uncertainties by scaling the formal parameter errors reported by the least-squares analysis by a uniform factor, F. This is numerically equivalent to uniformly scaling the uncertainty of each normal point by F. Essentially, we uniformly down-weight the data.

The F factor is empirically determined by holding the SME parameter values at zero but allowing the PPN values β and γ to vary (see Ref. 4 for an explanation of β and γ). We know from existing experiments[6] that these parameters are consistent with their General Relativity values ($\beta = \gamma = 1$) to within a part in 10^3 or better. We find that we require $F = 20$ to ensure that we are in accord with these earlier results.

4. SME Parameter Constraints and Verification

We constrain the SME parameters under the assumption that General Relativity is not violated (we set $\beta = \gamma = 1$). The resulting parameter estimates and their realistic errors (the formal errors scaled by $F = 20$) are reported in Table 2. All SME parameters are within $1.5F\sigma$ of zero. There is no evidence for Lorentz violation in the LLR data. The fit quality is shown in Fig. 1.

To verify our implementation of the partial derivatives of lunar range

Table 2. SME parameter estimates and their realistic (scaled) uncertainties ($F\sigma$) with $F = 20$.

Parameter	Estimate
$\bar{s}^{11} - \bar{s}^{22}$	$(1.3 \pm 0.9) \times 10^{-10}$
$\bar{s}^{12}$	$(6.9 \pm 4.5) \times 10^{-11}$
$\bar{s}^{02}$	$(-5.2 \pm 4.8) \times 10^{-07}$
$\bar{s}^{01}$	$(-0.8 \pm 1.1) \times 10^{-06}$
$\bar{s}_{\Omega_{\oplus}c}$	$(0.2 \pm 3.9) \times 10^{-07}$
$\bar{s}_{\Omega_{\oplus}s}$	$(-1.3 \pm 4.1) \times 10^{-07}$

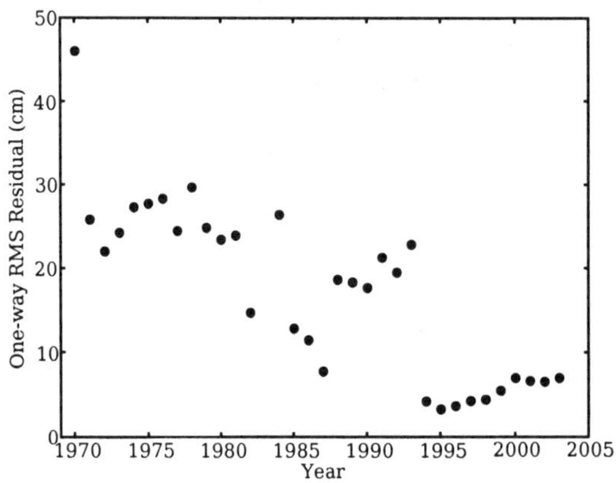

Figure 1. The agreement between model and data for the first 34 years of LLR.

with respect to the SME parameters, we generated, by hand, a perturbed LLR normal point data set by setting $\bar{s}^{11} - \bar{s}^{22} = 9 \times 10^{-10}$, a $10F\sigma$ deviation from the best-fit value of this parameter. A fit to this data recovers the perturbation: $\bar{s}^{11} - \bar{s}^{22} = [(1 + 9) \pm 0.9] \times 10^{-10}$ with the other SME parameters unchanged.

5. Conclusions and Future Prospects

We have analyzed 34+ years of LLR data and have derived constraints on six SME parameters combinations (see Table 2). We find no deviation from Lorentz Symmetry at the $10^{-6} - 10^{-11}$ level. This work provides the first LLR-based constraints of SME parameters.

There are several ways in which these constraints could be improved. First of all, by incorporating auxiliary solar system data (e.g., planetary radar ranging) model parameter correlations can be reduced, and F decreased. This would allow for tighter constraints on the SME parameters using the same LLR dataset. In addition, APOLLO data, which is about 10 times more precise than the archival data, will soon be ready for analysis. With this improved dataset, we will further tighten the SME parameter constraints.

Acknowledgments

This work developed from discussions with our colleagues including E. Adelberger, Q. Bailey, J. Davis, A. Kostelecký, J. Moran, T. Murphy, I. Shapiro and M. Zaldarriaga. We would also like to thank the staff observers at LLR stations, and the ILRS for the infrastructure that supports the normal point distribution. J.B.R.B. acknowledges financial support from the ASEE NDSEGF, the NSF GRFP and Harvard University. We also acknowledge the generous financial support of the National Science Foundation (grant PHY-0602507).

References

1. Q.G. Bailey and V.A. Kostelecký Phys. Rev. D **74**, 045001 (2006).
2. M.R. Pearlman *et al.*, Adv. Space Research **30**, 135 (2002).
3. J.G. Williams *et al.*, gr-qc/0507083.
4. C.M. Will and K.J. Nordtvedt, Astrophys. J. **177**, 757 (1972).
5. The Planetary Ephemeris Program is a solar system ephemeris and data analysis program that was developed at the Massachusetts Institute of Technology and its Lincoln Laboratory, beginning in 1964. The source code is currently maintained by John Chandler at the Center for Astrophysics in Cambridge, MA and is publicly available.
6. J.G. Williams *et al.*, Phys. Rev. Lett. **93**, 261101 (2004); B. Bertotti *et al.*, Nature (London) **425**, 374 (2003); S.S. Shapiro *et al.*, Phys. Rev. Lett. **92**, 121101 (2004); J.F. Chandler *et al.*, in R. Ruffini *et al.*, eds., *Proceedings of the Seventh Marcel Grossman Meeting*, World Scientific, Singapore, 1996.
7. See, for example, V.A. Kostelecký, ed., *CPT and Lorentz Symmetry III*, World Scientific, Singapore, 2005.
8. J. Lipa *et al.*, Phys. Rev. Lett. **90**, 060403 (2003); P.L. Stanwix *et al.*, Phys. Rev. D **74**, 081101 (2006); M.E. Tobar *et al.*, Phys. Rev. D **71**, 025004 (2005); M. Hohensee *et al.*, Phys. Rev. D **75**, 049902 (2007); J.P. Cotter and B. Varcoe, physics/0603111; H. Müller *et al.*, arXiv:0706.2031
9. T. Katori and R. Tayloe, in Ref. 7; LSND Collaboration, L.B. Auerbach *et al.*, Phys. Rev. D **72**, 076004 (2005); V.A. Kostelecký and M. Mewes, Phys. Rev. D **69**, 016005 (2004).

A THEORY OF GRAVITY FROM LORENTZ VIOLATION

R. POTTING
CENTRA and Physics Department
Faculdade de Ciências e Tecnologia
Universidade do Algarve
Faro, Portugal
E-mail: rpotting@ualg.pt

We construct a theory of gravity that arises through spontaneous Lorentz violation in a Lagrangian involving a Lorentz 2-tensor. Although at low energy the Einstein-Hilbert action is obtained, the theory differs in some key respects from General Relativity. We outline the construction of the model and discuss its key features.

1. Introduction

For vector particles masslessness is generally taken to arise from gauge invariance of the action. Another possibility that has been considered is that masslessness can arise from the spontaneous symmetry breaking of Lorentz symmetry[1] in the bumblebee model discussed in the talk by Robert Bluhm.

In this talk I show how this idea can be applied to the case of Lorentz 2-tensors. In the resulting cardinal model, gravitons arise by spontaneous breaking of local Lorentz invariance, and we obtain linearized general relativity in a fixed axial-like gauge. In this talk, a report on ongoing work with Alan Kostelecký,[2,3] we show this model can be extended to a nonlinear form in a self-consistent way, yielding the Einstein-Hilbert action of general relativity at low energies and temperatures.

Note that this construction goes beyond incorporating Lorentz violation into a theory of gravity.[4]

2. Linearized cardinal theory

The starting point for the linearized cardinal model is the lagrangian[2]

$$\mathcal{L} = \frac{1}{2} C^{\mu\nu} K_{\mu\nu\alpha\beta} C^{\alpha\beta} + V(C^{\mu\nu}, \eta_{\mu\nu}). \tag{1}$$

Here $C^{\mu\nu}$ is a 2-tensor, $K_{\mu\nu\alpha\beta}$ the quadratic kinetic operator for a massless a spin-2 field given by

$$K_{\mu\nu\alpha\beta} = -\partial^2(\eta_{\mu\alpha}\eta_{\nu\beta} - \tfrac{1}{2}\eta_{\mu\nu}\eta_{\alpha\beta}) + \partial_\mu\eta_{\nu\alpha}\partial_\beta + \partial_\nu\eta_{\mu\alpha}\partial_\beta \tag{2}$$

and $\eta_{\mu\nu}$ a flat background metric. V is a scalar potential constructed using the 4 independent scalars that can be built out of $C^{\mu\nu}$ and $\eta_{\mu\nu}$, namely $C_1 = C^{\mu\nu}\eta_{\nu\mu}$, $C_2 = (C\cdot\eta\cdot C\cdot\eta)^\mu_\mu$, $C_3 = (C\cdot\eta\cdot C\cdot\eta\cdot C\cdot\eta)^\mu_\mu$ and $C_4 = (C\cdot\eta\cdot C\cdot\eta\cdot C\cdot\eta\cdot C\cdot\eta)^\mu_\mu$. Assuming V has its minimum for $C^{\mu\nu} = c^{\mu\nu} \neq 0$, this will force $C^{\mu\nu}$ to acquire an expectation value $\langle C^{\mu\nu}\rangle = \Lambda^\mu_\rho\Lambda^\nu_\sigma c^{\rho\sigma}$ with Λ^μ_ρ some Lorentz matrix (which we will assume to be unity), thus spontaneously breaking Lorentz symmetry.

At low energies, fluctuations in the directions normal to the minimum of V will be negligible, and V can be approximated by sum of delta-functions that fix the 4 independent scalars multiplied by Lagrange multipliers λ_i $(i = 0,..,3)$. The low-energy equation of motion becomes

$$K_{\mu\nu\alpha\beta}h^{\alpha\beta} - \lambda_0\eta_{\mu\nu} - \lambda_1(\eta c\eta)_{\mu\nu} - \lambda_2(\eta c\eta c\eta)_{\mu\nu} - \lambda_3(\eta c\eta c\eta c\eta)_{\mu\nu} = 0 \tag{3}$$

with the constraints $h^\mu_\mu = c^{\mu\nu}h_{\mu\nu} = (c\eta c)^{\mu\nu}h_{\mu\nu} = (c\eta c\eta c)^{\mu\nu}h_{\mu\nu} = 0$. The resulting low-energy dynamics of fluctuations around the vacuum expectation value equals to linearized general relativity in the gauge defined by the constraints.[2]

It is instructive to count the propagating massless degrees of freedom. They can be parametrized as: $h_{\mu\nu} = \mathcal{E}_\mu{}^\alpha c_{\alpha\nu} + \mathcal{E}_\nu{}^\alpha c_{\mu\alpha}$, with the $\mathcal{E}_{\mu\nu} = -\mathcal{E}_{\nu\mu}$ corresponding to the 6 Nambu-Goldstone modes of the spontaneously broken Lorentz generators. Moreover, it can be checked that the equations of motion imply the masslessness condition $\partial^2 h_{\mu\nu} = 0$ and the 4 Lorenz conditions $\partial^\mu h_{\mu\nu} = 0$. This yields thus $6 - 4 = 2$ massless propagating degrees of freedom, corresponding to the graviton helicities.

3. Exponentiation

A requirement to be demanded of any theory of gravity is that it couple correctly to the total energy-momentum tensor. Inclusion of a coupling to matter can be accomplished by adding the term $h^{\mu\nu}\tau^m_{\mu\nu}$ to Eq. (1), where $\tau^m_{\mu\nu}$ is the trace-inversed energy-momentum tensor. We obtain the linearized Einstein equation

$$K_{\mu\nu\alpha\beta}h^{\alpha\beta} \equiv R^L_{\mu\nu} = \tau^m_{\mu\nu}. \tag{4}$$

However, the total energy-momentum tensor not only consists of contributions of matter. There is a contribution of the gravitons themselves as well, quadratic in $h_{\mu\nu}$. As a consequence, the inclusion of a cubic term in Eq. (1) is required. This, in turn, implies a cubic contribution to the energy-momentum tensor, corresponding to a quartic term in the Lagrangian. This process continues indefinitely, yielding in the limit the full Einstein-Hilbert action $R_{\mu\nu}(C^{\alpha\beta})$.[5]

This process of "exponentiation" can be carried out in one step by using a procedure used by Deser for general relativity.[6] Here one rewrites the kinetic term in Eq. (1) in the first-order form

$$h^{\mu\nu}(\Gamma^{\alpha}_{\mu\nu,\alpha} - \Gamma_{\mu,\nu}) + c^{\mu\nu}(\Gamma^{\alpha}_{\mu\nu}\Gamma_{\alpha} - \Gamma^{\alpha}_{\beta\mu}\Gamma^{\beta}_{\alpha\nu}) \tag{5}$$

where the connection coefficients $\Gamma^{\alpha}_{\mu\nu}$ are to be considered as independent additional variables. It can be shown that the exponentiation of the kinetic term then terminates after one step, yielding the Einstein-Hilbert action $R_{\mu\nu}(C^{\alpha\beta})$. We conclude that $C^{\alpha\beta}$ naturally assumes the role of curved-space metric density.

It can be shown that this exponentiation process can also be applied in a consistent way to the scalar potential and the matter energy-momentum tensor coupling. The exponentiation of the matter energy-momentum tensor coupling yields a curved-space matter lagrangian with $C^{\mu\nu}$ playing the role of curved-space metric density. The full exponentiated Lagrangian is given by

$$C^{\mu\nu}R_{\mu\nu}(\Gamma) - V(C^{\mu\nu}, \eta_{\mu\nu}) + L_{matter}(C, \eta, \phi_i, \partial_\mu\phi_i) \tag{6}$$

where the ϕ_i represent the matter fields. Consistent exponentiation of the scalar potential requires that the latter be built only of $C^{\mu\nu}$ and $\eta_{\mu\nu}$, which serves as a nice consistency check on our construction.

4. Vacuum energy-momentum tensor

Let us consider once again the linearized equations of motion of the cardinal model:

$$K_{\mu\nu\alpha\beta}h^{\alpha\beta} = (\eta_{\mu\nu}\partial_1 + 2\eta_{\mu\alpha}c^{\alpha\beta}\eta_{\beta\nu}\partial_2 + \ldots)V + \tau^{(m)}_{\mu\nu}(\eta, \phi_i, \partial_\mu\phi_i) \tag{7}$$

where we defined

$$\partial_n \equiv \frac{\partial}{\partial C_n} \qquad C_n = (C\cdot\eta)^n \qquad (n = 1\ldots 4). \tag{8}$$

Note that in Eq. (7) we did not use the Lagrange multiplier approximation for the scalar potential employed in Eq. (3).

We see from Eq. (7) that the first term on the right-hand side naturally takes the form of a (trace-reversed) energy momentum tensor. Explicitly, we can identify a "vacuum energy-momentum tensor"

$$T_{\mu\nu}^{(vac)} = \tau_{\mu\nu}^{(vac)} - \frac{1}{2}\eta_{\mu\nu}\left(\tau^{(vac)}\right)_{\alpha}^{\alpha} \tag{9}$$

where

$$\tau_{\mu\nu}^{(vac)} = (\eta_{\mu\nu}\partial_1 + 2\eta_{\mu\alpha}C^{\alpha\beta}\eta_{\beta\nu}\partial_2 + ...)V. \tag{10}$$

It takes nonzero values whenever the scalar potential takes values away from the minimum.

Explicit solutions of the linearized equations of motion can be obtained with nonzero vacuum energy-momentum tensor. In the latter case, independent initial/boundary field values can be defined on maximally 4 suitably defined timelike/spacelike spacetime slices. If the matter energy-momentum tensor is known to be independently conserved (e.g., by symmetry arguments), the same has to be true for the vacuum energy-momentum tensor. In such a case, choosing $T_{\mu\nu}^{(vac)}$ to be zero at the initial value spacetime slices ensures it is zero throughout spacetime. The vacuum energy-momentum tensor defined in Eq. (9) is analogous to the vacuum charge density that appears in bumblebee models (see Robert Bluhm's contribution to these proceedings, as well as Ref. 7).

5. Fluctuations in the non-flat directions

Leaving for a moment aside the issue of the vacuum energy-momentum tensor, we expect that, at low energies, the potential constrains $C^{\mu\nu}$-fluctuations to be strictly along the flat directions of its minimum. However, at high energies or temperatures, the fluctuations in the perpendicular directions, corresponding to positive curvature, contribute to and affect the solutions to the equations of motion.

These effects can be investigated systematically, for instance, by integrating out the fluctuations in the perpendicular directions, leaving an effective action for the remaining degrees of freedom. It can be shown that this yields the following R^2 correction to the effective action for the remaining degrees of freedom:

$$\mathcal{L}_{eff} \supset R(\Gamma)_{\mu\nu} M^{\mu\nu\rho\sigma} R(\Gamma)_{\rho\sigma} \tag{11}$$

for some coefficient matrix $M^{\mu\nu\rho\sigma}$ of which the order of magnitude is proportional to the inverse of the elements of $V_{\mu\nu\rho\sigma}$.[3]

Also significantly affected by, and sensitive to, the precise form of the cardinal potential will be the quantum properties of the model. This may shed light on the perennial problem of the consistent quantization of gravity.

6. Conclusions and outlook

We described the construction of an alternative theory of gravity, the cardinal model, with massless gravitons arising from spontaneous breaking of Lorentz symmetry by a scalar potential. The full nonlinear form of the Lagrangian is fixed by consistent coupling of gravity to the total energy-momentum tensor. At low energy, the Lagrangian corresponds to the Einstein-Hilbert action, with the caveat that the presence of the scalar potential can give rise to a nonzero "vacuum energy-momentum tensor." We also saw that at high energies and temperatures, the dynamics, both on the classical and the quantum level, becomes modified by the potential.

With regard to the vacuum energy-momentum tensor, it would be interesting to try and see if physical sense could be made of such an object, maybe in relation to dark matter/energy. Another issue that deserves attention is what counterpart, if any, solutions of General Relativity that include singularities have in the cardinal model.

We will report on these issues in the near future.[3]

Acknowledgments

It is a pleasure to thank Alan Kostelecký for collaboration, and Quentin Bailey, Don Colladay and Ted Jacobson for discussions. Financial support by the Fundação para a Ciência e a Tecnologia is gratefully acknowledged.

References

1. R. Bluhm and V.A. Kostelecký, Phys. Rev. D **71**, 065008 (2004).
2. V.A. Kostelecký and R. Potting, Gen. Rel. Grav. **37**, 1675 (2005).
3. V.A. Kostelecký and R. Potting, in preparation.
4. V.A. Kostelecký, Phys. Rev. D **69**, 105009 (2004).
5. R. Kraichnan, MIT thesis, 1947; Phys. Rev. **98**, 1118 (1955); A. Papapetrou, Proc. Roy. Irish Acad. **52A**, 11 (1948); S.N. Gupta, Proc. Phys. Soc. London **A65**, 608 (1952); R.P. Feynman, Chapel Hill Conference (1956).
6. S. Deser, Gen. Rel. Grav. **1**, 9 (1970).
7. J. L. Chkareuli, C. D. Froggatt and H. B. Nielsen, arXiv:hep-th/0610186.

SEARCH FOR LORENTZ VIOLATION IN A HIGH-FREQUENCY GRAVITATIONAL EXPERIMENT BELOW 50 MICRONS

W.A. JENSEN, S.M. LEWIS, AND J.C. LONG

Indiana University Cyclotron Facility
2401 Milo B Sampson Lane
Bloomington IN 47408, U.S.A.
E-mail: wajensen@indiana.edu

An experimental test of the Newtonian inverse square law below 50 μm is in progress at Indiana University. The experiment uses 1 kHz planar oscillators as test masses with a metal membrane stretched between them to suppress backgrounds, a technique showing potential for probing exceptionally small distances and operation at the limit of instrumental thermal noise. Previous data from this experiment, which set new constraints on short-range phenomena motivated by string models, are being re-analyzed for possible signals of Lorentz violation in the Standard-Model Extension.

1. Introduction

Laboratory–scale tests of the Newtonian inverse square law have received a revival of attention in the last decade. This is especially true of experiments in the regime below a few hundred microns, where gravity is poorly measured. Predictions of new effects in this range have arisen in several contexts, including attempts to describe gravity and the other fundamental interactions in the same theoretical framework.

Deviations from the inverse square law are commonly parameterized by a Yukawa interaction, the potential for which is given by:

$$V(r) = \frac{Gm_1m_2}{r}\alpha e^{-r/\lambda}. \tag{1}$$

Here, G is the Newtonian Gravitational Constant, m_1 and m_2 are test masses, r is their separation, α is the strength of a new interaction relative to gravity and λ is the range. Figure 1, adapted from Refs. 1 and 2, shows the limits in the range between 1 μm and 1 cm. The published limits on new forces above 10 μm are defined by classical gravity measurements (curves labeled "Irvine"[3] and "Washington"), including the Eöt-Wash torsion pendulum experiment.[4] Below a few microns the best limits are derived primarily from Casimir force measurements.[5] The limits in the intermediate range are the result of two high-frequency (kilohertz–range) experiments ("Stanford",[6] "Colorado"). The latter experiment, published by one of the authors in 2003,[7] has been re-located to Indiana University and is described below. The limits allow for new forces in nature several million times stronger than gravity at ranges resolvable by the unaided eye.

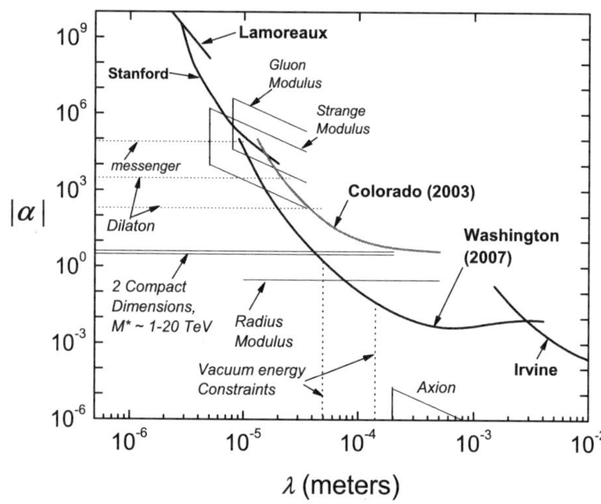

Figure 1. Parameter space for Yukawa-type forces in which the strength relative to gravity (α) is plotted versus the range (λ). The experimentally excluded region is above and to the right of the bold, solid curves. Fine and dashed lines are theoretical predictions.

In addition to this very general motivation for new experiments, a number of specific theoretical predictions of new physics have arisen in the short-range regime, represented by the fine and dashed lines in Fig. 1. Most notable is the line indicating Yukawa corrections to the inverse square law which arise from compact extra dimensions.[8] Corrections are also predicted from massive scalars in string theories, such as moduli and dilatons.[9,10,11] The other predictions shown are motivated by the cosmological constant problem[12] and the strong CP problem of QCD.[13]

2. The Indiana Short–Range Experiment

For an experiment to be sensitive to gravity or possible new forces in nature at short range, the test masses must in general be scaled to that range, so as not to be overwhelmed by Newtonian forces at larger scales. The highly sensitive, linear response of thin fibers under tension has made the torsion pendulum the instrument of choice for laboratory gravity measurements. However, it tends to be limited by backgrounds including vibrations, tilts and mechanical relaxation—which would be expected to be exacerbated at small test mass separations—and it is to the credit of the Eöt-Wash group for having made that technology successful below 100 μm. The Indiana experiment takes a different approach, showing potential for greater sensitivity at shorter ranges.

Figure 2, adapted from Ref. 7, shows the central components of the experiment. The detector is driven by the source mass on resonance near 1 kHz, placing a heavy burden on vibration isolation. The 1 kHz operation is chosen since at this frequency it is possible to construct a simple, passive vibration isolation system. The entire apparatus is enclosed in a vacuum chamber and operated at 10^{-7} torr to further reduce the acoustic coupling. Detector oscillations are read out with a capacitive transducer and lock-in amplifier. This design has proven effective for suppressing all background forces to the extent that the only effect observed is thermal noise due to dissipation in the detector mass. At the time of publication, this experiment set the strongest limits on new forces of nature between 10 and 100 μm, and improved on previous searches by up to a factor of 1000.

The 100 μm test mass separation in the 2003 experiment was determined largely by the shield thickness (60 μm) and not the amplitude of the test mass oscillations; test mass separations as small as $\sim$30 μm could be achieved with a thin enough shield. Since the relocation of the experiment to Indiana, shield prototypes 10 μm thick have been made using stretched metal membranes, reducing the test mass gap to below 50 μm.

First data with the membrane shield in 2007 revealed a large signal that was traced to electronic pick-up. This signal has been suppressed below the detector thermal noise and more sensitive force data are now being obtained with the smaller test mass gap. Assuming the backgrounds can be controlled, the projected limits attainable with the current experiment are about two orders of magnitude more sensitive than the best limits in Fig. 1 at 10 μm. A cryogenic version of the experiment could attain sensitivity to gravitational strength forces at 20 μm.

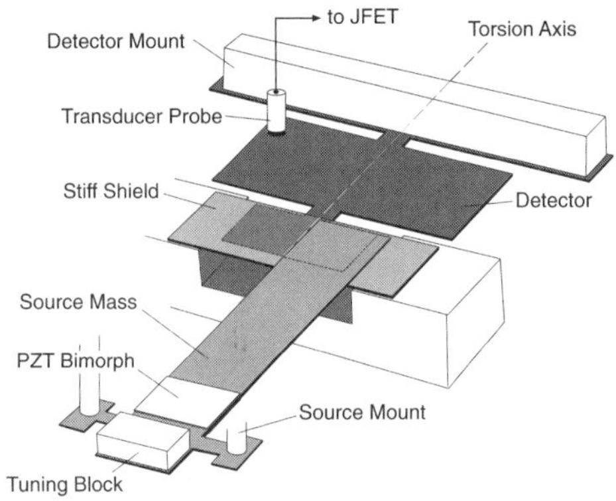

Figure 2. Central components of Indiana short–range experiment, to scale (width of detector mount is approximately 2.5 cm). Test masses consist of 250 μm thick planar tungsten oscillators, a geometry which concentrates as much mass as possible at the scale of interest. It is nominally null with respect to $1/r^2$ forces, which is effective in suppressing the Newtonian background relative to new short–range effects. A stiff conducting shield is placed between the test masses to suppress electrostatic and acoustic coupling.

2.1. *Search for Lorentz Violation in the 2002 Data Set*

Recently, a new analysis was begun to look for evidence of the violation of Lorentz symmetry, based on the data set used to extract the Yukawa limits published in 2003. Lorentz violation would manifest itself in the Indiana short–range experiment as a sidereal modulation of the force signal.

Figure 3 is a first look at the 2002 data set as a function of time. While it may hardly be suggestive of any periodicity in the data set, a sidereal fit could yield useful limits on Lorentz violating effects.

Derivation of a fitting function, based on the most general Lorentz violating force compatible with the Standard Model and the known behavior of gravity, has been made possible by the recent extension of the Standard-Model Extension (SME) into the gravitational sector.[14] The fitting function has the form

$$F = C_1 + C_2 \sin(\omega_\oplus T_\oplus) + C_3 \cos(\omega_\oplus T_\oplus) + C_4 \sin(2\omega_\oplus T_\oplus) \\ + C_5 \cos(2\omega_\oplus T_\oplus), \tag{2}$$

where the C_i are linear combinations of the coefficients $\bar{s}^{JK}$ controlling Lorentz violation in the SME gravity sector, $\omega_\oplus$ is the sidereal angular frequency of the Earth, and $T_\oplus$ is the time as measured in the Sun–centered

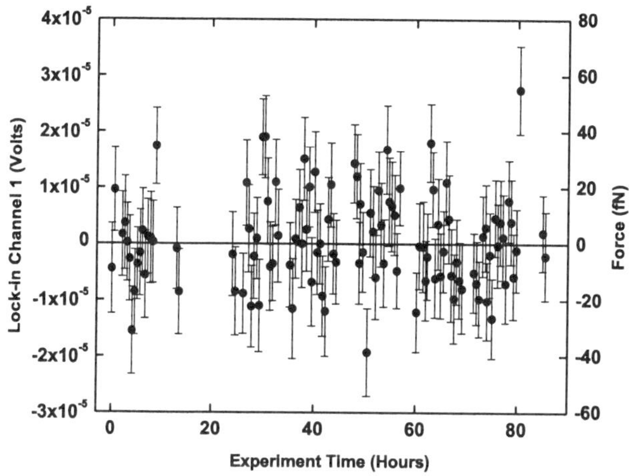

Figure 3. Data from the 2002 run of the Indiana short–range experiment as a function of time. Force data were collected in 12 minute sets at a 1 Hz rate, with comparable intervals in between each set during which diagnostic data were taken to monitor the experiment for gain and frequency drifts. Each data point represents the mean of a 12 minute set, where the error bar is the 1σ standard deviation of the mean.

celestial equatorial frame.[15]

The SME expression for the gravitational potential between two test masses m_1 and m_2 is given by:

$$V = -\frac{Gm_1m_2}{|\vec{x}_1 - \vec{x}_2|}\left(1 + \frac{1}{2}\hat{x}^{\hat{j}}\hat{x}^{\hat{k}}\bar{s}^{\hat{j}\hat{k}}\right), \tag{3}$$

where $\vec{x}_1 - \vec{x}_2$ is the vector separating m_1 and m_2, $\hat{x}^{\hat{i}}$ is the projection of the unit vector along $\vec{x}_1 - \vec{x}_2$ in the ith direction, and $\bar{s}^{\hat{j}\hat{k}}$ is a set of 9 dimensionless coefficients of Lorentz violation in the standard laboratory frame.[16] As pointed out in Ref. 16, Lorentz violation leads to a misalignment of the force associated with Eq. (3) relative to the vector $\vec{x}_1 - \vec{x}_2$, however, the inverse–square behavior is preserved. Therefore it is expected that the sensitivity of the Indiana experiment to the $\bar{s}^{\hat{j}\hat{k}}$ will be poor (at least relative to the anticipated sensitivity of dedicated torsion pendulum experiments, which the authors of Ref. 16 estimate at $10^{-4} - 10^{-15}$). However, it is possible that the Indiana experiment may be able to set fairly long–standing, if weak, limits on unique combinations of the $\bar{s}^{\hat{j}\hat{k}}$ due to the specific geometry.

The force corresponding to Eq. (3) has been integrated over the experimental test mass geometry using a Monte Carlo technique. The next step, in progress as of this writing, is to transform the results into the Sun-centered celestial equatorial frame to calculate expressions for the C_i in Eq. (2) in terms of the $\bar{s}^{JK}$. In the meantime, the approximate sensitivity of the experiment to the $\bar{s}^{\hat{j}\hat{k}}$ can be estimated from the laboratory frame calculation.

The result of the calculation can be expressed as the sum of terms each consisting of a numerical term in Newtons times an individual coefficient of Lorentz violation:

$$\begin{aligned} F_{LV} = &(-8.7 \times 10^{-16}\mathrm{N})\bar{s}^{11} + (-8.7 \times 10^{-16}\mathrm{N})\bar{s}^{22} \\ &+(9.4 \times 10^{-16}\mathrm{N})\bar{s}^{33} + (-4.2 \times 10^{-16}\mathrm{N})\bar{s}^{23}. \end{aligned} \tag{4}$$

The value of each numerical term is, unsurprisingly, approximately equal to the Fourier amplitude of the (unmodified) Newtonian force on the detector mass at the resonant frequency. The numerical terms proportional to $\bar{s}^{12}$ and $\bar{s}^{13}$ are about 1-2 orders of magnitude smaller than the terms in Eq. (4), and appear to be very sensitive to various run control parameters in the integration. The Monte Carlo is being further optimized to make these results more robust.

Sensitivity to the $\bar{s}^{\hat{j}\hat{k}}$ can be estimated by comparison of the Lorentz violating signal to the limiting thermal noise. The thermal noise force is found from the mechanical Nyquist theorem to be:

$$F_T = \sqrt{\frac{4kT}{\tau}\left(\frac{m\omega}{Q}\right)}, \tag{5}$$

where m is detector mass, ω is the resonant frequency, Q is the detector quality factor, T is the temperature and τ is the measurement integration time. For typical Q values (2.5×10^4), a temperature of 300 K, and an integration time of 720 s (equal to the averaging time of a single point in Fig. 3), this force is about 1×10^{-14} N.

Keeping only the diagonal terms in Eq. (4), setting the ratio $SNR = |F_{LV}/F_T|$ to unity results in the surfaces shown in Fig. 4. The basic structure of the surfaces in Fig. 4 is a consequence of the signs of the terms in Eq. (4), and indicative in part of the combination of the $\bar{s}^{\hat{j}\hat{k}}$ to which the Indiana short–range experiment will be sensitive. The $\bar{s}^{11}$ and $\bar{s}^{22}$ terms are of the same sign, and opposite in sign to the $\bar{s}^{33}$ term. In the case when the signs of the $\bar{s}^{11}$ and $\bar{s}^{22}$ coefficients themselves are opposite, all of the signal is in the $\bar{s}^{33}$ term, the limits on which might be constrained to order unity or better.

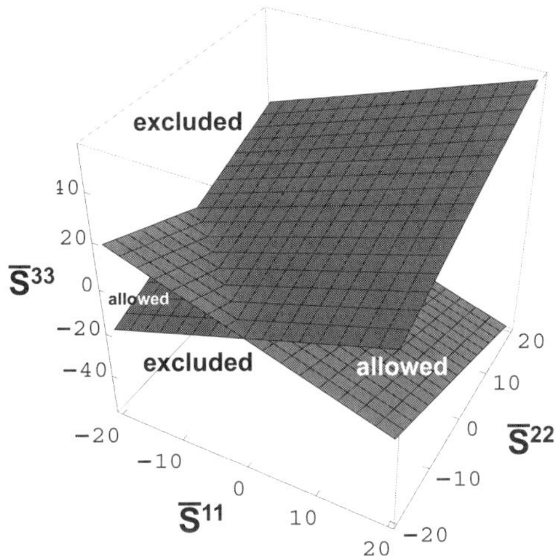

Figure 4. Relationship between diagonal elements of $\bar{s}^{\hat{j}\hat{k}}$ matrix in the case where $|F_{LV}/F_T| = 1$. Allowed and excluded values of the $\bar{s}^{\hat{j}\hat{j}}$ are shown assuming no evidence of sidereal variation in the force signal.

References

1. J. C. Long, J. C. Price, C. R. Physique 4, 337 (2003).
2. J.C. Long, in J. Hewett *et al.*, eds., *Proceedings of the SLAC Summer Institute on Particle Physics*, Menlo Park, CA, 2005.
3. J. H. Hoskins *et al.*, Phys. Rev. D **32**, 3084 (1985).
4. D. Kapner *et al.*, Phys. Rev. Lett. **98**, 021101 (2007).
5. S. K. Lamoreaux, Phys. Rev. Lett. **78**, 5 (1997).
6. S. J. Smullin *et al.*, Phys. Rev. D **72**, 122001 (2005).
7. J. C. Long *et al.*, Nature **421**, 922 (2003).
8. N. Arkani-Hamed, S. Dimopoulos, G. Dvali, Phys. Lett. B **429**, 263 (1998); A. Kehagias, K. Sfetsos, Phys. Lett. B **472**, 39 (2000).
9. S. Dimopoulos, A. Geraci, Phys. Rev. D **68**, 124021 (2003).
10. I. Antoniadis, S. Dimopoulos, G. Dvali, Nucl. Phys. B **516**, 70 (1998).
11. D. B. Kaplan, M. B. Wise, J. High Energy Phys. **0008**, 037 (2000).
12. R. Sundrum, J. High Energy Phys. **7**, 1 (1999); S. R. Beane, Gen. Rel. Grav. **29**, 945 (1997).
13. J. E. Moody, F. Wilczek, Phys. Rev. D **30**, 130 (1984); L. J. Rosenberg, K. A. Van Bibber, Phys. Rep. **325**, 1 (2000).
14. V. A. Kostelecký, Phys. Rev. D **69**, 105009 (2004).
15. V. A. Kostelecký and M. Mewes, Phys. Rev. D **66**, 056005 (2002).
16. Q. G. Bailey and V. A. Kostelecký, Phys. Rev. D **74**, 045001 (2006).

LORENTZ VIOLATION AND TORSION

NEIL RUSSELL
Physics Department, Northern Michigan University,
Marquette, MI 49855, U.S.A.

Similarities between the structure of theories with Lorentz violation and theories with constant torsion in flat spacetime are exploited to place bounds on torsion components. An example is given showing the analysis leading to bounds on the axial-vector and mixed-symmetry components of torsion, based on a dual-maser experiment.

1. Introduction

In conventional Riemann-Cartan theory, torsion is minimally coupled to a fermion. Several nonminimal generalizations are possible, including nonminimal couplings to a fermion, nonminimal couplings involving a single particle of another species (e.g., the photon), and couplings involving more than one particle. We limit consideration to signals of torsion effects arising from nonminimal couplings of the torsion tensor to one or more fermions only. This is ongoing work and a more complete analysis is available elsewhere.[1]

We focus on the analysis of effects relevant for laboratory experiments that can be, or have been, done. For simplicity, torsion is taken as a fixed background field in a Sun-centered inertial frame. To consider fluctuations, the Nambu-Goldstone and massive sectors[2] have to be incorporated. Other approaches, which include ones seeking sensitivity to effects in dynamical torsion theories,[3,4] are not considered here. The literature on torsion includes reviews by Hehl *et al.*,[5] Shapiro,[6] and Hammond.[7] Possible bounds from Hughes-Drever experiments have been discussed by Lämmerzahl.[8]

Studies have shown that Lorentz violation and the associated CPT violation can arise in string theory.[9,10] However, it is possible to describe all the effects at the level of effective field theory.[11] A systematic framework that encompasses global Lorentz violation[12] and local Lorentz violation[13] exists, known as the Standard-Model Extension, or SME.

The similarities between lagrangian terms coupling fermions to torsion and ones coupling fermions to Lorentz-violating backgrounds means that experimental sensitivity to Lorentz-violation effects cannot easily be decoupled from sensitivity to torsion effects. Here, we assume no Lorentz violation, and interpret experimental results entirely in terms of torsion.

2. Basics

We adopt the conventions of Ref. 13. The general metric $g_{\mu\nu}$ has diagonal entries $(-1,1,1,1)$ in the flat-space limit, and the antisymmetric tensor $\epsilon^{\mu\nu\alpha\beta}$ is defined so that $\epsilon^{0123}=-1$.

The Riemann-Cartan curvature tensor, denoted by $R_{\mu\nu\alpha\beta}$, consists of the usual Riemann curvature tensor $\widetilde{R}_{\mu\nu\alpha\beta}$ and added terms involving the contortion. We are interested in the limit of spacetime with diagonal metric $(-1,1,1,1)$, in which case the Christoffel symbols are zero and the usual curvature tensor $\widetilde{R}_{\mu\nu\alpha\beta}$ vanishes. We refer to this as 'flat spacetime.' In flat spacetime, the Riemann-Cartan curvature tensor $R_{\mu\nu\alpha\beta}$ does not necessarily vanish.

The torsion tensor $T^{\alpha}{}_{\mu\nu}$ is antisymmetric in the second and third indices, and so has 24 independent components. We define the trace part T^{μ} and the antisymmetric part A^{μ} of the torsion tensor as follows: $T_{\mu}\equiv g^{\alpha\beta}T_{\alpha\beta\mu}$ and $A^{\nu}\equiv\frac{1}{6}\epsilon^{\alpha\beta\mu\nu}T_{\alpha\beta\mu}$. The torsion tensor can be decomposed into irreducible components

$$T_{\alpha\mu\nu}=\frac{1}{3}(g_{\mu\alpha}T_{\beta}-g_{\mu\beta}T_{\alpha})-\epsilon_{\alpha\beta\mu\nu}A^{\nu}+M_{\mu\alpha\beta}\,, \tag{1}$$

where $M_{\mu\alpha\beta}$ is unique and is called the mixed-symmetry component.

3. Fermions in flat spacetime with torsion

The lagrangian for an electron of mass m in flat spacetime with all possible independent torsion couplings up to dimension five is:[1,13]

$$\begin{aligned}\mathcal{L}^{T,5}\approx{}&\tfrac{1}{2}i\bar\psi\gamma^{\mu}\overleftrightarrow{\partial_{\mu}}\psi-m\bar\psi\psi\\&+\xi_1^{(4)}T_{\mu}\bar\psi\gamma^{\mu}\psi+\xi_2^{(4)}T_{\mu}\bar\psi\gamma_5\gamma^{\mu}\psi+\xi_3^{(4)}A_{\mu}\bar\psi\gamma^{\mu}\psi+\xi_4^{(4)}A_{\mu}\bar\psi\gamma_5\gamma^{\mu}\psi\\&+\tfrac{1}{2}i\xi_1^{(5)}T^{\mu}\bar\psi\overleftrightarrow{\partial_{\mu}}\psi+\tfrac{1}{2}\xi_2^{(5)}T^{\mu}\bar\psi\gamma_5\overleftrightarrow{\partial_{\mu}}\psi+\tfrac{1}{2}i\xi_3^{(5)}A^{\mu}\bar\psi\overleftrightarrow{\partial_{\mu}}\psi\\&+\tfrac{1}{2}\xi_4^{(5)}A^{\mu}\bar\psi\gamma_5\overleftrightarrow{\partial_{\mu}}\psi+\tfrac{1}{2}i\xi_5^{(5)}M^{\lambda}{}_{\mu\nu}\bar\psi\overleftrightarrow{\partial_{\lambda}}\sigma^{\mu\nu}\psi+\tfrac{1}{2}i\xi_6^{(5)}T_{\mu}\bar\psi\overleftrightarrow{\partial_{\nu}}\sigma^{\mu\nu}\psi\\&+\tfrac{1}{2}i\xi_7^{(5)}A_{\mu}\bar\psi\overleftrightarrow{\partial_{\nu}}\sigma^{\mu\nu}\psi+\tfrac{1}{2}i\xi_8^{(5)}\epsilon^{\lambda\kappa\mu\nu}T_{\lambda}\bar\psi\overleftrightarrow{\partial_{\kappa}}\sigma_{\mu\nu}\psi\\&+\tfrac{1}{2}i\xi_9^{(5)}\epsilon^{\lambda\kappa\mu\nu}A_{\lambda}\bar\psi\overleftrightarrow{\partial_{\kappa}}\sigma_{\mu\nu}\psi\,.\end{aligned} \tag{2}$$

This expression includes four coupling constants with dimension m^0: $\xi_1^{(4)}$, $\ldots$, $\xi_4^{(4)}$, and nine with dimension m^{-1}: $\xi_1^{(5)}$, $\ldots$, $\xi_9^{(5)}$. These terms can be arranged so as to match the Minkowski-spacetime limit of the fermion-sector Lorentz-violating lagrangian as given in Ref. 13. In making this match, we assume the torsion components are constants, as are the SME coefficients. Using Eqs. (12) to (14) of Ref. 13, and assuming zero torsion and no electromagnetic field, we have:

$$\begin{aligned}\mathcal{L}^{SME} &= \tfrac{1}{2} i\bar{\psi}\gamma^\mu \overleftrightarrow{\partial_\mu} \psi - m\bar{\psi}\psi \\ &\quad -a_\mu\bar{\psi}\gamma^\mu\psi - b_\mu\bar{\psi}\gamma_5\gamma^\mu\psi \\ &\quad -\tfrac{1}{2}H_{\mu\nu}\bar{\psi}\sigma^{\mu\nu}\psi - \tfrac{1}{2}ic_{\mu\nu}\bar{\psi}\gamma^\mu \overleftrightarrow{\partial^\nu} \psi - \tfrac{1}{2}id_{\mu\nu}\bar{\psi}\gamma_5\gamma^\mu \overleftrightarrow{\partial^\nu} \psi \\ &\quad -\tfrac{1}{2}ie_\mu\bar{\psi} \overleftrightarrow{\partial^\mu} \psi + \tfrac{1}{2}f_\mu\bar{\psi}\gamma_5 \overleftrightarrow{\partial^\mu} \psi \\ &\quad -\tfrac{1}{4}ig_{\mu\nu\lambda}\bar{\psi}\sigma^{\mu\nu} \overleftrightarrow{\partial^\lambda} \psi . \end{aligned} \tag{3}$$

The SME coefficient $g_{\lambda\mu\nu}$ appearing in the last term can be decomposed in the same way as in Eq. (1):

$$g_{\mu\nu}{}^\lambda = \tfrac{1}{3}(g_\mu^{(T)}\delta_\nu^\lambda - g_\nu^{(T)}\delta_\mu^\lambda) - \epsilon_{\mu\nu}{}^{\lambda\kappa}g_\kappa^{(A)} + g_{\mu\nu}^{(M)\lambda} , \tag{4}$$

where $g_\mu^{(T)}$, $g_\mu^{(A)}$ and $g_{\lambda\mu\nu}^{(M)}$ are suitably-defined trace, axial-vector, and mixed-symmetry components. If we make this substitution and match Eqs. (2) and (3) term by term, a number of identities result, including, for example:

$$b_\mu = -\xi_2^{(4)}T_\mu - \xi_4^{(4)}A_\mu , \tag{5}$$

$$g_\kappa^{(A)} = 2(\xi_8^{(5)}T_\kappa + \xi_9^{(5)}A_\kappa) , \tag{6}$$

$$g_{\mu\nu\lambda}^{(M)} = 2\xi_5^{(5)}M_{\lambda\mu\nu} . \tag{7}$$

A variety of experiments are sensitive to Lorentz-violation coefficients,[14] including b_μ and $g_{\lambda\mu\nu}$, and these equations show they must also be sensitive to torsion effects.

4. Connecting with experiments

Experimental sensitivities in the case of ordinary matter are to 40 tilde coefficients, defined in Appendix B of Ref. 15. As an example, we'll consider two of these, $\tilde{b}_X$ and $\tilde{g}_{DX}$. Using the decomposition in Eq. (4), they can be expressed in terms of the irreducible components of $g_{\lambda\mu\nu}$:

$$\tilde{b}_X = b_X - mg_X^{(A)} + mg_{YZT}^{(M)} , \tag{8}$$

$$\tilde{g}_{DX} = -b_X + mg_X^{(A)} + 2mg_{YZT}^{(M)} . \tag{9}$$

If we now use Eqs. (5), (6), and (7), we obtain relationships between experimental tilde observables in the SME and irreducible components of the torsion tensor:

$$\tilde{b}_X = -(\xi_2^{(4)} + 2m\xi_8^{(5)})T_X - (\xi_4^{(4)} + 2m\xi_9^{(5)})A_X + 2m\xi_5^{(5)} M_{TYZ}\,, \quad (10)$$

$$\tilde{g}_{DX} = +(\xi_2^{(4)} + 2m\xi_8^{(5)})T_X + (\xi_4^{(4)} + 2m\xi_9^{(5)})A_X + 4m\xi_5^{(5)} M_{TYZ}\,. \quad (11)$$

It follows that any experiment with sensitivity to $\tilde{b}_X$ or $\tilde{g}_{DX}$ is also sensitive to the torsion components T_X, A_X, and M_{TYZ}. To investigate the expected sensitivity that can be expected, we next consider a specific experiment.

5. Dual-maser experiment

As an example, consider the result for a combination of coefficients[16] given in the second line of Table II of Ref. 17, which reports on a recent He-Xe dual-maser experiment:

$$-\tilde{b}_X + 0.0034\tilde{d}_X - 0.0034\tilde{g}_{DX} < (2.2 \pm 7.9) \times 10^{-32}\ \text{GeV}\,. \quad (12)$$

We note by inspection of Eqs. (2) and (3) that the $\tilde{d}_X$ coefficient is not relevant for torsion bounds, and may therefore be taken as zero. To introduce the experimental result, we substitute Eqs. (10) and (11), with the mass taken as that for the neutron, $m_n \simeq 0.938$ GeV. This gives one particular torsion combination bounded by this experiment. Some simplification occurs because the substitution involves combinations of the same torsion terms, and so we may neglect the third term since the factor of 0.0034 is small. We extract a bound with a confidence level of about 90% by doubling the one-sigma uncertainty:

$$|(\xi_2^{(4)} + 2m_n\xi_8^{(5)})T_X + (\xi_4^{(4)} + 2m_n\xi_9^{(5)})A_X \ -2m_n\xi_5^{(5)} M_{TYZ}| < 1.6 \times 10^{-31}\ \text{GeV}\,. \quad (13)$$

In the case of minimally-coupled torsion, we take $\xi_4^{(4)} = 1$ and assume all the others are zero. Then, Eq. (13) yields the result

$$|A_X| < 10^{-31}\ \text{GeV}\,. \quad (14)$$

To extract additional torsion results, we look at each term in Eq. (13) under the assumption that the other terms vanish. We find, for example, that

$$|\xi_5^{(5)} M_{TYZ}| < 10^{-31}\,. \quad (15)$$

Additional bounds on components of the torsion tensor can be extracted by considering the other terms in Eq. (13), and by considering the other bounds reported in this particular experiment. More generally, the method adopted here can be used to seek torsion bounds from other experiments.

References

1. V.A. Kostelecký and N. Russell, in preparation.
2. R. Bluhm and V.A. Kostelecký, Phys. Rev. D **71**, 065008 (2005); B. Altschul and V.A. Kostelecký, Phys. Lett. B **628**, 106 (2005); V.A. Kostelecký and R. Potting, Gen. Rel. Grav. **37**, 1675 (2005); R. Bluhm *et al.*, in preparation.
3. S.M. Carroll and G.B. Field, Phys. Rev. **D 50**, 3867 (1994).
4. A.S. Belyaev, I.L. Shapiro, and M.A.B. do Vale, Phys. Rev. **D 75**, 034014 (2007).
5. F.W. Hehl *et al.*, Rev. Mod. Phys. **48**, 393 (1974).
6. I.L. Shapiro, Phys. Rep. **357**, 113 (2002).
7. R.T. Hammond, Rep. Prog. Phys. **65** 599 (2002).
8. C. Lämmerzahl, Phys. Lett. A **228**, 223 (1997).
9. V.A. Kostelecký and S. Samuel, Phys. Rev. D **39**, 683 (1989); Phys. Rev. D **40**, 1886 (1989); Phys. Rev. Lett. **63**, 224 (1989).
10. V.A. Kostelecký and R. Potting, Nucl. Phys. B **359**, 545 (1991).
11. V.A. Kostelecký and R. Potting, Phys. Rev. D **51**, 3923 (1995).
12. D. Colladay and V.A. Kostelecký, Phys. Rev. D **55**, 6760 (1997); Phys. Rev. D **58**, 116002 (1998); V.A. Kostelecký and R. Lehnert, Phys. Rev. D **63**, 065008 (2001).
13. V.A. Kostelecký, Phys. Rev. D **69**, 105009 (2004).
14. For a tabulation, see V.A. Kostelecký and N. Russell, these proceedings.
15. R. Bluhm *et al.*, Phys. Rev. D **68**, 125008 (2003).
16. V.A. Kostelecký and C.D. Lane, Phys. Rev. D **60**, 116010 (1999).
17. F. Canè *et al.*, Phys. Rev. Lett. **93**, 230801 (2004).

PROBING LORENTZ INVARIANCE USING COHERENT OPTICAL PHENOMENA

J. P. COTTER, M. P. HILL AND B. T. H. VARCOE

Department of Physics and Astronomy
The University of Leeds, Leeds, U.K.
E-mail: joe.cotter@quantuminfo.org

We have demonstrated an experimental method for detecting extremely small frequency shifts which can be used to access the least well known of the photon sector parameters κ_{tr}. Using an interferometric coherent double resonance a spectral feature of $\sim$ 50Hz in width can be generated in an Ives-Stilwell like apparatus. The feature is robust against Doppler broadening and can be observed in a vapour cell.

Ives-Stilwell experiments are leading order sensitive to κ_{tr}, the least well known of the photon sector parameters.[1,2] In a typical experiment a fast beam of atoms, or ions, is used to measure the Doppler shifted frequencies of two optical fields counter propagating co-linearly with the beam. It has recently been proposed that the dramatically reduced effective linewidths which result from coherent interactions could be used improve the sensitivity of experiments of this type.[4]

An intense, near resonant laser can modify the internal structure of an ensemble of atoms, generating pairs of dressed states. The transition amplitudes of a perturbatively weak probing field between these unpopulated dressed states and a populated ground state will then encounter interference between the two excitation pathways of this doublet. For near resonant fields this restricts the absorption of probe photons, trapping the ensemble in a dark state. The spectral feature which results is critically dependent on the detunings of the coupling and probing fields from their respective transitions, making them extremely sensitive to small frequency shifts. Counter propagating probe and coupling fields can be used to form an Ives-Stilwell experiment where the absorption of probe field photons can be related to its detuning from resonance:[5]

$$\alpha = \Im\left[\chi^{(1)}\right] \propto \left[\frac{8\delta^2 + 2\gamma_{31} + 2\gamma_{21}(|\Omega_c|^2 + \gamma_{21}\gamma_{31})}{||\Omega_c|^2 + (\gamma_{31} + 2i\Delta_1)(\gamma_{21} + 2i\delta)|^2}\right]. \qquad (1)$$

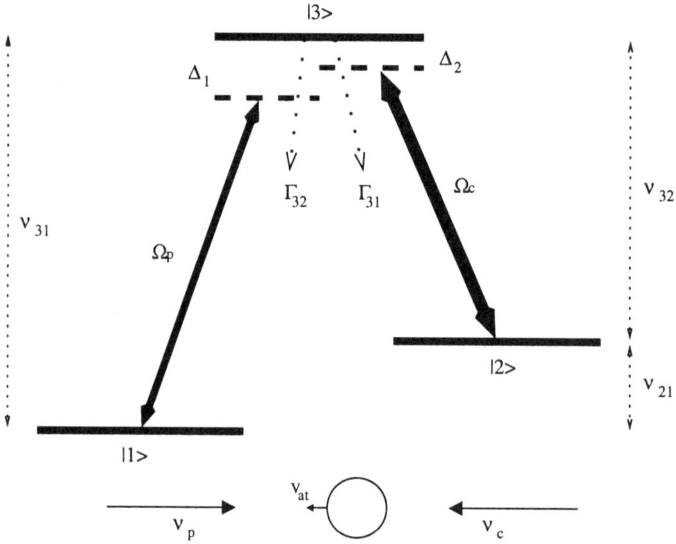

Figure 1. Three atomic levels in a Λ configuration are coupled using one strong and one perturbatively weak probing fields counter propagating with respect to one another with Rabi frequencies Ω_c and Ω_p respectively. The coupling and probing fields also form the parallel and anti-parallel fields of an Ives-Stilwell experiment with frequencies ν_c and ν_p in the atomic frame.

A folded Mach-Zender interferometer,[6] which is inherently resistant to vibrational noise, provides an ideal apparatus for this kind of experiment. Both counter propagating fields are derived from the same laser with rest frame frequency ν_0. For $\nu_0 = (\nu_{31} + \nu_{32})/2$ there exists a critical atomic velocity class such that the atoms see both counter-propagating fields Doppler shifted into resonance with their respective transitions.[3] In a vapour cell there are two atomic velocity classes which satisfy these resonance conditions. Frame dependencies in the frequencies observed by the different velocity classes alters their resonance condition. This causes a splitting between the absorption spectra of the two systems which is sensitive to κ_{tr}, also providing additional common mode noise rejection. If $|1\rangle$, $|2\rangle$ and $|3\rangle$ are hyperfine states then an external magnetic field can be used to scan across the resonance, removing the need to dither the laser. Figure 2 shows a coherent double resonance generated using a diode laser locked to a frequency $\nu_0 = (\nu_{31} + \nu_{32})/2$ in the D1 line of ^{85}Rb using an optical frequency comb.[7] An external magnetic field was applied using a precision solenoid.

Both the cell and solenoid were encased in four layers of μ-metal to reduce background fields. The resonance has a width of 20nT, or $\sim$ 50Hz FWHM.

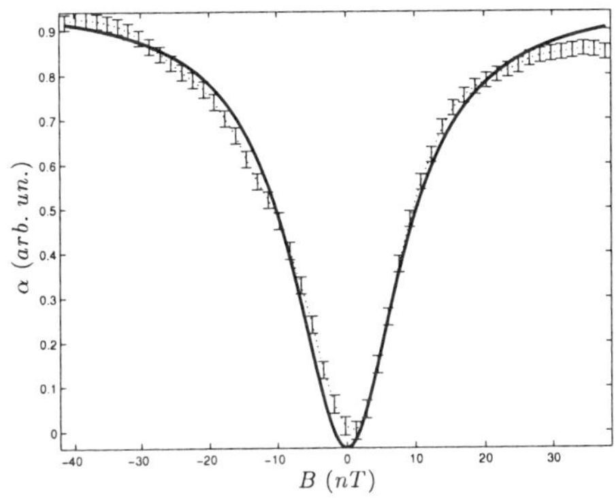

Figure 2. Experimental observation of coherent resonance in an effective beam of ^{85}Rb atoms with a velocity of 1200 m/s. The solid line represents the best fit of Eq. (1) to the data. The coupling field had an intensity of $64\mu W$ and a $1/e^2$ beam radius of 3.31mm. The width of the resonance is 20nT FWHM, which corresponds to an effective frequency width of $\sim$ 50Hz.

This corresponds to a sensitivity of $\delta\nu/\nu_0 = 1 \times 10^{-14}$ which translates to a sensitivity to κ_{tr} on the order of a part in 1×10^7 via sidereal modulations alone.

References

1. V. A. Kostelecký and M. Mewes, Phys. Rev. D. **66**, 056005 (2002).
2. M. E. Tobar *et al.*, Phys. Rev. D. **71**, 025004 (2005); M. Hohensee *et al.*, Phys. Rev. D. **75**, 049902(E) (2007).
3. J. P. Cotter, M. P. Hill and B. T. H. Varcoe, to be submitted (2007).
4. B. T. H. Varcoe, Cont. Phys. **47**, 25 (2006).
5. M. Fleischauer *et al.*, Rev. Mod. Phys. **77**, 633 (2005).
6. G. Jundt *et al.*, Euro. Phys. D. **27**, 273 (2003).
7. R. Holzwart *et al.*, Phys. Rev. Lett. **85**, 2264 (2000).

CONSTRAINT ANALYSIS OF BUMBLEBEE MODELS

N. GAGNE[1] AND A. VRUBLEVSKIS[1,2]
[1] *Physics Department, Colby College, Waterville, ME 04901, U.S.A.*
[2] *Physics Department, M.I.T., Cambridge, MA 02139, U.S.A.*

Dirac's Hamiltonian constraint analysis is used to study vector theories with spontaneous Lorentz violation known as bumblebee models. In certain of these models, the Nambu-Goldstone sector has properties similar to those of photons in classical electromagnetism. An analysis of the different types of constraints and resulting number of degrees of freedom is presented here for models with different kinetic and potential terms, and the results are compared with electromagnetism.

1. Introduction

Bumblebee models are vector field theories in which spontaneous Lorentz violation is dynamically induced by a potential term in the action.[1,2] They are of interest as potential alternative theories of gravity and as models illustrating features of spontaneous Lorentz breaking, such as the appearance of Nambu-Goldstone (NG) and massive (Higgs) modes.[3] In flat spacetime, certain bumblebee models have solutions that are equivalent to those describing photons in classical electromagnetism. In these models, the photons arise as massless NG modes.

The goal of this work is to use Dirac's constraint analysis[4] to study the extent of the equivalence of bumblebee models with classical electromagnetism. This approach is useful since many bumblebee models as well as electromagnetism have singular Lagrangians. By analyzing the constraint structure, the number of physical degrees of freedom and their behavior can be determined and compared with the case of the photon.

2. Bumblebee Models and Electromagnetism

Bumblebee models can be defined in different geometries and with a variety of kinetic and potential terms. In this analysis, the models are restricted to Minkowski spacetime, and different examples of kinetic and potential terms are used to study the types of constraint structures that can arise.

The Lagrangian is defined to have the form $\mathcal{L} = \mathcal{L}_0 - V(B_\mu B^\mu \pm b^2)$. The kinetic terms $\mathcal{L}_0$ are chosen to have derivatives up to second-order, with a general form parameterized as

$$\mathcal{L}_0 = -\frac{1}{4}\tau_1\, B_{\mu\nu}B^{\mu\nu} + \frac{1}{2}\tau_2\, \partial_\mu B_\nu \partial^\mu B^\nu + \frac{1}{2}\tau_3\, \partial_\mu B^\mu \partial_\nu B^\nu, \tag{1}$$

where $B_{\mu\nu} = \partial_\mu B_\nu - \partial_\nu B_\mu$. This form is equivalent to the flat-spacetime limit of the kinetic terms in Will-Nordvedt vector-tensor theories.[5]

The potential is chosen to have the functional form $V(B_\mu B^\mu \pm b^2)$ considered by Kostelecký and Samuel.[1] Two specific examples are considered. The first introduces a Lagrange-multiplier field λ and has a linear form, $V = \lambda(B_\mu B^\mu \pm b^2)$. The Lagrange-multipler imposes the constraint $B_\mu B^\mu \pm b^2 = 0$. The second is a smooth quadratic potential, $V = \frac{1}{2}\kappa(B_\mu B^\mu \pm b^2)^2$, where κ is a constant. It has a minimum when $B_\mu B^\mu \pm b^2 = 0$. In both cases, this condition is satisfied when the vector field has a nonzero vacuum value $\langle B_\mu \rangle - b_\mu$.

Notice that the Lagrangian for classical electromagnetism can be obtained from $\mathcal{L}$ by setting the potential V to zero and choosing $\tau_1 = 1$, $\tau_2 = \tau_3 = 0$. The kinetic term then has the standard Maxwell form, $-\frac{1}{4}B_{\mu\nu}B^{\mu\nu}$, where $B_{\mu\nu}$ is the field strength. Conventional notation in this case would be $B_\mu = A_\mu$ and $B_{\mu\nu} = F_{\mu\nu}$, describing a photon field.

In total, four bumblebee models comprised of different choices for the kinetic and potential terms are analyzed. The fields entering in these models along with those in electromagnetism are summarized in Table 1.

Table 1. Models, field content, and total number of fields.

Model	Kinetic Term	Potential V	Fields	Fields
EM	$-\frac{1}{4}F_{\mu\nu}F^{\mu\nu}$		A_μ, Π^μ	8
BB	$\tau_1 \neq 0$ and	$\lambda(B_\mu B^\mu \pm b^2)$	B_μ, Π^μ, λ, Π^λ	10
	$\tau_2 + \tau_3 \neq 0$	$\frac{1}{2}\kappa(B_\mu B^\mu \pm b^2)^2$	B_μ, Π^μ	8
BB	$-\frac{1}{4}B_{\mu\nu}B^{\mu\nu}$	$\lambda(B_\mu B^\mu \pm b^2)$	B_μ, Π^μ, λ, Π^λ	10
		$\frac{1}{2}\kappa(B_\mu B^\mu \pm b^2)^2$	B_μ, Π^μ	8

3. Constraint Analysis

Dirac's constraint analysis is useful in identifying the physical degrees of freedom in a theory. Each constraint removes one degree of freedom. In addition, Dirac showed that every primary first-class constraint is associated with an unphysical degree of freedom, such as a gauge freedom. He conjectured that this is true as well for secondary first-class constraints. From this a counting rule for degrees of freedom follows. In a theory with n total field and conjugate-momentum components, if there are n_1 first-class constraints and n_2 second-class constraints, there will be a total of $(n - 2n_1 - n_2)$ physical degrees of freedom. Table 2 shows the result of this counting argument applied to the four bumblebee models and electromagnetism.

Table 2. Number of primary (1°), secondary (2°), first-class (FC), and second-class (SC) constraints and independent degrees of freedom (DF).

Model	Kinetic Term	Potential V	1°	2°	FC	SC	DF
EM	$-\frac{1}{4}F_{\mu\nu}F^{\mu\nu}$	–	1	1	2	0	4
BB	$\tau_1 \neq 0$ and	$\lambda(B_\mu B^\mu \pm b^2)$	1	3	0	4	6
	$\tau_2 + \tau_3 \neq 0$	$\frac{1}{2}\kappa(B_\mu B^\mu \pm b^2)^2$	0	0	0	0	8
BB	$-\frac{1}{4}B_{\mu\nu}B^{\mu\nu}$	$\lambda(B_\mu B^\mu \pm b^2)$	2	2	0	4	6
		$\frac{1}{2}\kappa(B_\mu B^\mu \pm b^2)^2$	1	1	0	2	6

This analysis reveals that the bumblebee models considered here have different constraint structures in comparison with each other and with electromagnetism. As a result, the total number of physical degrees of freedom depends on the choice of model. Each of the bumblebee models has one or more additional pairs of degrees of freedom beyond those of the two photon fields and their corresponding conjugate momenta. These are associated with massive modes and/or additional Nambu-Goldstone modes. The behavior of these extra modes can be studied using the canonical equations of motion obtained in terms of Poisson brackets. This analysis is currently being carried out, and the results will be presented in a future work.

Acknowledgments

This work was supported by Professor R. Bluhm's NSF Research at an Undergraduate Institution (RUI) grant PHY-0554663.

References

1. V.A. Kostelecký and S. Samuel, Phys. Rev. D **40**, 1886 (1989).
2. V.A. Kostelecký, Phys. Rev. D **69**, 105009 (2004).
3. R. Bluhm and V.A. Kostelecký, Phys. Rev. D **71**, 0065008 (2005); R. Bluhm, S.-H. Fung, and V.A. Kostelecký, to appear.
4. P.A.M. Dirac, *Lectures on Quantum Mechanics*, Yeshiva University, New York, 1964.
5. C.M. Will, *Theory and Experiment in Gravitational Physics*, Cambridge University Press, Cambridge, England, 1993.

IVES-STILWELL FOR THE NEW MILLENNIUM

MICHAEL A. HOHENSEE, DAVID F. PHILLIPS,
AND RONALD L. WALSWORTH
Harvard-Smithsonian Center for Astrophysics
Cambridge, MA 02138, U.S.A.

Ives-Stilwell measurements have been used to measure relativistic time dilation effects and place constraints upon directional and frame-dependent anisotropies of the speed of light c. We propose using coherent two-photon processes and coherent pulsed spectroscopic techniques to move beyond existing scaling limits.

The construction of the Standard-Model Extension has greatly aided the development of low energy experiments seeking to detect Lorentz violating physics by enabling direct comparison of a wide variety of effects.[1] In the photon sector of the SME, all but one of the Lorentz-violating coefficients have been constrained to be less (in many cases far less) than 10^{-11}.[1,2,3] The exception, $\tilde{\kappa}_{tr}$, has been limited to be no larger than the relatively loose bound of 2.2×10^{-7}.[4] Thus far, Ives-Stilwell experiments have offered the best measurements of $\tilde{\kappa}_{tr}$.[4]

Elements of the Ives-Stilwell experiment have seen significant improvement over the past seventy years,[4,5] but the core has remained unchanged. All Ives-Stilwell experiments use continuous wave spectroscopy to measure anomalous Doppler shifts. Low-velocity experiments (where the atomic velocity β_{at} is smaller than the lab velocity β_{lab}) have been limited by the minimum linewidths afforded by optical transitions, while high velocity experiments (where β_{at} exceeds β_{lab}) are approaching the fundamental and practical limits of such techniques. Here, we outline a means by which low velocity experiments ($\beta_{at} \leq \beta_{lab}$) can approach their theoretical peak sensitivities, and propose the use of pulsed techniques to allow both high and low velocity experiments to beat the single-particle Fourier transform limit.

The precision to which $\tilde{\kappa}_{tr}$ can be measured with cw-spectroscopy scales as $\delta\tilde{\kappa}_{tr} \propto \Gamma / \left(\beta_{at}^2 + 2\vec{\beta}_{at} \cdot \vec{\beta}_{lab}\right)$, where Γ is the measured linewidth. For all experiments, the precision $\delta\tilde{\kappa}_{tr}$ improves with increasing β_{at} until they

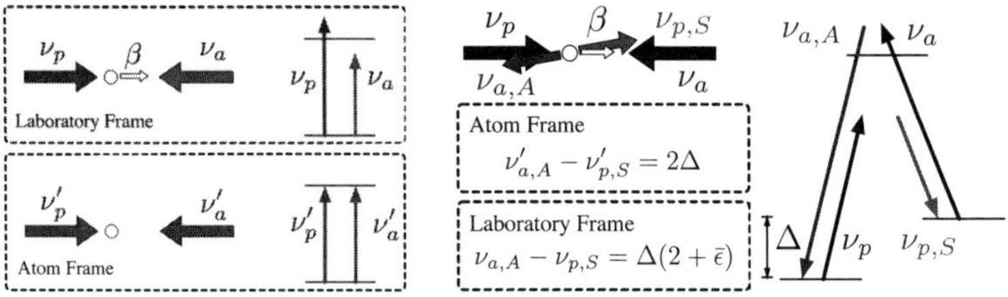

(a) Schematic of an Ives-Stilwell Experiment

(b) Ives-Stilwell Using Four Wave Mixing

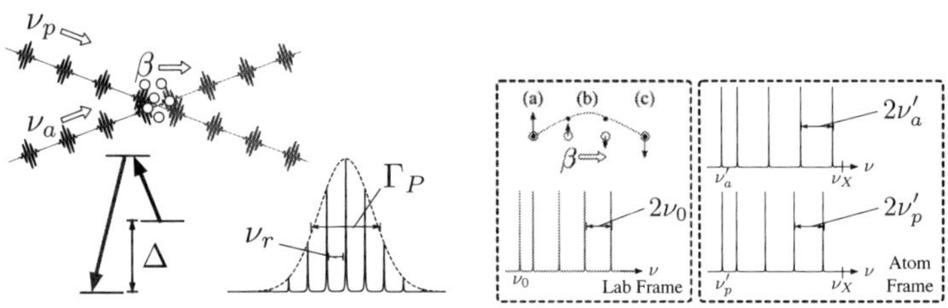

(c) Pulse Raman Comb Ives-Stilwell

(d) High Harmonic Ives-Stilwell

Figure 1. (a) In the lab frame, atoms are illuminated with light propagating parallel (with freq. ν_p) and anti-parallel (freq. ν_a) to their motion ($v = \beta c$). The frequencies of the applied fields are tuned to simultaneous resonance $\nu'_p = \nu'_a$ in the atoms' frame. (b) As in (a), counterpropagating fields with frequencies ν_p and ν_a are applied to moving atoms. The Λ-system generates an anti-Stokes field $\nu_{a,A}$ propagating anti-parallel, and a Stokes field $\nu_{p,S}$ propagating parallel to the atoms (angled for clarity). Generated fields are compared in the lab frame. (c) Atoms in a long-lived metastable state move at β_{at}, and interact with a frequency comb, generating a chain of anti-stokes Raman pulses. Anomalous Doppler shifts of the applied field from the lab frame to the boosted frame are preserved in the frequency of the generated field in the lab frame. (d) When intense (nearly ionizing) fields are applied, bound electrons can tunnel into the ionization band (d-a), acquire energy from the electric field (d-b), and recombine (d-c), releasing the gained energy as a photon at an odd-harmonic of the applied frequency. All odd-harmonics of the incident field are produced, up to a fixed "cutoff" energy, determined by the interaction of the free charge with the applied field. HHG from a frequency comb has recently been demonstrated.[8]

reach the interaction time limit $\Gamma \simeq \frac{\beta_{at}c}{L}$, where L is the length of the interaction region. For low-velocity experiments the sensitivity scales as $1/L$ for $\beta_{at} \geq \gamma L/c$ (with γ the natural or minimally broadened linewidth), while for high-velocity experiments the sensitivity scales as $1/\beta_{at}L$. Low-velocity experiments require smaller linewidths to reach their peak sensitivities. Both low and high velocity experiments are limited by the scaling with size of the interaction region.

Two-Photon Tests with Metastable Atoms. Smaller linewidths for low-velocity experiments can be found in two-photon transitions. In Figure 1, we outline an example using four-wave mixing. The generated fields' linewidth is set by the long-lived coherence between metastable states. The sensitivity to $\tilde{\kappa}_{tr}$, for $\nu_a = \nu_p$ is

$$\nu_{a,A} - \nu_{p,S} = \frac{2\Delta}{\sqrt{1-\beta_{at}^2}} + \Delta\left(2\beta_{at}^2 - 4\vec{\beta}_{at}\cdot\vec{\beta}_{lab}\right)\tilde{\kappa}_{tr}. \tag{1}$$

Pulsed Spectroscopy. For moving atoms, cw-spectroscopy is effectively single-pulse spectroscopy, and cannot yield spectra with features narrower than the inverse interaction time. This Fourier transform limit can be beaten by coherently combining the results of multiple pulsed interactions, as in frequency combs.[6] The bandwidth of the resulting spectrum is independent of the size of the interaction region.

One possible experiment involves using a frequency comb to generate another, shifted comb using Raman scattering. The sensitivity of such an experiment would be similar to that for the four-wave mixing test.

Another method involves using the high harmonic generation cutoff energy[7] to detect anomalous Doppler (and thus energy) shifts in the moving atomic frame. This test is of particular interest because it can be applied to detect Lorentz violation for other species of charged particles (i.e., muons) and not merely electrons. A more detailed analysis will be presented in future publications.

References

1. V.A. Kostelecký and M. Mewes, Phys. Rev. D **66**, 056005 (2002).
2. M.E. Tobar, P. Wolf, A. Fowler and J.G. Hartnett, Phys. Rev. D **71**, 025004 (2005).
3. H. Müller *et al.*, Phys. Rev. Lett **99**, 050401 (2007).
4. G. Saathoff *et al.*, Phys. Rev. Lett **91**, 190403 (2003).
5. H.E. Ives and G.R. Stilwell, J. Opt. Soc. Am. **28**, 215 (1938).
6. J.L. Hall, Rev. Mod. Phys. **78** 1279 (2006).
7. M. Lewenstein *et al*, Phys. Rev. A **49** 2117 (1994).
8. R.J. Jones, K.D. Moll, M.J. Thorpe and J. Ye, Phys. Rev. Lett **94** 193201 (2005).

TEST FOR LORENTZ VIOLATION IN THE MINIBOONE NEUTRINO OSCILLATION EXPERIMENT

T. KATORI AND R. TAYLOE FOR THE MINIBOONE COLLABORATION
Indiana University Cyclotron Facility
2401 Milo B. Sampson Lane
Bloomington, IN 47408, U.S.A.
E-mail: katori@fnal.gov

The small signal of Lorentz and CPT violation is the interest of neutrino oscillation experiments, such as MiniBooNE. MiniBooNE shows unexpected data excess of ν_e neutrino oscillation candidate events at low energy region. From the sidereal time dependence of these events, we performed several statistical tests to find any statistically significant structures in the sidereal time distribution. We found that there is no statistically significant sidereal time dependence in the low energy ν_e excess events.

1. MiniBooNE initial oscillation result

The MiniBooNE neutrino oscillation experiment is designed to confirm or reject the oscillation signal observed in the LSND experiment at LANL.[1] Since the LSND signal cannot be reconciled with the standard three neutrino massive model picture, the confirmation of the LSND signal would suggest physics beyond the Standard Model.

MiniBooNE recently showed first oscillation results.[2] The data show an anomalously high number of ν_e events in the low reconstructed ν_e energy region, and these signals cannot be understood from the standard three-neutrino massive model. However, this is a type of signal predicted by Lorentz violation,[3] and especially, a specific neutrino oscillation model based on Lorentz violation predicted a low energy event excess for MiniBooNE.[4] If they are the signals of Lorentz violating neutrino oscillation, they may have sidereal time dependence.

2. Data quality test

Figure 1 shows the time distribution of MiniBooNE low energy ν_e oscillation candidate events. Ideally, Fig. 1b should be completely filled out, then any structure in day-night (Greenwich Mean, or GM, time) distribution would be averaged out over a solar year. This is not the case here. The MiniBooNE data show a band in Fig. 1b that corresponds to a summer shutdown, as can be seen from Fig. 1a.

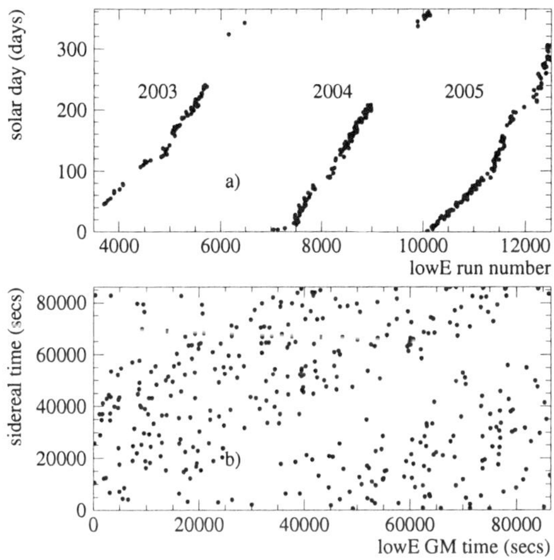

Figure 1. Distribution of low energy ν_e oscillation candidate events. a) Run number vs solar day (day 1 = January 1). b) GM time vs sidereal time. The year of each set of runs is indicated in a).

3. Statistical tests

Figure 2 shows the sidereal and GM time distribution of the low energy region. Here, "low energy" is defined from the reconstructed electron neutrino spectrum E_ν^{QE}, $300 < E_\nu^{QE}(MeV) < 475$. To avoid possible effects in sidereal time distribution due to the structure of the day-night (GM time) distribution, we also applied the same tests to the GM time distribution.

The results of compatibility of flatness from Pearson's χ^2 (P-χ^2) and the unbinned Kolmogorov-Smirnov test (K-S) are shown in Table 1. This table has results from both MiniBooNE and LSND.[5] Compared with LSND, MiniBooNE shows a sidereal time distribution consistent with no time-varying terms. Therefore, one can expect only a weak signal in sidereal dependent terms after the fit using SME parameters.[6]

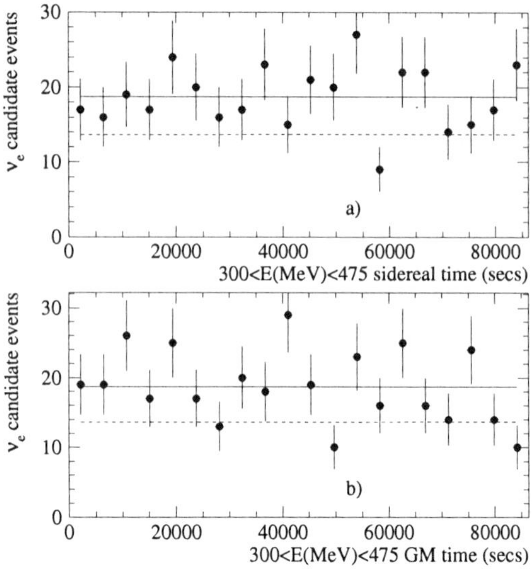

Figure 2. a) Sidereal and b) GM time distributions of the low energy ν_e oscillation candidate events. The solid line shows the expected underlying distribution with no sidereal time variation, and the dashed line indicates the estimated background contribution.

Table 1. The results of compatibility of flatness from Pearson's χ^2 (P-χ^2) test and the unbinned Kolmorogorov-Smirnov (K-S) test from the MiniBooNE and LSND experiments. (MCD=maximum cumulative deviation)

		P-χ^2			K-S	
	# of events	# of bins	χ^2	P(χ^2)	MCD	P(K-S)
MiniBooNE						
sidereal	374	74	79.5	0.28	0.027	1.00
GM	374	74	72.8	0.49	0.045	0.84
LSND[5]						
sidereal	186	37	44.8	0.15	0.076	0.23
GM	186	37	27.6	0.84	0.066	0.39

References

1. A. A. Aguilar-Arevalo *et al.* (LSND Collaboration), Phys. Rev. D **64**, 112007 (2001).
2. A. A. Aguilar-Arevalo *et al.* (MiniBooNE Collaboration), Phys. Rev. Lett. **98**, 231801 (2007); T. Katori and R. Tayloe, these proceedings.
3. V.A. Kostelecký and M. Mewes, Phys. Rev. D **69**, 016005 (2004).
4. T. Katori, R. Tayloe, and V.A. Kostelecký, Phys. Rev. D **74**, 105009 (2006).
5. L.B. Auerbach *et al.* (LSND Collaboration), Phys. Rev. D **72**, 076004 (2005).
6. V.A. Kostelecký and M. Mewes, Phys. Rev. D **70**, 076002 (2004).

DOPPLER-EFFECT EXPERIMENTS AND LORENTZ VIOLATION

CHARLES D. LANE

Berry College Physics Department
2277 Martha Berry Hwy
Mount Berry, GA 30149-5004, U.S.A.
E-mail: clane@berry.edu

Within the context of the Lorentz-violating Standard-Model Extension, I present results of a study of precision Doppler-effect experiments. While other types of experiments, such as clock-comparison and spin-torsion experiments, already provide sensitive tests of Lorentz symmetry in atomic systems, they do not cover all possible types of Lorentz-violating interactions. Doppler-effect experiments provide tests of some sectors of the Standard-Model Extension that are unprobed by most other classes of experiments. I present analysis of already-completed Doppler experiments from the Heidelberg Storage Ring.

Special relativity predicts specific relations between measurements of transition frequencies of atoms in relative motion. If Lorentz symmetry is violated,[1] however, the conventional relations may not be precisely correct. In this article, I present analysis[2] of sensitive Doppler-shift experiments[3] within the framework of the Lorentz-violating Standard-Model Extension[4] (SME).

The basic setup for a Doppler-effect measurement requires a set of stationary atoms and a beam of atoms, 2 reference frames (S and $\bar{S}$, respectively corresponding to the rest frames of the two sets of atoms), and 3 transition-frequency measurements. The same transition should be studied in both sets of atoms. The transition-frequency measurements are as follows: (1) ν_{lab} is the frequency of the lab atoms in their own rest frame, (2) ν_{beam} is the frequency of the beam atoms as measured in the lab frame, and (3) $\bar{\nu}_{\text{beam}}$ is the frequency of the beam atoms should their transition be measured in their own rest frame.

Relations between these frequencies may be discovered by considering some general Lorentz-transformation properties. First: ν_{beam} and $\bar{\nu}_{\text{beam}}$ denote the same quantity measured in different frames, so they are related by an observer Lorentz transformation, and hence obey the conventional Doppler relationship. Second: $\nu_{\rm lab}$ and $\bar{\nu}_{\rm beam}$, while not the same quantity, denote the same type of quantity (namely, the same transition) in atoms moving with respect to each other. They are related by a particle Lorentz transformation, and hence in the context of the SME don't obey their conventional relationship (equality).

The key experimental quantity is the ratio

$$\frac{\bar{\nu}^2_{\rm beam}}{\nu^2_{\rm lab}} = 1 + \varepsilon \quad , \tag{1}$$

where ε must be much less than one because Lorentz symmetry is approximately valid. The Heidelberg experiment[3] reported that $|\varepsilon| \lesssim 2 \times 10^{-9}$ for particular lithium-ion transitions.

In the SME, new interactions cause shifts in atomic-energy levels: every frequency can be expressed as $\nu = \nu_{\rm SM} + \delta\nu$, where $\nu_{\rm SM}$ is the frequency as predicted by conventional Standard-Model physics and $\delta\nu$ is a linear combination of SME coefficients for Lorentz violation. These components, such as b_3 and c_{11}, depend on the frame in which they are evaluated. In general, $b_3 \neq b_{\bar{3}}$ and $c_{11} \neq c_{\bar{1}\bar{1}}$, so $\delta\bar{\nu}_{\rm beam} \neq \delta\nu_{\rm lab}$. In fact, the quantity of interest $\varepsilon = \frac{2}{\nu_{\rm SM}}[\delta\bar{\nu}_{\rm beam} - \delta\nu_{\rm lab}]$, so ε itself can be expressed as a linear combination of SME Lorentz-violation coefficients.

Moreover, the components of the coefficients vary with time as Earth rotates (since each basis used to define them rotates along with Earth's surface). The quantity of interest ε can then be split into a constant term, a term that varies at Earth's sidereal frequency $\omega_\oplus$, and a term that varies at twice Earth's sidereal frequency:

$$\varepsilon = {\rm const.} + (\ldots)\cos\omega_\oplus T + (\ldots)\sin\omega_\oplus T + (\ldots)\cos 2\omega_\oplus T + (\ldots)\sin 2\omega_\oplus T \quad . \tag{2}$$

Even without timing information, interesting bounds may result. For example, in the Heidelberg experiment, the frequency measurements were effectively time averaged. In this situation, the SME predicts that ε is a constant depending on certain combinations of Lorentz-violation coefficients. The tight bound on ε then leads to bounds on those coefficients:

$$\begin{aligned} |c^p_{XX} + c^p_{YY} - 2c^p_{ZZ}| &\lesssim 10^{-11} \quad , \\ |c^p_{TJ} + c^p_{JT}| &\lesssim 10^{-8} \quad (J = X, Y, Z) \quad , \\ |c^e_{XX} + c^e_{YY} - 2c^e_{ZZ}| &\lesssim 10^{-5} \quad , \\ |c^e_{TJ} + c^e_{JT}| &\lesssim 10^{-2} \quad (J = X, Y, Z) \quad . \end{aligned} \tag{3}$$

These expressions give bounds on Lorentz-violation coefficients defined in a nonrotating Sun-centered frame.[5] The superscripts indicate whether the coefficients are associated with protons or electrons.

To summarize: Doppler-effect experiments provide a good example of the differences between observer and particle transformations. Already completed experiments provide bounds on Lorentz-violating SME coefficients that nicely complement the results of other types of atomic experiments.

References

1. For recent reviews of various theoretical and experimental aspects of Lorentz violation, see, for example, V.A. Kostelecký, ed., *CPT and Lorentz Symmetry III*, World Scientific, Singapore, 2005.
2. C.D. Lane, Phys. Rev. D **72**, 016005 (2005).
3. G. Saathoff *et al.*, Phys. Rev. Lett. **91**, 190403 (2003); G. Gwinner, these proceedings.
4. D. Colladay and V.A. Kostelecký, Phys. Rev. D **55**, 6760 (1997); Phys. Rev. D **58**, 116002 (1998).
5. V.A. Kostelecký and M. Mewes, Phys. Rev. D **66**, 056005 (2002).

COMPLEMENTARY MICHELSON-MORLEY EXPERIMENTS: COORDINATE AND FIELD REDEFINITIONS

H. MÜLLER,[1] P.L. STANWIX,[2] M.E. TOBAR,[2] E. IVANOV,[2] P. WOLF,[3] S. HERRMANN,[4] A. SENGER,[4] E. KOVALCHUK,[4] AND A. PETERS[4]

[1] *Physics Dept., Varian 226, 382 Via Pueblo Mall, Stanford, CA94305, U.S.A.*
E-mail: holgerm@stanford.edu

[2] *Univ. of Western Australia, 35 Stirling Hwy., Crawley 6009 WA, Australia*

[3] *LNE-SYRTE, 61 Av. de l'Observatoire, 75014 Paris, France*

[4] *Humboldt-Universität zu Berlin, Hausvogteiplatz 5-7, 10117 Berlin, Germany*

We illustrate the role of coordinate and field redefinitions in complementary Michelson-Morley experiments that determine separate limits on Lorentz violation in the photon and electron sector. We present a simplified but explicit model of how the electron (or photon) terms can be moved to the proton sector.

Our complementary Michelson-Morley experiments were built and operated in Berlin, Germany and Perth, Australia.[1] The Berlin setup compares the resonance frequencies $\omega_r = 2\pi c/(2L)$ (near a wavelength of 1064 nm) of Fabry-Perot resonators of length L, where c is the velocity of light along the cavity direction. One cavity rotating on a turntable at a rate of 1/43 s is compared to a similar, but non-rotating, one.

The Perth setup compares two whispering-gallery resonators (WGR), in which a standing microwave (at $\sim$ 10 GHz) resonates inside the perimeter of a sapphire cylinder. In a simplified but sufficient model, the resonance frequency of the WGR is proportional to $\bar{c}/(nL)$, where $\bar{c}$ is the velocity of light integrated along the circular path, n is the index of refraction, and L the circumference of the cylinder. Both WGRs co-rotate on the same turntable at a rate of 1/18 s.

Müller[2] has calculated in detail how the cavity lengths of both experiments, as well as the index of refraction of the WGR, are both subject to changes due to Lorentz violation in the matter sector. Each experiment can be analyzed for the modulation of the cavities' frequency difference as a function of rotation (due to Earth's rotation, Earth's orbit, and the turntables). By combining the data of both experiments, separate limits on Lorentz violation as given by the electron c-coefficients and the photon $\tilde{\kappa}_{e-}$ and $\tilde{\kappa}_{o+}$-coefficients can be deduced.

It is well known that by suitably defining coordinates and fields, one can set the electron (or photon) coefficients that are relevant in this experiment to zero.[3,4,5] In doing so, however, one also modifies the other sectors (like the photon, neutron, and the proton sector) in such a way that measurable quantities will not be affected by the redefinition. For our analysis,[1] we took the nucleon sectors to be conventional, but an analysis that assumes a Lorentz violating nucleon sector should be possible. While a general analysis along these lines is beyond the scope of this proceedings, we will gain insight by transforming three photon parameters to the proton sector.

We use a simplified version of the SME that describes photons by the field tensor $F^{\mu\nu}$; electrons and protons are described as scalar fields ϕ^e and ϕ^p. We start by assuming a conventional proton sector. Lorentz violation for the electron is given by $k_{\mu\nu}$, in analogy to the $c_{\mu\nu}$ of the full SME.[3,4] We assume the existence of a frame of reference in which this has only one nonzero component $k_{00} = k^2 - 1$, where $k \approx 1$. The Lagrangian is

$$\begin{aligned}\mathcal{L}^\phi &= (D_\mu\phi^e)^\dagger D^\mu\phi^e + (k^2-1)|D_0\phi^e|^2 - m_e^2(\phi^e)^\dagger\phi^e + (D_\mu\phi^p)^\dagger D^\mu\phi^p \\ &\quad -(m^p)^2(\phi^p)^\dagger\phi^p \quad \frac{1}{4}F^{\mu\nu}F_{\mu\nu} - \frac{1}{4}(k_F)_{\kappa\lambda\mu\nu}F^{\kappa\lambda}F^{\mu\nu} . \qquad (1)\end{aligned}$$

As usual, the covariant derivative is given by $D_\mu\phi^{e,p} = (\partial_\mu + iq^{e,p}A_\mu)\phi^{e,p}$, where $q^{e,p}$ is the particle's electric charge.

The Lorentz violation for the electron acts via the term $(k^2-1)|D_0\phi^e|^2$. A modified coupling of the electron to the Coulomb potential causes a geometry change of the crystal. Thus, a combination of $k^2 - 1$ and the modified velocity of light given by $(k_F)_{\kappa\lambda\mu\nu}$ are measurable. Using the existing theory of the matter sector[2] it is straightforward to derive the detailed signals in cavity experiments in all frames of reference.

By the transformations $t \to tk$, $\vec{x} \to \vec{x}$, $A_0 \to A_0$, $A_i \to A_i k$, and $q \to q/k$, one obtains

$$\begin{aligned}\mathcal{L}^\phi &= (D_\mu\phi^e)^\dagger D^\mu\phi^e - m^2(\phi^e)^\dagger\phi^e + (D_\mu\phi^p)^\dagger D^\mu\phi^p + (k^{-2}-1)|D_0\phi^p|^2 \\ &\quad -(m^p)^2(\phi^p)^\dagger\phi^p - \frac{1}{4}F^{\mu\nu}F_{\mu\nu} - \frac{1}{4}(k_F)_{\kappa\lambda\mu\nu}F^{\kappa\lambda}F^{\mu\nu} + \frac{1-k^2}{2}B^2 , \qquad (2)\end{aligned}$$

where B is the magnetic field. Thus, the Lorentz violation in the electron sector has been moved to the proton and photon sectors. It now acts via a rescaled coupling of the proton to the electric field, and a modified velocity of light given by $(k_F)_{\kappa\lambda\mu\nu}$ and $k^2 - 1$.

Here, we considered only a single parameter analogous to c_{00}, that causes a rotation-invariant scaling of the solid. We can easily generalize this model by applying a Lorentz boost to $k_{\mu\nu}$ that is characterized by a velocity $0 \leq v < 1$ into an arbitrary direction and subsequent rotation by an angle β around an axis normal to $\vec{v}$ and an angle γ around the axis needed to complete a dreibein. The measurable electron coefficients are given by a quantity[2] $E'_{jk} = -k_{jk} - \frac{1}{2}\delta_{jk}k_{00}$, for which our model gives

$$E' = \frac{k^2-1}{1-v^2} \times \begin{pmatrix} -\frac{1}{2} - v^2\cos^2\beta\cos^2\gamma & v^2\cos\beta\cos\gamma\sin\gamma & v^2\cos\beta\cos^2\gamma\sin\beta \\ v^2\cos\beta\cos\gamma\sin\gamma & -\frac{1}{2} - v^2\sin^2\gamma & -v^2\cos\gamma\sin\beta\sin\gamma \\ v^2\cos\beta\cos^2\gamma\sin\beta & -v^2\cos\gamma\sin\beta\sin\gamma & -\frac{1}{2} - v^2\cos^2\gamma\sin^2\beta \end{pmatrix}. \quad (3)$$

The Lorentz violating degrees of freedom are $k^2 - 1$, v, β, and γ. In our cavity experiments,[1] the diagonal elements of E' enter only by the combinations[2] $E'_{11} + E'_{22} - 2E'_{33}$ and $E'_{11} - E'_{22}$. Thus, no independent measurements of $k^2 - 1$ and v^2 are possible using cavities. We end up with three measurable proton parameters that can be chosen as $(k^2-1)v^2/(1-v^2)$, β, and γ.

We have shown how to describe the influence of three proton parameters on cavity experiments. As a corollary, at least 12 out of 18 parameters from the photon and electron sector are separately measurable. They can, for example, be chosen as three $c_{\mu\nu}$ and nine $(k_F)_{\kappa\lambda\mu\nu}$. A more detailed model may allow to analyze the effect of further proton parameters.

References

1. H. Müller *et al.*, Phys. Rev. Lett. **99,** 050401 (2007).
2. H. Müller *et al.*, Phys. Rev. D **67,** 056006 (2003); Phys. Rev. D **68,** 116006 (2003); H. Müller, Phys. Rev. D **71,** 045004 (2005); H. Müller and C. Lämmerzahl in V.A. Kostelecký, ed., *CPT and Lorentz Symmetry III*, World Scientific, Singapore, 2005.
3. D. Colladay and V.A. Kostelecký, Phys. Rev. D **55,** 6760 (1997); Phys. Rev. D **58**, 116002 (1998).
4. V.A. Kostelecký and M. Mewes, Phys. Rev. D **66,** 056005 (2002).
5. V.A. Kostelecký and A.G.M. Pickering, Phys. Rev. Lett. **91,** 031801, (2003).

A POEM PROGRESS REPORT

ROBERT D. REASENBERG AND JAMES D. PHILLIPS

Smithsonian Astrophysical Observatory
Harvard-Smithsonian Center for Astrophysics
60 Garden Street
Cambridge, MA 02138, U.S.A.
E-mail: reasenberg@cfa.harvard.edu

1. Introduction

The POEM instrument[1] directly examines the relative motion of a pair of test mass assemblies (TMA) that are freely falling in a comoving vacuum chamber. A single "toss" lasts 0.8 s as the chamber moves up and down ≈ 90 cm of vertical path. Recycling between tosses requires an additional 0.5 s. Each TMA contains $\approx 30\%$ test substance. The scientific goal of POEM is to determine the EP parameter, η, in a short experiment with an uncertainty under 5×10^{-14}. The POEM approach was derived from the successful EP test at JILA[2] that reached $\sigma(\eta) = 5 \times 10^{-10}$.

The primary measurement is made with our tracking frequency laser gauge (TFG), and the transverse motion and rotation of the TMA are measured by a multi-channel capacitance gauge (see both below). In the next version of POEM, which will reach the above sensitivity goal, there will be two pairs of TMA. These will be laterally separated by ≈ 7 cm, with test substances in reversed locations. This four-TMA configuration supports a double difference observable, which cancels systematic errors.

2. Why a Galilean Test Now?

Torsion-fiber experiments have succeeded in reducing $\sigma(\eta)$ 100-fold over the past four decades, even though they are sensitive to only $\approx 10^{-3}$ of the Earth's field. They have now encountered the limit imposed by thermal noise in the suspension fiber. Galilean tests such as POEM, which are sensitive to the full Earth field, will continue the improvement. Furthermore, unlike the torsion-balance technique, the Galilean technique lends itself naturally to a space-based version. A version on a sounding rocket, SR-POEM, will improve upon the current ground-based limit by a further 1000-fold, to $\sigma(\eta) = 10^{-16}$, in the duration of a coffee break.

Technology has advanced in important ways since the pioneering work at JILA. Better lasers are available and the TFG has been developed. There has been an explosion in the motion-control industry that has resulted in improved motor controllers, position sensors, and air bearings. Computers are faster and cheaper. Similarly, the data acquisition devices and related software available for these computers are much improved.

3. Multi-channel Capacitance Gauge

In collaboration with Winfield Hill (Rowland Institute at Harvard), we are developing a capacitance-gauge system that will measure five degrees of freedom per TMA. Each degree of freedom is assigned a unique frequency and complementary signals are applied to facing plates. The signal on the TMA at each frequency measures the TMA displacement from the center of the corresponding pair of plates. That signal is capacitively coupled to an amplifier-digitizer and correlated in a PC to extract the position measurements. Some parts of the system have been received and tested.[3]

4. Tracking Frequency Laser Gauge

We developed the TFG[4] more than 15 years ago for POINTS. The original version used a HeNe laser and both an ADM (frequency shifter) and an EOM (phase modulator). The present development is with a semiconductor laser and no external frequency shifter. The TFG offers six advantages over the classical heterodyne gauge: 1) Intrinsically free of the nm-scale cyclic bias characteristic of the heterodyne laser gauge (e.g., Hewlett-Packard, Zygo). 2) Uses one beam, not two: simplifies the beam launcher. 3) Distance changes measured by a radio frequency: more stable and more easily measured than RF phase. 4) Able to operate in a resonant cavity: improves precision and suppresses misalignment error. 5) Suppress (nm scale) polarization error from non-normal incidence in a cornercube. 6) Measures absolute distance with a minimum of added cost or complexity.

In any TFG realization, there is a limit to the range of optical frequency shift available for following incremental distance smoothly. The TFG controller has a nonlinear aspect such that when it reaches the limit of the available range, a "hop" is introduced. The interferometer jumps from one null to another. If the hop is precise and faster than the closed-loop response time, it goes "unnoticed," and yields minimal disruption.

5. Motion System

The motion system must meet two requirements. First, the vacuum chamber needs to move smoothly in a vertical free-fall trajectory during the TMA launch and the subsequent measurement interval. Second, the energy of the falling chamber must be absorbed and used later to send the chamber moving upward with minimal delay and energy loss. In the version now under construction, the smooth motion is ensured by an air-bearing cart running on a massive granite beam with lapped ways. All components of the new slide have been ordered and some have been delivered. The vertical motion is determined by a linear motor with position feedback and a digital controller-amplifier. The key requirement for the figure of the granite beam is that the slope along the ways not deviate from the mean slope by more than 2×10^{-6}. We employ commercially available porous graphite air bearings, which provide adjustability and guarantee bearing stability.

The "bouncer" stores the energy of the falling chamber in two torsion bars (working size: 1 by 75 inch, custom made of 4340 steel). Unlike coil springs, the torsion bars will not have their internal modes excited during a bounce. The pulley under the chamber encounters a cable stretched between levers that pivot on the torsion bars. This provides minimal "initial-contact shock" and a smooth onset of force. The effective mass of the cable (0.05 kg) is small compared to the mass of the chamber and associated moving parts, which is now ≈ 40 kg and will be > 55 kg with the air-bearing.

Acknowledgments

This work has been supported by NASA-UG (fundamental physics) through grant NNC04GB30G.

References

1. R.D. Reasenberg and J.D. Phillips, in H. Dittus, C. Lämmerzahl, and S.G. Turyshev, eds., *Lasers, Clocks, and Drag-Free Control*, Springer, New York, 2007.
2. T.M. Niebauer, G.S. Sasagawa, J.E. Faller, R. Hilt, and F. Klopping, Metrologia **32**, 159 (1995).
3. R.D.G. Hickman, M.S. thesis, Southampton, 2007.
4. J.D. Phillips and R.D. Reasenberg, Rev. Sci. Instrum. **76**, 064501 (2005).

DATA TABLES FOR LORENTZ AND CPT VIOLATION

V. ALAN KOSTELECKÝ
Physics Department, Indiana University
Bloomington, IN 47405, U.S.A.

NEIL RUSSELL
Physics Department, Northern Michigan University
Marquette, MI 49855, U.S.A.

This work tabulates experimental results for coefficients for Lorentz and CPT violation in the minimal Standard-Model Extension. Summary tables are provided listing attained sensitivities in the matter and photon sectors.

1. Introduction

Recent years have seen a renewed interest in experimental tests of Lorentz and CPT symmetry. Observable signals of Lorentz and CPT violation can be described in a model-independent way using effective field theory.[1]

The general realistic effective field theory for Lorentz violation is called the Standard-Model Extension (SME).[2,3] It includes the Standard Model coupled to General Relativity along with all possible operators for Lorentz violation. Both global[2] and local[3] Lorentz violation are incorporated. Since CPT violation in realistic field theories is accompanied by Lorentz violation,[4] the SME also describes general CPT violation. Reviews of the SME can be found in Refs. 5, 6 and these proceedings.

In the SME, each Lorentz-violating operator is controlled by a coefficient for Lorentz violation. Any experimental signal for Lorentz violation can be expressed in terms of one or more of these coefficients. The Lorentz-violating operators are systematically classified according to their mass dimension. A limiting case of particular interest is the minimal SME, defined as the restriction of the SME to include only Lorentz-violating operators of mass dimension four or less.

The results summarized here concern coefficients for Lorentz violation in the minimal SME. We compile data tables for these SME coefficients, including both existing experimental measurements and some theory-derived limits. We also extract summary tables listing attained sensitivities in two sectors: ordinary matter (electrons, protons, and neutrons), and photons. The tables include results announced at the CPT'07 meeting and in these proceedings, along with a few others released shortly thereafter.

2. Summary tables

Table 1 lists the maximum attained sensitivities involving protons, neutrons, and electrons. For each distinct massive spin-half fermion in the minimal SME, there are 44 independent observable combinations of coefficients for Lorentz violation. Of these, 20 also control CPT violation. The 44 combinations are conventionally chosen as the tilde coefficients shown. The definitions of 40 of these in terms of minimal SME coefficients are given in Appendix B of Ref. 7. In the same notation, the remaining four are defined as

$$\tilde{b}_J^* = b_J + \tfrac{1}{2}\varepsilon_{JKL}H_{KL} + m(d_{JT} + \tfrac{1}{2}\varepsilon_{JKL}g_{KLT}), \quad \tilde{c}_{TT} = mc_{TT}. \quad (1)$$

In Table 1, all tilde coefficients have dimensions of GeV in natural units. A dash indicates that no measurement of the coefficient exists to date.

Table 2 displays the maximum attained sensitivities to SME coefficients for Lorentz violation in the photon sector. There are 23 observable coefficient combinations for photons, of which four also control CPT violation. The first 19 listed in the table are dimensionless and are defined in Section II B of Ref. 8. The remaining four have dimensions of GeV in natural units and are defined in Ref. 9.

In both summary tables, each sensitivity value displayed represents our best estimate of a definitive two-σ limit on the modulus of the corresponding coefficient. In a few cases, tighter results may exist when suitable theoretical assumptions are adopted; these results can be found in the data tables that follow. Also, in circumstances where observations involve a linear combination of coefficients in the tables, the displayed sensitivity for each coefficient assumes for definiteness that no other coefficient contributes. Some caution is therefore advisable in adapting the results in these summary tables to situations involving two or more nonzero coefficient values.

3. Data tables

The remaining 10 data tables compile existing results for various sectors of the minimal SME: the electron sector (Table 3), the proton sector (Table

4), the neutron sector (Table 5), the photon sector (Table 6), the charged-lepton sector (Table 7), the neutrino sector (Table 8), the meson sector (Table 9), the electroweak sector (Table 10), the gluon sector (Table 11), and the gravity sector (Table 12). Each table lists the relevant coefficient or combination of coefficients, the result in natural units, the physical system involved, and the reference.

In these data tables, a theoretical deduction can be distinguished from a direct experimental measurement by an asterisk placed after the reference. Certain results involve combinations of coefficients across more than one sector; each of these has been listed only once in the table deemed most appropriate. Some minor changes in notation or format have been introduced as needed, but for the most part the results are quoted as they appear in the cited references.

Acknowledgments

This work was supported in part by DoE grant DE-FG02-91ER40661 and NASA grant NAG3-2914.

References

1. V.A. Kostelecký and R. Potting, Phys. Rev. D **51**, 3923 (1995).
2. D. Colladay and V.A. Kostelecký, Phys. Rev. D **58**, 116002 (1998).
3. V.A. Kostelecký, Phys. Rev. D **69**, 105009 (2004).
4. O.W. Greenberg, Phys. Rev. Lett. **89**, 231602 (2002).
5. V.A. Kostelecký, ed., *CPT and Lorentz Symmetry I, II, III*, World Scientific, Singapore 1999, 2002, 2005.
6. R. Bluhm, Lect. Notes Phys. **702**, 191 (2006) [hep-ph/0506054].
7. R. Bluhm *et al.*, Phys. Rev. D **68**, 125008 (2003).
8. V.A. Kostelecký and M. Mewes, Phys. Rev. D **66**, 056005 (2002).
9. V.A. Kostelecký and M. Mewes, Phys. Rev. Lett. **99**, 011601 (2007).
10. B.R. Heckel *et al.*, Phys. Rev. Lett. **97**, 021603 (2006).
11. L.-S. Hou, W.-T. Ni, and Y.-C.M. Li, Phys. Rev. Lett. **90**, 201101 (2003).
12. T.W. Kornack, G. Vasilakis, and M.V. Romalis, these proceedings.
13. L. Hunter *et al.*, in Ref. 5, vol. *I*.
14. H. Dehmelt *et al.*, Phys. Rev. Lett. **83**, 4694 (1999).
15. R.K. Mittleman *et al.*, Phys. Rev. Lett. **83**, 2116 (1999).
16. V.A. Kostelecký and C.D. Lane, Phys. Rev. D **60**, 116010 (1999).
17. H. Müller *et al.*, Phys. Rev. Lett. **99**, 050401 (2007).
18. H. Müller *et al.*, Phys. Rev. D **68**, 116006 (2003).
19. C.D. Lane, Phys. Rev. D **72**, 016005 (2005).
20. B. Altschul, Phys. Rev. D **74**, 083003 (2006).
21. B. Altschul, Astropart. Phys. **28**, 380 (2007).

22. B. Altschul, Phys. Rev. D **75**, 041301(R) (2007).
23. M.A. Humphrey *et al.*, Phys. Rev. **A 68**, 063807 (2003).
24. D.F. Phillips *et al.*, Phys. Rev. D **63**, 111101 (2001).
25. P. Wolf *et al.*, Phys. Rev. Lett. **96**, 060801 (2006).
26. G. Gabrielse *et al.*, Phys. Rev. Lett. **82**, 3198 (1999).
27. F. Canè *et al.*, Phys. Rev. Lett. **93**, 230801 (2004).
28. D. Bear *et al.*, Phys. Rev. Lett. **85**, 5038 (2000).
29. B. Altschul, Phys. Rev. D **75**, 023001 (2007).
30. S. Herrmann *et al.*, these proceedings.
31. P.L. Stanwix *et al.*, Phys. Rev. D **74**, 081101 (R) (2006).
32. S. Herrmann *et al.*, Phys. Rev. Lett. **95**, 150401 (2005).
33. P.L. Stanwix, *et al.*, Phys. Rev. Lett. **95**, 040404 (2005).
34. P. Wolf *et al.*, Phys. Rev. D **70**, 051902 (2004).
35. H. Müller *et al.*, Phys. Rev. Lett. **91** 020401 (2003).
36. J.A. Lipa *et al.*, Phys. Rev. Lett. **90**, 060403 (2003).
37. P. Antonini *et al.*, Phys. Rev. A **72**, 066102 (2005).
38. S. Reinhardt *et al.*, Nature Physics **3**, 861 (2007).
39. M. Hohensee *et al.*, Phys. Rev. D **71**, 025004 (2007).
40. C.D. Carone, M. Sher and M. Vanderhaeghen, Phys. Rev. D **74**, 077901 (2006).
41. J.P. Cotter and B.T.H. Varcoe, arXiv:physics/0603111.
42. V.A. Kostelecký and M. Mewes, Phys. Rev. Lett. **97**, 140401 (2006).
43. V.A. Kostelecký and M. Mewes, Phys. Rev. Lett. **99**, 011601 (2007).
44. B. Feng *et al.*, Phys. Rev. Lett. **96**, 221302 (2006).
45. S.M. Carroll, G.B. Field, and R. Jackiw, Phys. Rev. D **41**, 1231 (1990).
46. G.W. Bennett *et al.*, Muon g–2 Collaboration, arXiv:0709.4670.
47. V.W. Hughes *et al.*, Phys. Rev. Lett. **87**, 111804 (2001).
48. M. Deile *et al.*, Muon g–2 Collaboration, in Ref. 5, vol. *II* [hep-ex/0110044].
49. R. Bluhm, V.A. Kostelecký and C.D. Lane, Phys. Rev. Lett. **84**, 1098 (2000).
50. L.B. Auerbach *et al.*, LSND Collaboration, Phys. Rev. D **72**, 076004 (2005).
51. M.D. Messier, in Ref. 5, vol. *III*.
52. A. Di Domenico, KLOE Collaboration, these proceedings.
53. H. Nguyen, KTeV Collaboration, in Ref. 5, vol. *II* [hep-ex/0112046].
54. V.A. Kostelecký, Phys. Rev. Lett. **80**, 1818 (1998).
55. J. Link *et al.*, FOCUS Collaboration, Phys. Lett. B **556**, 7 (2003).
56. B. Aubert *et al.*, BaBar Collaboration, arXiv:0711.2713.
57. B. Aubert *et al.*, BaBar Collaboration, hep-ex/0607103.
58. B. Altschul, arXiv:0712.1579.
59. D.L. Anderson, M. Sher, and I. Turan, Phys. Rev. D **70**, 016001 (2004).
60. H. Müller *et al.*, arXiv:0710.3768.
61. J.B.R. Battat, J.F. Chandler and C.W. Stubbs, Phys. Rev. Lett. **99**, 241103 (2007).
62. Q.G. Bailey and V.A. Kostelecký, Phys. Rev. D **74**, 045001 (2006).

Table 1. Matter-sector summary

Coefficient	**Proton**	**Neutron**	**Electron**
$\tilde{b}_X$	10^{-27} GeV	10^{-31} GeV	10^{-31} GeV
$\tilde{b}_Y$	10^{-27} GeV	10^{-31} GeV	10^{-30} GeV
$\tilde{b}_Z$	–	–	10^{-29} GeV
$\tilde{b}_T$	–	10^{-27} GeV	–
$\tilde{b}^*_J$ $(J = X, Y, Z)$	–	–	–
$\tilde{c}_-$	10^{-25} GeV	10^{-27} GeV	10^{-19} GeV
$\tilde{c}_Q$	10^{-22} GeV	–	10^{-19} GeV
$\tilde{c}_X$	10^{-25} GeV	10^{-25} GeV	10^{-19} GeV
$\tilde{c}_Y$	10^{-25} GeV	10^{-25} GeV	10^{-19} GeV
$\tilde{c}_Z$	10^{-24} GeV	10^{-27} GeV	10^{-19} GeV
$\tilde{c}_{TX}$	10^{-20} GeV	–	10^{-18} GeV
$\tilde{c}_{TY}$	10^{-20} GeV	–	10^{-18} GeV
$\tilde{c}_{TZ}$	10^{-21} GeV	–	10^{-20} GeV
$\tilde{c}_{TT}$	–	–	10^{-18} GeV
$\tilde{d}_+$	–	10^{-27} GeV	10^{-17} GeV
$\tilde{d}_-$	–	10^{-27} GeV	10^{-17} GeV
$\tilde{d}_Q$	–	10^{-27} GeV	10^{-17} GeV
$\tilde{d}_{XY}$	–	10^{-27} GeV	10^{-18} GeV
$\tilde{d}_{YZ}$	–	10^{-26} GeV	10^{-18} GeV
$\tilde{d}_{ZX}$	–	–	10^{-17} GeV
$\tilde{d}_X$	10^{-25} GeV	10^{-29} GeV	10^{-22} GeV
$\tilde{d}_Y$	10^{-25} GeV	10^{-28} GeV	10^{-22} GeV
$\tilde{d}_Z$	–	–	10^{-19} GeV
$\tilde{H}_{XT}$	–	10^{-26} GeV	–
$\tilde{H}_{YT}$	–	10^{-27} GeV	–
$\tilde{H}_{ZT}$	–	10^{-27} GeV	–
$\tilde{g}_T$	–	10^{-27} GeV	–
$\tilde{g}_c$	–	10^{-27} GeV	–
$\tilde{g}_Q$	–	–	–
$\tilde{g}_-$	–	–	–
$\tilde{g}_{TJ}$ $(J = X, Y, Z)$	–	–	–
$\tilde{g}_{XY}$	–	–	–
$\tilde{g}_{YX}$	–	–	–
$\tilde{g}_{ZX}$	–	–	–
$\tilde{g}_{XZ}$	–	–	–
$\tilde{g}_{YZ}$	–	–	–
$\tilde{g}_{ZY}$	–	–	–
$\tilde{g}_{DX}$	10^{-25} GeV	10^{-29} GeV	10^{-22} GeV
$\tilde{g}_{DY}$	10^{-25} GeV	10^{-28} GeV	10^{-22} GeV
$\tilde{g}_{DZ}$	–	–	–

Table 2. Photon-sector summary

Coefficient	**Sensitivity**
$(\tilde{\kappa}_{e+})^{XY}$	10^{-32}
$(\tilde{\kappa}_{e+})^{XZ}$	10^{-32}
$(\tilde{\kappa}_{e+})^{YZ}$	10^{-32}
$(\tilde{\kappa}_{e+})^{XX} - (\tilde{\kappa}_{e+})^{YY}$	10^{-32}
$(\tilde{\kappa}_{e+})^{ZZ}$	10^{-32}
$(\tilde{\kappa}_{o-})^{XY}$	10^{-32}
$(\tilde{\kappa}_{o-})^{XZ}$	10^{-32}
$(\tilde{\kappa}_{o-})^{YZ}$	10^{-32}
$(\tilde{\kappa}_{o-})^{XX} - (\tilde{\kappa}_{e+})^{YY}$	10^{-32}
$(\tilde{\kappa}_{o-})^{ZZ}$	10^{-32}
$(\tilde{\kappa}_{e-})^{XY}$	10^{-16}
$(\tilde{\kappa}_{e-})^{XZ}$	10^{-16}
$(\tilde{\kappa}_{e-})^{YZ}$	10^{-16}
$(\tilde{\kappa}_{e-})^{XX} - (\tilde{\kappa}_{e-})^{YY}$	10^{-15}
$(\tilde{\kappa}_{e-})^{ZZ}$	10^{-14}
$(\tilde{\kappa}_{o+})^{XY}$	10^{-12}
$(\tilde{\kappa}_{o+})^{XZ}$	10^{-12}
$(\tilde{\kappa}_{o+})^{YZ}$	10^{-12}
$\tilde{\kappa}_{\rm tr}$	10^{-7}
$k^{(3)}_{(V)00}$	10^{-42} GeV
$k^{(3)}_{(V)10}$	10^{-42} GeV
$\mathrm{Re}\, k^{(3)}_{(V)11}$	10^{-42} GeV
$\mathrm{Im}\, k^{(3)}_{(V)11}$	–

Table 3. Electron sector

Combination	Result	System	Ref.
$\tilde{b}_X$	$(0.1 \pm 2.4) \times 10^{-31}$ GeV	Torsion pendulum	10
$\tilde{b}_Y$	$(-1.7 \pm 2.5) \times 10^{-31}$ GeV	"	10
$\tilde{b}_Z$	$(-29 \pm 39) \times 10^{-31}$ GeV	"	10
$\tilde{b}_\perp$	$< 3.1 \times 10^{-29}$ GeV	"	11
$\|\tilde{b}_Z\|$	$< 7.1 \times 10^{-28}$ GeV	"	11
$\tilde{b}_X$	$(2.8 \pm 6.1) \times 10^{-29}$ GeV	K/He magnetometer	12
$\tilde{b}_Y$	$(6.8 \pm 6.1) \times 10^{-29}$ GeV	"	12
r_e	$< 3.2 \times 10^{-24}$	Hg/Cs comparison	13
$\|\vec{b}\|$	< 20 radians/s	Penning trap	14
$r_{\omega_a^-,\mathrm{diurnal}}$	$< 1.6 \times 10^{-21}$	"	15
$\|\tilde{b}_J\|$ $(J = X, Y)$	$< 10^{-27}$ GeV	Hg/Cs comparison	16*
$c_{XX} - c_{YY}$	$(-2.9 \pm 6.3) \times 10^{-16}$	Optical, microwave	17
$\frac{1}{2}c_{(XY)}$	$(2.1 \pm 0.9) \times 10^{-16}$	resonators	17
$\frac{1}{2}c_{(XZ)}$	$(-1.5 \pm 0.9) \times 10^{-16}$	"	17
$\frac{1}{2}c_{(YZ)}$	$(-0.5 \pm 1.2) \times 10^{-16}$	"	17
$c_{XX} + c_{YY} - 2c_{ZZ}$	$(-106 \pm 147) \times 10^{-16}$	"	17
λ^{ZZ}	$(13.3 \pm 9.8) \times 10^{-16}$	"	17
$\|\frac{1}{2}c_{(XY)}\|$	$< 8 \times 10^{-15}$	Optical resonators	18
$\|c_{XX} - c_{YY}\|$	$< 1.6 \times 10^{-15}$	"	18
$\|c_{XX} + c_{YY} - 2c_{ZZ}\|$	$< 10^{-5}$	Heavy-ion storage ring	19*
$\|c_{(TX)}\|, \|c_{(TY)}\|, \|c_{(TZ)}\|$	$< 10^{-2}$	"	19*
c_{XX}	$(-3$ to $5) \times 10^{-15}$	Astrophysics	20*
c_{YY}	$(-0.7$ to $2.5) \times 10^{-15}$	"	20*
c_{ZZ}	$(-1.6$ to $2.5) \times 10^{-15}$	"	20*
$c_{(XY)}, c_{(XZ)}$	$(-3$ to $3) \times 10^{-15}$	"	20*
$c_{(YZ)}$	$(-2.5$ to $1.8) \times 10^{-15}$	"	20*
c_{0X}	$(-7$ to $4) \times 10^{-15}$	"	20*
c_{0Y}	$(-0.5$ to $1.5) \times 10^{-15}$	"	20*
c_{0Z}	$(-4$ to $2) \times 10^{-17}$	"	20*
$\|0.05c_{XX} + 0.55c_{YY} + 0.41c_{ZZ}$ $+0.16c_{(XY)} - 0.14c_{(XZ)} - 0.47c_{(YZ)}$ $+0.22c_{(0X)} + 0.74c_{(0Y)} - 0.64c_{(0Z)}$ $+c_{00}\|$	$< 1.3 \times 10^{-15}$	"	21*
$\|0.58c_{XX} + 0.04c_{YY} + 0.38c_{ZZ}$ $-0.14c_{(XY)} - 0.47c_{(XZ)} + 0.12c_{(YZ)}$ $+0.76c_{(0X)} - 0.19c_{(0Y)} - 0.62c_{(0Z)}$ $+c_{00}\|$	$< 2.5 \times 10^{-15}$	"	21*

Table 3. Electron sector (continued)

Combination	Result	System	Ref.
$\lvert d_{XX}\rvert$	$< 2 \times 10^{-14}$	Astrophysics	22*
$\lvert d_{YY}\rvert, \lvert d_{ZZ}\rvert$	$< 3 \times 10^{-15}$	"	22*
$\lvert d_{(XY)}\rvert$	$< 2 \times 10^{-15}$	"	22*
$\lvert d_{(XZ)}\rvert$	$< 2 \times 10^{-14}$	"	22*
$\lvert d_{(YZ)}\rvert$	$< 7 \times 10^{-15}$	"	22*
$\lvert d_{TX}\rvert$	$< 5 \times 10^{-14}$	"	22*
$\lvert d_{TY}\rvert$	$< 5 \times 10^{-15}$	"	22*
$\lvert d_{TZ}\rvert$	$< 8 \times 10^{-17}$	"	22*
$\lvert \tilde{d}_J\rvert, \lvert \tilde{g}_{D,J}\rvert\ (J = X, Y)$	$< 10^{-22}$ GeV	Hg/Cs comparison	16*

Table 4. Proton sector

Combination	Result	System	Ref.
$\tilde{b}_X$	$(6.0 \pm 1.3) \times 10^{-31}$ GeV	K/He magnetometer	12
$\tilde{b}_Y$	$(1.5 \pm 1.2) \times 10^{-31}$ GeV	"	12
$\sqrt{(\tilde{b}^e_X + \tilde{b}^p_X)^2 + (\tilde{b}^e_Y + \tilde{b}^p_Y)^2}$	$(3 \pm 2) \times 10^{-27}$ GeV	H maser	23
$\lvert \tilde{b}_J\rvert\ (J = X, Y)$	$< 2 \times 10^{-27}$ GeV	"	24
$\lvert \tilde{b}_J\rvert\ (J = X, Y)$	$< 10^{-27}$ GeV	Hg/Cs comparison	16*
$\tilde{c}_Q$	$(-0.3 \pm 2.2) \times 10^{-22}$ GeV	Cs fountain	25
$\tilde{c}_-$	$(-1.8 \pm 2.8) \times 10^{-25}$ GeV	"	25
c_X	$(0.6 \pm 1.2) \times 10^{-25}$ GeV	"	25
$\tilde{c}_Y$	$(-1.9 \pm 1.2) \times 10^{-25}$ GeV	"	25
$\tilde{c}_Z$	$(-1.4 \pm 2.8) \times 10^{-25}$ GeV	"	25
$\tilde{c}_{TX}$	$(-2.7 \pm 3.0) \times 10^{-21}$ GeV	"	25
$\tilde{c}_{TY}$	$(-0.2 \pm 3.0) \times 10^{-21}$ GeV	"	25
$\tilde{c}_{TZ}$	$(-0.4 \pm 2.0) \times 10^{-21}$ GeV	"	25
$\lvert c_{XX} + c_{YY} - 2c_{ZZ}\rvert$	$< 10^{-11}$	Doppler shift	19*
$\lvert c_{(TX)}\rvert, \lvert c_{(TY)}\rvert, \lvert c_{(TZ)}\rvert$	$< 10^{-8}$	"	19*
$r^{H^-}_{\omega_c}$	$< 4 \times 10^{-26}$	Penning trap	26
$\lvert \tilde{d}_J\rvert, \lvert \tilde{g}_{D,J}\rvert\ (J = X, Y)$	$< 10^{-25}$ GeV	Hg/Cs comparison	16*

Table 5. Neutron sector

Combination	Result	System	Ref.
$\tilde{b}_X$	$(-3.7 \pm 8.1) \times 10^{-32}$ GeV	K/He magnetometer	12
$\tilde{b}_Y$	$(-9.0 \pm 7.5) \times 10^{-32}$ GeV	"	12
$\tilde{b}_Y - 0.0034\tilde{d}_Y + 0.0034\tilde{g}_{DY}$	$(8.0 \pm 9.5) \times 10^{-32}$ GeV	Xe/He maser	27
$-\tilde{b}_X - 0.0034\tilde{d}_X - 0.0034\tilde{g}_{DX}$	$(2.2 \pm 7.9) \times 10^{-32}$ GeV	"	27
$-\cos\eta(\frac{1}{2}\tilde{b}_T + \frac{1}{2}\tilde{d}_- - \tilde{g}_c - \frac{1}{2}\tilde{g}_T)$ $-\cos\eta(\tilde{g}_T - 2\tilde{d}_+ + \frac{1}{2}\tilde{d}_Q)$ $+\sin\eta(\tilde{d}_{YZ} - \tilde{H}_{XT})$	$(-1.1 \pm 1.0) \times 10^{-27}$ GeV	"	27
$-\tilde{H}_{ZT}$	$(0.2 \pm 1.8) \times 10^{-27}$ GeV	"	27
$(\frac{1}{2}\tilde{b}_T + \frac{1}{2}\tilde{d}_- - \tilde{g}_c - \frac{1}{2}\tilde{g}_T)$ $-(\tilde{g}_T - 2\tilde{d}_+ + \frac{1}{2}\tilde{d}_Q)$	$(-1.8 \pm 1.9) \times 10^{-27}$ GeV	"	27
$\cos\eta(\tilde{H}_{ZT} - \tilde{d}_{XY}) - \sin\eta\tilde{H}_{YT}$	$(-1.1 \pm 0.8) \times 10^{-27}$ GeV	"	27
$\sqrt{(\tilde{b}_X)^2 + (\tilde{b}_Y)^2}$	$(4.0 \pm 3.3) \times 10^{-31}$ GeV	"	28
r_n	$< 1.5 \times 10^{-30}$	Hg/Cs comparison	13
$\lvert\tilde{b}_J\rvert$ $(J = X, Y)$	$< 10^{-30}$ GeV	"	16*
$\min(\lvert c_{11} - c_{22}\rvert, \lvert c_{11} - c_{33}\rvert, \lvert c_{22} - c_{33}\rvert)$	$< 1.7 \times 10^{-8}$	Pulsar timing	29*
$\lvert\tilde{c}_J\rvert$ $(J = X, Y)$	$< 10^{-25}$ GeV	Be/H comparison	16*
$\lvert\tilde{c}_-\rvert, \lvert\tilde{c}_Z\rvert$	$< 10^{-27}$ GeV	Hg/Hg & Ne/He comparison	16*
$\lvert\tilde{d}_J\rvert, \lvert\tilde{g}_{D,J}\rvert$ $(J = X, Y)$	$< 10^{-28}$ GeV	Hg/Cs comparison	16*

Table 6. Photon sector

Combination	Result	System	Ref.
$(\tilde{\kappa}_{e-})^{XY}$	$(-0.1 \pm 0.6) \times 10^{-17}$	Rotating optical resonators	30
"	$(-7.7 \pm 4.0) \times 10^{-16}$	Optical, microwave resonators	17
"	$(2.9 \pm 2.3) \times 10^{-16}$	Rotating microwave resonators	31
"	$(-3.1 \pm 2.5) \times 10^{-16}$	Rotating optical resonators	32
"	$(-0.63 \pm 0.43) \times 10^{-15}$	Rotating microwave resonators	33
"	$(-5.7 \pm 2.3) \times 10^{-15}$	Microwave resonator, maser	34
"	$(1.7 \pm 2.6) \times 10^{-15}$	Optical resonators	35
"	$(1.4 \pm 1.4) \times 10^{-13}$	Microwave resonators	36
$(\tilde{\kappa}_{e-})^{XZ}$	$(-2.0 \pm 0.9) \times 10^{-17}$	Rotating optical resonators	30
"	$(-10.3 \pm 3.9) \times 10^{-16}$	Optical, microwave resonators	17
"	$(-6.9 \pm 2.2) \times 10^{-16}$	Rotating microwave resonators	31
"	$(5.7 \pm 4.9) \times 10^{-16}$	Rotating optical resonators	32
"	$(0.19 \pm 0.37) \times 10^{-15}$	Rotating microwave resonators	33
"	$(-3.2 \pm 1.3) \times 10^{-15}$	Microwave resonator, maser	34
"	$(-6.3 \pm 12.4) \times 10^{-15}$	Optical resonators	35
"	$(-3.5 \pm 4.3) \times 10^{-13}$	Microwave resonators	36
$(\tilde{\kappa}_{e-})^{YZ}$	$(-0.3 \pm 1.4) \times 10^{-17}$	Rotating optical resonators	30
"	$(0.9 \pm 4.2) \times 10^{-16}$	Optical, microwave resonators	17
"	$(2.1 \pm 2.1) \times 10^{-16}$	Rotating microwave resonators	31
"	$(-1.5 \pm 4.4) \times 10^{-16}$	Rotating optical resonators	32
"	$(-0.45 \pm 0.37) \times 10^{-15}$	Rotating microwave resonators	33
"	$(-0.5 \pm 1.3) \times 10^{-15}$	Microwave resonator, maser	34
"	$(3.6 \pm 9.0) \times 10^{-15}$	Optical resonators	35
"	$(1.7 \pm 3.6) \times 10^{-13}$	Microwave resonators	36
$(\tilde{\kappa}_{e-})^{XX} - (\tilde{\kappa}_{e-})^{YY}$	$(-2.0 \pm 1.7) \times 10^{-17}$	Rotating optical resonators	30
"	$(-12 \pm 16) \times 10^{-16}$	Optical, microwave resonators	17
"	$(-5.0 \pm 4.7) \times 10^{-16}$	Rotating microwave resonators	31
"	$(5.4 \pm 4.8) \times 10^{-16}$	Rotating optical resonators	32
"	$(-1.3 \pm 0.9) \times 10^{-15}$	Rotating microwave resonators	33
"	$(-3.2 \pm 4.6) \times 10^{-15}$	Microwave resonator, maser	34
"	$(8.9 \pm 4.9) \times 10^{-15}$	Optical resonators	35
"	$(-1.0 \pm 2.1) \times 10^{-13}$	Microwave resonators	36
$(\tilde{\kappa}_{e-})^{ZZ}$	$(-0.2 \pm 3.1) \times 10^{-17}$	Rotating optical resonators	30
"	$(223 \pm 290) \times 10^{-16}$	Optical, microwave resonators	17
"	$(143 \pm 179) \times 10^{-16}$	Rotating microwave resonators	31
"	$(-1.9 \pm 5.2) \times 10^{-15}$	Rotating optical resonators	32
"	$(21 \pm 57) \times 10^{-15}$	Rotating microwave resonators	33
"	$(-2.9 \pm 2.2) \times 10^{-14}$	Optical resonators	37

Table 6. Photon sector (continued)

Combination	Result	System	Ref.
$\beta(\tilde{\kappa}_{o+})^{XY}$	$(-2.5 \pm 2.5) \times 10^{-17}$	Rotating optical resonators	30
$(\tilde{\kappa}_{o+})^{XY}$	$(1.7 \pm 2.0) \times 10^{-12}$	Optical, microwave resonators	17
"	$(-0.9 \pm 2.6) \times 10^{-12}$	Rotating microwave resonators	31
"	$(-2.5 \pm 5.1) \times 10^{-12}$	Rotating optical resonators	32
"	$(0.20 \pm 0.21) \times 10^{-11}$	Rotating microwave resonators	33
"	$(-1.8 \pm 1.5) \times 10^{-11}$	Microwave resonator, maser	34
"	$(14 \pm 14) \times 10^{-11}$	Optical resonators	35
$\beta(\tilde{\kappa}_{o+})^{XZ}$	$(1.5 \pm 1.7) \times 10^{-17}$	Rotating optical resonators	30
$(\tilde{\kappa}_{o+})^{XZ}$	$(-3.1 \pm 2.3) \times 10^{-12}$	Optical, microwave resonators	17
"	$(-4.4 \pm 2.5) \times 10^{-12}$	Rotating microwave resonators	31
"	$(-3.6 \pm 2.7) \times 10^{-12}$	Rotating optical resonators	32
"	$(-0.91 \pm 0.46) \times 10^{-11}$	Rotating microwave resonators	33
"	$(-1.4 \pm 2.3) \times 10^{-11}$	Microwave resonator, maser	34
"	$(-1.2 \pm 2.6) \times 10^{-11}$	Optical resonators	35
$\beta(\tilde{\kappa}_{o+})^{YZ}$	$(-1.0 \pm 1.5) \times 10^{-17}$	Rotating optical resonators	30
$(\tilde{\kappa}_{o+})^{YZ}$	$(-2.8 \pm 2.2) \times 10^{-12}$	Optical, microwave resonators	17
"	$(-3.2 \pm 2.3) \times 10^{-12}$	Rotating microwave resonators	31
"	$(2.9 \pm 2.8) \times 10^{-12}$	Rotating optical resonators	32
"	$(0.44 \pm 0.46) \times 10^{-11}$	Rotating microwave resonators	33
"	$(2.7 \pm 2.2) \times 10^{-11}$	Microwave resonator, maser	34
"	$(0.1 \pm 2.7) \times 10^{-11}$	Optical resonators	35
$(\tilde{\kappa}_{o+})^{YX} - 0.432(\tilde{\kappa}_{o+})^{ZX}$	$(4.0 \pm 8.4) \times 10^{-9}$	Microwave resonators	36
$(\tilde{\kappa}_{o+})^{XY} - 0.209(\tilde{\kappa}_{o+})^{YZ}$	$(4.0 \pm 4.9) \times 10^{-9}$	"	36
$(\tilde{\kappa}_{o+})^{XZ} - 0.484(\tilde{\kappa}_{o+})^{YZ}$	$(1.6 \pm 1.7) \times 10^{-9}$	"	36
$(\tilde{\kappa}_{o+})^{YZ} + 0.484(\tilde{\kappa}_{o+})^{XZ}$	$(0.6 \pm 1.9) \times 10^{-9}$	"	36
$\lvert\tilde{\kappa}_{\rm tr}\rvert$	$< 8.4 \times 10^{-8}$	Optical atomic clocks	38
"	$< 2.2 \times 10^{-7}$	Heavy-ion storage ring	39*
"	$< 2 \times 10^{-14}$	Astrophysics	40*
"	$< 3 \times 10^{-8}$	$g_e - 2$	40*
"	$< 1.6 \times 10^{-5}$	Sagnac interferometer	41*
$\lvert k^a\rvert$ for some a	$< 2 \times 10^{-37}$	Cosmological birefringence	42*
$\lvert k^a\rvert$ for $a = 1, \ldots, 10$	$< 2 \times 10^{-32}$	"	8*
$k^{(3)}_{(V)10}$	$\pm(3 \pm 1) \times 10^{-42}$ GeV	CMB polarization	43*
$k^{(3)}_{(V)11}$	$\pm(21^{+7}_{-9}) \times 10^{-43}$ GeV	"	43*
$k^{(3)}_{(V)00}$	$(12 \pm 7) \times 10^{-43}$ GeV	"	43*
$k^{(3)}_{(V)00}$	$(6 \pm 4) \times 10^{-43}$ GeV	"	44*
m	$< 1.7 \times 10^{-42} h_0$ GeV	Cosmological birefringence	45*

Table 7. Charged-lepton sector

Combination	Result	System	Ref.
b_Z^μ	$-(8.7 \pm 8.9) \times 10^{-24}$ GeV	BNL $g_\mu - 2$	46
$\sqrt{(\check{b}_X^{\mu^+})^2 + (\check{b}_Y^{\mu^+})^2}$	$< 1.3 \times 10^{-24}$ GeV	"	46
$\sqrt{(\check{b}_X^{\mu^-})^2 + (\check{b}_Y^{\mu^-})^2}$	$< 2.5 \times 10^{-24}$ GeV	"	46
$\sqrt{(\tilde{b}_X^\mu)^2 + (\tilde{b}_Y^\mu)^2}$	$< 2 \times 10^{-23}$ GeV	Muonium spectroscopy	47
$b_Z^\mu - 1.19(m_\mu d_{Z0}^\mu + H_{XY}^\mu)$	$(-1.4 \pm 1.0) \times 10^{-22}$ GeV	BNL, CERN $g_\mu - 2$ data	48
b_Z^μ	$(-2.3 \pm 1.4) \times 10^{-22}$ GeV	CERN $g_\mu - 2$ data	49*, 48
$m_\mu d_{ZT}^\mu + H_{XY}^\mu$	$(1.6 \pm 5.6) \times 10^{-23}$ GeV	BNL $g_\mu - 2$	46
$\lvert c^\mu \rvert$	$< 10^{-11}$	Astrophysics	21*
$\lvert c^\tau \rvert$	$< 10^{-8}$	"	21*

Table 8. Neutrino sector

Combination	Result	System	Ref.
$\lvert(\mathcal{C})_{\bar{e}\bar{\mu}}\rvert^2$	$(10.7 \pm 2.6 \pm 1.3) \times (10^{-19}\text{ GeV})^2$	Accelerator	50
$\lvert(\mathcal{C})_{\bar{e}\bar{\mu}}\rvert^2 + \frac{1}{2}\lvert(\mathcal{A}_s)_{\bar{e}\bar{\mu}}\rvert^2 + \frac{1}{2}\lvert(\mathcal{A}_c)_{\bar{e}\bar{\mu}}\rvert^2$	$(9.9 \pm 2.3 \pm 1.4) \times (10^{-19}\text{ GeV})^2$	"	50
$\lvert(\mathcal{C})_{\bar{e}\bar{\mu}}\rvert^2 + \frac{1}{2}\lvert(\mathcal{A}_s)_{\bar{e}\bar{\mu}}\rvert^2 + \frac{1}{2}\lvert(\mathcal{A}_c)_{\bar{e}\bar{\mu}}\rvert^2 + \frac{1}{2}\lvert(\mathcal{B}_s)_{\bar{e}\bar{\mu}}\rvert^2 + \frac{1}{2}\lvert(\mathcal{B}_c)_{\bar{e}\bar{\mu}}\rvert^2$	$(10.5 \pm 2.4 \pm 1.4) \times (10^{-19}\text{ GeV})^2$	"	50
b	$< 1.6 \times 10^{-23}$ GeV	Atmospheric	51
c	$< 1.4 \times 10^{-26}$	"	51
$\check{a}/\mathring{c}$	< 5 GeV	"	51

Table 9. Meson sector

Combination	Result	System	Ref.
Δa_0^K	$(0.4 \pm 1.8) \times 10^{-17}$ GeV	K oscillations	52
Δa_Z^K	$(-1 \pm 4) \times 10^{-17}$ GeV	"	52
$\|\Delta a_1^K\|$	$< 9.2 \times 10^{-22}$ GeV	"	53
$\|\Delta a_2^K\|$	$< 9.2 \times 10^{-22}$ GeV	"	53
$\|\Delta a_0^K - 0.6\Delta a_{\|\|}^K\|$	$< 10^{-20} m_K$	K oscillations, E773 data	54*
$N^D(\Delta a_0^D - 0.6\Delta a_Z^D)$	$(-2.8$ to $4.8) \times 10^{-16}$ GeV	D oscillations	55
$N^D \Delta a_X^D$	$(-7$ to $3.8) \times 10^{-16}$ GeV	"	55
$N^D \Delta a_X^D$	$(-7$ to $3.8) \times 10^{-16}$ GeV	"	55
$N^B(\Delta a_0^B - 0.30\Delta a_Z^B)$	$(-3.0 \pm 2.4) \times 10^{-15}$ GeV	B_d oscillations	56
$N^B \Delta a_X$	$(-22 \pm 7) \times 10^{-15}$ GeV	"	56
$N^B \Delta a_Y$	$(-27$ to $-4) \times 10^{-15}$ GeV	"	56
$N^B(\Delta a_0^B - 0.3\Delta a_Z^B)$	$-(5.2 \pm 4.0) \times 10^{-15}$ GeV	"	57
$N^B\sqrt{(\Delta a_X^B)^2 + (\Delta a_Y^B)^2}$	$(37 \pm 16) \times 10^{-15}$ GeV	"	57
δ^π	$(-1.5$ to $200) \times 10^{-11}$	Astrophysics	58*
$\|c^\pi\|$	$< 10^{-10}$	"	21*
$\|c^K\|$	$< 10^{-9}$	"	21*
$\|c^D\|$	$< 10^{-8}$	"	21*
$\|c^{B_d}\|, \|c^{B_s}\|$	$< 10^{-7}$	"	21*

Table 10. Electroweak sector

Combination	Result	System	Ref.
$\|(k_{\phi\phi}^A)_{\mu\nu}\|$	$< 3 \times 10^{-16}$	Cosmological birefringence	59*
$\|(k_{\phi B})_{\mu\nu}\|$	$< 0.9 \times 10^{-16}$	"	59*
$\|(k_{\phi W})_{\mu\nu}\|$	$< 1.7 \times 10^{-16}$	"	59*
$\|(k_{\phi\phi}^S)_{XX}\|, \|(k_{\phi\phi}^S)_{YY}\|, \|(k_{\phi\phi}^S)_{ZZ}\|$	$< 10^{-27}$	Clock comparisons	59*
$\|(k_{\phi\phi}^S)_{XY}\|$	$< 10^{-27}$	"	59*
$\|(k_{\phi\phi}^S)_{XZ}\|, \|(k_{\phi\phi}^S)_{YZ}\|$	$< 10^{-25}$	"	59*
$\|(k_{\phi\phi}^S)_{TT}\|$	$< 4 \times 10^{-13}$	H^- ion, $\bar{p}$ comparison	59*
$\|(k_\phi)_X\|, \|(k_\phi)_Y\|$	$< 10^{-31}$	Xe-He maser	59*
$\|(k_\phi)_Z\|, \|(k_\phi)_T\|$	$< 2.8 \times 10^{-27}$	"	59*
$\|k_W\|$	$< 10^{-5}$	Astrophysics	21*

Table 11. Gluon sector

Combination	Result	System	Ref.
$\lvert\tilde{\kappa}_{\rm tr}^{\rm QCD}\rvert$	$< 2 \times 10^{-13}$	Astrophysics	40*

Table 12. Gravity sector

Combination	Result	System	Ref.
$\sigma^{XX} - \sigma^{YY}$	$(-5.6 \pm 2.1) \times 10^{-9}$	Atom interferometry	60
σ^{XY}	$(-0.09 \pm 79) \times 10^{-9}$	"	60
σ^{XZ}	$(-13 \pm 37) \times 10^{-9}$	"	60
σ^{YZ}	$(-61 \pm 38) \times 10^{-9}$	"	60
σ^{TY}	$(-2.0 \pm 4.4) \times 10^{-5}$	"	60
σ^{TX}	$(5.4 \pm 4.5) \times 10^{-5}$	"	60
σ^{TZ}	$(1.1 \pm 26) \times 10^{-5}$	"	60
$\overline{s}^{11} - \overline{s}^{22}$	$(1.3 \pm 0.9) \times 10^{-10}$	Lunar laser ranging	61
$\overline{s}^{12}$	$(6.9 \pm 4.5) \times 10^{-11}$	"	61
$\overline{s}^{02}$	$(5.2 \pm 4.8) \times 10^{-7}$	"	61
$\overline{s}^{01}$	$(-0.8 \pm 1.1) \times 10^{-6}$	"	61
$\overline{s}_{\Omega_\oplus c}$	$(0.2 \pm 3.9) \times 10^{-7}$	"	61
$\overline{s}_{\Omega_\oplus s}$	$(-1.3 \pm 4.1) \times 10^{-7}$	"	61
$\lvert\overline{s}_{\rm Mercury}\rvert$	$\leq 10^{-9}$	Perihelion precession	62*
$\lvert\overline{s}_\oplus\rvert$	$\leq 10^{-8}$	"	62*
$\lvert\overline{s}_{\rm SSP}\rvert$	$\leq 10^{-13}$	Solar-spin precession	62*